KV-638-324

Mathematics for Computing

Macmillan Computer Science Series

Consulting Editor
Professor F. H. Sumner, University of Manchester

S. T. Allworth, *Introduction to Real-time Software Design*

Ian O. Angell, *A Practical Introduction to Computer Graphics*

G. M. Birtwistle, *Discrete Event Modelling on Simula*

T. B. Boffey, *Graph Theory in Operations Research*

Richard Bornat, *Understanding and Writing Compilers*

J. K. Buckle, *The ICL 2900 Series*

Robert Cole, *Computer Communications*

Derek Coleman, *A Structured Programming Approach to Data**

Andrew J. T. Colin, *Fundamentals of Computer Science*

Andrew J. T. Colin, *Programming and Problem solving in Algol 68**

S. M. Deen, *Fundamentals of Data Base Systems**

J. B. Gosling, *Design of Arithmetic Units for Digital Computers*

David Hopkin and Barbara Moss, *Automata**

Roger Hutty, *Fortran for Students*

H. Kopetz, *Software Reliability*

A. M. Lister, *Fundamentals of Operating Systems, second edition**

G. P. McKeown and V. J. Rayward-Smith, *Mathematics for Computing*

Brian Meek, *Fortran, PL/I and the Algols*

Derrick Morris and Roland N. Ibbett, *The MU5 Computer System*

John Race, *Case Studies in Systems Analysis*

Peter Wallis, *Portable Programming*

I. R. Wilson and A. M. Addyman, *A Practical Introduction to Pascal*

* The titles marked with an asterisk were prepared during the Consulting Editorship of Professor J. S. Rohl, University of Western Australia.

18-00

Mathematics for Computing

G. P. McKeown
V. J. Rayward-Smith

*School of Computing Studies and Accountancy,
University of East Anglia*

First published 1982 by
THE MACMILLAN PRESS LTD
London and Basingstoke
Companies and representatives throughout the world

Filmset in 10/12 Monophoto Times by
MID-COUNTY PRESS, LONDON SW15
Printed in Hong Kong

ISBN 0 333 29169 7
ISBN 0 333 29170 0 pbk

To Mary and Sheila

Contents

Preface

The problems that early computers solved were mostly mathematical. Since then, as everyone knows, the use of computers has greatly diversified such that today the majority of applications are non-numerical. Thus, while it is immediately clear that a good mathematical background is essential for the obviously mathematical areas of computing, such as numerical techniques, computer simulation or the theory of computation, the question arises as to whether mathematics is now important in computing as a whole. We believe the answer to this question to be an unqualified yes. The last decade or so has seen the development of computing from little more than a 'bag of tricks' into a science. As in any scientific or engineering discipline, mathematics is the medium through which the underlying concepts and principles of computer science may be understood. Mathematical notation is now found in virtually every branch of computing, from the mathematical areas mentioned above, through compiling techniques, data structures and algorithm design and analysis, to data base systems. Without the ability to appreciate and apply mathematical concepts and techniques, the aspiring computer scientist cannot hope to grasp the fundamental principles of computing, principles that will still be relevant even if the particular programming skills that he has learnt become obsolete.

This book is designed for two types of user. The first is the student starting tertiary education in computing, who will need to develop a reasonable mathematical maturity in order to cope with the use of mathematical notation in subsequent computer science courses. We assume that no such student will be without a reasonable 'A'-level (or equivalent) in mathematics. The book provides a basis for a course equivalent to about one-third of the first year of study in a degree programme.

The second category of reader for whom the book is designed is the practising computer scientist who needs a reference book on his shelf to which he can go when he needs a definition and an example of some concept only vaguely remembered. For this reason, after most definitions in the book, one or more examples are given to illustrate the new term.

While the material in this book is essentially traditional mathematics, it has been given a computer science flavour through the use of algorithms.

The algorithm is the core concept in computing, but one which normally has little place in traditional mathematics. Nevertheless, an algorithm is often the best way of describing well-known mathematical techniques.

Since this book is intended primarily to be of use on a general first mathematics course in a computer science degree programme, it is not constrained to just discrete mathematics. We believe that calculus contains a wealth of results with applications in computing, and for this reason a substantial amount of calculus is presented in the third chapter. Although this chapter is called 'Calculus' it does, in fact, contain some finite mathematics. In particular, the section on series includes both the finite and the infinite cases.

G. P. McKEOWN
V. J. RAYWARD-SMITH

ACKNOWLEDGEMENTS

The authors wish to thank Miss Gillian Hall, who produced all of the computer-generated figures in the book. The figures were generated using the GINO-F general-purpose graphics package on a PRIME 400 computer.

The authors also wish to thank Mr E. W. Haddon for his constructive criticism of the text.

Finally, the authors offer their warmest thanks to Mrs Jane Copeman and to Mrs Jane Loughlin for their excellent and patient typing of a difficult manuscript.

1 Foundations

1.1 PROPOSITIONAL LOGIC

Throughout this book, the mathematical reasoning is presented in the English language, suitably augmented by a collection of special symbols. These symbols are defined not only as a shorthand tool but also for the sake of clarity and precision. In this opening section, notation is introduced to show how a complex statement in English is constructed from simple statements and how, given the truth value of these simple statements, the truth value of the complex statement can be determined.

A simple statement may be represented by a *statement letter*, either an upper-case letter of the roman alphabet (A, B, C, ...) or such a letter with an integer subscript (A_1, A_2, ...). The simple statement

It is raining

might thus be represented by the statement letter, R, while H might be chosen to represent

Today is a holiday

Complex statements are constructed from simple statements using connectives such as: 'not', 'and', 'or', 'implies'. For example, a complex statement constructed from the above simple statements using the connectives 'not' and 'and' is

It is not raining and today is a holiday

Whether this statement is true or not will depend on the truth values of the simple statements used in its construction. For this example, the complex statement constructed from R and H is true if and only if R is false and H is true.

Connectives, Statement Forms and Truth Tables

A statement is either a simple or a complex statement and is represented by a capital, script letter of the roman alphabet ($\mathscr{A}$, $\mathscr{B}$, $\mathscr{C}$, ...).

Given a statement $\mathscr{A}$, $(\sim\mathscr{A})$ [read: not $\mathscr{A}$] represents the statement that is true if $\mathscr{A}$ is false and is false if $\mathscr{A}$ is true. If true is denoted by T and false by F, then this definition can be summarised in a tabular form

$\mathscr{A}$	$(\sim\mathscr{A})$
T	F
F	T

Such a table is known as a *truth table.*

Given statements $\mathscr{A}$ and $\mathscr{B}$, $(\mathscr{A}\wedge\mathscr{B})$ [read: $\mathscr{A}$ and $\mathscr{B}$] represents the statement that is true if and only if both $\mathscr{A}$ and $\mathscr{B}$ are true. In terms of a truth table

$\mathscr{A}$	$\mathscr{B}$	$(\mathscr{A}\wedge\mathscr{B})$
T	T	T
T	F	F
F	T	F
F	F	F

Further connectives, $\vee$ [or], $\supset$ [implies] and $\equiv$ [if and only if, sometimes written iff], can also be defined using truth tables

$\mathscr{A}$	$\mathscr{B}$	$(\mathscr{A}\vee\mathscr{B})$	$(\mathscr{A}\supset\mathscr{B})$	$(\mathscr{A}\equiv\mathscr{B})$
T	T	T	T	T
T	F	T	F	F
F	T	T	T	F
F	F	F	T	T

The definition of $(\mathscr{A}\supset\mathscr{B})$ may seem a little strange at first — the idea that $(\mathscr{A}\supset\mathscr{B})$ should be true whenever $\mathscr{A}$ is false can be puzzling. However, consider the following example. Let $\mathscr{A}$ be the statement

It is a sunny day

and let $\mathscr{B}$ be the statement

It is daytime

No one would dispute that statement $\mathscr{A}$ implies statement $\mathscr{B}$, even though $\mathscr{A}$ is, in fact, often false. However, $\mathscr{A}$ can never be true when $\mathscr{B}$ is false. In general, therefore, $(\mathscr{A}\supset\mathscr{B})$ is taken to be true when either $\mathscr{A}$ is false or both $\mathscr{A}$ and $\mathscr{B}$ are true. If $\mathscr{A}$ is true and $\mathscr{B}$ is false then $(\mathscr{A}\supset\mathscr{B})$ is taken to be false.

The five connectives $(\sim, \vee, \wedge, \supset, \equiv)$ are used to combine simple statements into complex statements whose truth values will usually depend on the truth values of the constituent simple statements. A statement may thus be represented by an expression composed of statement letters and connectives. Such an expression is called a statement form. For example, the statement

It is not raining and today is a holiday

can be represented by the statement form $((\sim R) \wedge H)$.

A *statement form* can be formally defined as follows.

Definition 1.1.1

(a) All statement letters (capital roman letters and such letters with numerical subscripts) are statement forms.
(b) If $\mathscr{A}$ and $\mathscr{B}$ are statement forms, then so are $(\sim \mathscr{A})$, $(\mathscr{A} \wedge \mathscr{B})$, $(\mathscr{A} \vee \mathscr{B})$, $(\mathscr{A} \supset \mathscr{B})$ and $(\mathscr{A} \equiv \mathscr{B})$.
(c) Only those expressions that are determined by (a) and (b) are statement forms.

In ordinary algebra, $a+b \times c+d$ may be used to represent $((a+(b \times c))+d)$ since there is a convention for the restoration of parentheses based on the priorities of operators. There is a similar convention in propositional logic which works according to the following rules.

(a) The connectives

$$\sim, \wedge, \vee, \supset, \equiv$$

are ranked from left to right, with $\sim$ having the highest priority.
(b) If there is only one connective in a statement form then parentheses can be omitted by association to the left.
(c) The outer parentheses can be omitted.

Example 1.1.1

Removing parentheses from $((A \supset (\sim B)) \supset (C \vee (\sim A)))$ results in the following steps

$$(A \supset (\sim B)) \supset (C \vee (\sim A))$$

$$A \supset (\sim B) \supset (C \vee (\sim A))$$

$$A \supset (\sim B) \supset C \vee (\sim A)$$

$$A \supset \sim B \supset C \vee \sim A$$

Of course, it is not always possible to remove all the parentheses from a statement form and indeed, for the sake of readability, it is often not desirable to remove the maximum number of parentheses.

Given any statement form, $\mathscr{A}$, and an assignation of truth values to the individual statement letters occurring in $\mathscr{A}$, a truth value for $\mathscr{A}$ can be deduced. It is possible to tabulate all possible such assignations of truth values to the individual statement letters, giving the resulting truth values of the statement form.

Example 1.1.2 The truth table for $A \supset \sim B \supset C \vee \sim A$

A	B	C	$\sim A$	$C \vee \sim A$	$\sim B$	$A \supset \sim B$	$A \supset \sim B \supset C \vee \sim A$
T	T	T	F	T	F	F	T
T	T	F	F	F	F	F	T
T	F	T	F	T	T	T	T
T	F	F	F	F	T	T	F
F	T	T	T	T	F	T	T
F	T	F	T	T	F	T	T
F	F	T	T	T	T	T	T
F	F	F	T	T	T	T	T

A more compact form of the truth table can be achieved by writing the truth values of statement letters immediately beneath them and writing the truth values constructed using a particular connective immediately beneath that connective. The final result is indicated by means of the symbol ↑. Using this technique for the above table results in a table of the following form.

A	$\supset$	$\sim$	B	$\supset$	C	$\vee$	$\sim$	A
T	F	F	T	T	T	T	F	T
T	F	F	T	T	F	F	F	T
T	T	T	F	T	T	T	F	T
T	T	T	F	F	F	F	F	T
F	T	F	T	T	T	T	T	F
F	T	F	T	T	F	T	T	F
F	T	T	F	T	T	T	T	F
F	T	T	F	T	F	T	T	F
				↑				

Tautologies and Contradictions

Definition 1.1.2
A statement form that is always true no matter what truth values are assigned to its statement letters is called a *tautology*. The truth table for a tautology thus has only T occurring in the final column calculated.

A *contradiction* is the opposite of a tautology, with only F appearing in the final column calculated.

An immediate consequence of these definitions is the result that $\mathscr{A}$ is a tautology if and only if $\sim \mathscr{A}$ is a contradiction.

Example 1.1.3
$(A \wedge B) \supset A$ is a tautology since its truth value yields a column consisting entirely of T values.

(A	$\wedge$	B)	$\supset$	A
T	T	T	T	T
T	F	F	T	T
F	F	T	T	F
F	F	F	T	F
			↑	

A complex statement in English that can be derived from a tautology by substituting English statements for statement letters, such that each occurrence of a particular statement letter is replaced by the same English statement, is said to be *logically true* (*according to propositional logic*). Similarly, a complex English statement arising from substitution into a contradiction is said to be *logically false* (*according to propositional logic*).

In example 1.1.3, it was shown that $(A \wedge B) \supset A$ is a tautology. If A is replaced by 'the sun shines' and B by 'the grass grows' then the result is the logically true statement, 'the sun shines and the grass grows implies the sun shines.'

Definition 1.1.3
Two statement forms, $\mathscr{A}$ and $\mathscr{B}$, are said to be *equivalent* if and only if $(\mathscr{A} \equiv \mathscr{B})$ is a tautology. $\mathscr{A}$ *implies* $\mathscr{B}$ if and only if $(\mathscr{A} \supset \mathscr{B})$ is a tautology.

Theorem 1.1.1
$\mathscr{A}$ and $\mathscr{B}$ are equivalent if and only if $\mathscr{A}$ implies $\mathscr{B}$ and $\mathscr{B}$ implies $\mathscr{A}$.
Proof There are two parts to this proof.

The first part is to show that $\mathscr{A}$ and $\mathscr{B}$ are equivalent if $\mathscr{A}$ implies $\mathscr{B}$ and $\mathscr{B}$ implies $\mathscr{A}$. If $\mathscr{A}$ is true then, since $(\mathscr{A} \supset \mathscr{B})$ is a tautology, it follows that $\mathscr{B}$ is true. If $\mathscr{A}$ is false then, since $(\mathscr{B} \supset \mathscr{A})$ is a tautology, $\mathscr{B}$ must be false. Hence, $(\mathscr{A} \equiv \mathscr{B})$ is a tautology.

The second part is to show that $\mathscr{A}$ and $\mathscr{B}$ are equivalent only if $\mathscr{A}$ implies $\mathscr{B}$ and $\mathscr{B}$ implies $\mathscr{A}$, that is, if $\mathscr{A}$ and $\mathscr{B}$ are equivalent then it follows that $\mathscr{A}$ implies $\mathscr{B}$ and $\mathscr{B}$ implies $\mathscr{A}$. If $(\mathscr{A} \equiv \mathscr{B})$ is a tautology then, if $\mathscr{A}$ is true, $\mathscr{B}$ must be true and, if $\mathscr{A}$ is false, then $\mathscr{B}$ must be false. In either case, $(\mathscr{A} \supset \mathscr{B})$ and $(\mathscr{B} \supset \mathscr{A})$ are tautologies.

A more concise proof of this theorem can be achieved by rewriting it in propositional logic. The theorem states that $(\mathscr{A} \equiv \mathscr{B}) \equiv (\mathscr{A} \supset \mathscr{B}) \wedge (\mathscr{B} \supset \mathscr{A})$ and it is left as an exercise for the reader to check that this is a tautology by using a truth table.

Definition 1.1.4
Two English statements represented by $\mathscr{A}$ and $\mathscr{B}$ are said to be *logically equivalent* if and only if $\mathscr{A}$ and $\mathscr{B}$ are equivalent. The statement represented by $\mathscr{A}$ is said to *logically imply* the statement represented by $\mathscr{B}$ if and only if $\mathscr{A}$ implies $\mathscr{B}$.

Proving two English sentences to be logically equivalent is particularly important.

Example 1.1.4
Show that 'It is not raining or snowing so it is sunny' is logically equivalent to 'It is not raining and it is not snowing so it is sunny'.

If A represents 'It is raining', B represents 'It is snowing' and C represents 'It is sunny', then the first sentence is represented by $\sim(A \vee B) \supset C$ and the second sentence is represented by $(\sim A \wedge \sim B) \supset C$. The truth table below shows the two statement forms to be equivalent.

$\sim$	$(A$	$\vee$	$B)$	$\supset$	C	$\equiv$	$(\sim$	A	$\wedge$	$\sim$	$B)$	$\supset$	C
F	T	T	T	T	T	T	F	T	F	F	T	T	T
F	T	T	T	T	F	T	F	T	F	F	T	T	F
F	T	T	F	T	T	T	F	T	F	T	F	T	T
F	T	T	F	T	F	T	F	T	F	T	F	T	F
F	F	T	T	T	T	T	T	F	F	F	T	T	T
F	F	T	T	T	F	T	T	F	F	F	T	T	F
T	F	F	F	T	T	T	T	F	T	T	F	T	T
T	F	F	F	F	F	T	T	F	T	T	F	F	F
						↑							

Adequate Sets of Connectives

A truth table involving n statement letters $A_1, \ldots, A_n$ will consist of 2^n rows. If a truth value is arbitrarily assigned to each row, the following question naturally arises: can a statement form involving $A_1, \ldots, A_n$ and the connectives $\sim$, $\vee$, $\wedge$, $\supset$ and $\equiv$ be found whose truth values correspond to the assigned column of truth values? For example, consider a truth table involving three statement letters A_1, A_2, A_3 and assume the last column arbitrarily chosen as below.

A_1	A_2	A_3	
T	T	T	T
T	T	F	F
T	F	T	F
T	F	F	F
F	T	T	T
F	T	F	T
F	F	T	F
F	F	F	T

What statement form would then give a truth table with this last column? One way to construct such a statement is to use algorithm 1.1.1.

Algorithm 1.1.1

```
if last column has no T values then the statement form required is
A_1 ∧ ~A_1
else for each row i where the last column has value T do
        for j from 1 in steps of 1 to n do
            if A_j has value T then
                U^i_j ← A_j
            else
                U^i_j ← ~A_j
            endif
        endfor
        C_i ← U^i_1 ∧ U^i_2 ∧ ... ∧ U^i_n
    endfor
    The required statement form is given by combining the various
    C_i s using the connective ∨.
endif
```

For the above example, $\mathscr{C}_1$ is $A_1 \wedge A_2 \wedge A_3$, $\mathscr{C}_5$ is $\sim A_1 \wedge A_2 \wedge A_3$, $\mathscr{C}_6$ is $\sim A_1 \wedge A_2 \wedge \sim A_3$ and $\mathscr{C}_8$ is $\sim A_1 \wedge \sim A_2 \wedge \sim A_3$. The required statement form is thus

$$(A_1 \wedge A_2 \wedge A_3) \vee (\sim A_1 \wedge A_2 \wedge A_3) \vee (\sim A_1 \wedge A_2 \wedge \sim A_3) \vee (\sim A_1 \wedge \sim A_2 \wedge \sim A_3)$$

The construction above shows that $\sim$, $\vee$ and $\wedge$ form an adequate set of connectives in the sense that every truth table corresponds to some statement form constructed using just these connectives.

A stronger result can be obtained by noting that $(\mathscr{A} \vee \mathscr{B})$ is equivalent to $\sim(\sim \mathscr{A} \wedge \sim \mathscr{B})$, so every occurrence of $\vee$ can be replaced using $\sim$ and $\wedge$. Hence just $\sim$ and $\wedge$ form an adequate set of connectives. Similarly, by noting the equivalence of $(\mathscr{A} \wedge \mathscr{B})$ and $\sim(\sim \mathscr{A} \vee \sim \mathscr{B})$ one can see that $\sim$ and $\vee$ also form an adequate set of connectives.

The connective $\vee$ is sometimes known as a *disjunction* and the connective $\wedge$ as a *conjunction.* If a statement form $\mathscr{C}$ can be written as $\mathscr{A} \vee \mathscr{B}$ then $\mathscr{A}$, $\mathscr{B}$ are known as *disjuncts.* Similarly, if $\mathscr{C}$ can be written as $\mathscr{A} \wedge \mathscr{B}$, $\mathscr{A}$ and $\mathscr{B}$ are known as *conjuncts.*

Definition 1.1.5

A statement form is in *disjunctive normal form* if it is a disjunction of one or more disjuncts, each of which is a conjunction of one or more statement letters and negations of statement letters.

Algorithm 1.1.1 shows that for every statement form there is an equivalent statement form in disjunctive normal form.

Exercise 1.1

1. Write the following as statement forms, using statement letters to stand for simple statements.

(a) If John is good, Mark is bad and if Mark is bad, John is good.
(b) A sufficient condition for John to be good is that Mark is bad.
(c) A necessary condition for John to be good is that Mark is bad.

2. Determine whether each of the following statement forms is a tautology, contradiction or neither.

(a) $A \equiv A \vee B$
(b) $(A \supset B) \wedge B \supset A$
(c) $\sim(A \supset (B \supset A))$
(d) $A \supset (B \supset C) \supset ((A \supset B) \supset (A \supset C))$
(e) $A \supset ((B \supset \sim A) \supset \sim B)$

3. Show that the following pairs of statements are equivalent.

(a) $\mathscr{A}$ and $\sim\sim\mathscr{A}$
(b) $\sim(\mathscr{A} \vee \mathscr{B})$ and $\sim\mathscr{A} \wedge \sim\mathscr{B}$
(c) $\sim(\mathscr{A} \wedge \mathscr{B})$ and $\sim\mathscr{A} \vee \sim\mathscr{B}$
(d) $\mathscr{A} \wedge (\mathscr{B} \vee \mathscr{C})$ and $(\mathscr{A} \wedge \mathscr{B}) \vee (\mathscr{A} \wedge \mathscr{C})$
(e) $\mathscr{A} \vee (\mathscr{B} \wedge \mathscr{C})$ and $(\mathscr{A} \vee \mathscr{B}) \wedge (\mathscr{A} \vee \mathscr{C})$

4. Find a statement form $\mathscr{A}$ constructed from the statement letters A, B, C corresponding to the following truth table.

A	B	C	$\mathscr{A}$
T	T	T	F
T	T	F	T
T	F	T	F
T	F	F	T
F	T	T	F
F	T	F	T
F	F	T	F
F	F	F	F

5. Define the connective | [alternative denial] by

$\mathscr{A}$	$\mathscr{B}$	$\mathscr{A} \mid \mathscr{B}$
T	T	F
T	F	T
F	T	T
F	F	T

Show that $|$, on its own, is an adequate set of connectives. (Hint: express $\sim \mathscr{A}$ and $\mathscr{A} \vee \mathscr{B}$ in terms of $\mathscr{A}$, $\mathscr{B}$ and $|$.)

6. Prove that algorithm 1.1.1 is correct.

7. Define conjunctive normal form and prove that every statement form is equivalent to some statement form in conjunctive normal form. (Hint: look at the rows of the truth table resulting in an F in the final column.)

1.2 SET THEORY

Defining a Set

In everyday English the word 'set' is used in phrases such as 'a set of spanners', 'a geometry set' or 'a set of rules' and similarly, in mathematics, a set is simply a collection of objects.

One way of defining a set is to list all members or *elements* of that set. For example,

{Monday, Tuesday, Wednesday, Thursday, Friday, Saturday, Sunday}

is the set of the days of the week. When defining a set by listing its elements, the elements themselves are separated by commas and the total collection of elements is surrounded by special parentheses, $\{$ and $\}$. The order of the elements is unimportant and each element must only be listed once. Two sets are equal only if they contain exactly the same elements.

Definition 1.2.1
If A is a set and x is an element, $x \in A$ signifies that x is an element of A and $x \notin A$ signifies that x is not an element of A.

Example 1.2.1

$$\{3, 2, 1, 4\} = \{1, 2, 3, 4\} = \{1, 4, 3, 2\}$$
$$2 \in \{3, 2, 1, 4\} \text{ but } 5 \notin \{3, 2, 1, 4\}$$

Definition 1.2.2
The number of elements in a set, A, is known as the *cardinality* of the set A and is denoted by $\#(A)$.

Having no elements, the *empty set* has cardinality zero and is denoted by $\{\ \}$ or $\emptyset$.

A set is *finite* if its cardinality is finite, otherwise it is an *infinite* set.

When a set is finite, it is possible to define it by listing all its elements. Clearly, this method of definition is not possible for infinite sets. An example of such a set is the set of all prime numbers, P. This set can only be defined by giving some property (that is, being a prime number) that each element of the set must satisfy. Thus P can be defined

$$P = \{x | x \text{ is a prime number}\}$$

that is, P is the set of elements x such that x is a prime number. Note that for any set, A

$$A = \{x | x \in A\}$$

Example 1.2.2

(1) The set with cardinality 0

$$\{\ \} = \emptyset$$

(2) Sets with cardinality 1, that is *singleton* sets

$$\{0\}, \{1\}, \{\emptyset\}, \{\{0,1\}\}$$

(3) Sets with cardinality seven

$$\{1, 2, 3, 4, 5, 6, 7\}$$

$$\{x | x \text{ is a positive integer less than } 8\}$$

$$\{d | d \text{ is a day of the week}\}$$

(4) Sets with infinite cardinality

$$\{x | x \text{ is an integer and } x \text{ is greater than } 2\}$$

$$\{y | y \text{ is a real number}\}$$

If the set, X, is defined to be $\{x | x \text{ is greater than } 2\}$ then the precise nature of X can only be determined given the values which x might take. For example, if x could only range over the positive integers, X would be a different set from the case where x could range over all numbers. In the former case, $2.1 \notin X$ but, in the latter, $2.1 \in X$. Whenever sets are being considered, it is important to specify the *universe* — the set of values over which elements can range. Possible universes in which $\{x | x \text{ is greater than } 2\}$ might occur are

the set of integers

the set of positive integers

the set of real numbers

the set of prime numbers.

The universe is normally denoted by $\mathscr{U}$.

Subsets

Definition 1.2.3
If every element of a set A is also a member of a set B then A is a *subset* of B, denoted by $A \subset B$.

A is a *proper subset* of B, denoted by $A \subsetneqq B$, if $A \subset B$ and $A \neq B$. Two sets, A and B, are *comparable* if either $A \subset B$ or $B \subset A$.

Note that every set is a subset of itself, that all sets are subsets of their universe and that the empty set is a subset of every set.

If $a \in A$ then $\{a\} \subset A$. The distinction between the element, a, and the singleton subset of A containing the element a is important.

Example 1.2.3

Let $X = \{3, 4, 5, 6, 7\}$ and $Y = \{6, 7, 8, 9\}$. Then X and Y are incomparable, and $\{6, 7\}$ is a proper subset of both X and Y.

Definition 1.2.4
Let A be any set. The set of subsets of A is denoted by 2^A and is called the *power set* of A.

The elements of a power set are themselves sets and the power set of a set always includes the empty set and the set itself.

Example 1.2.4

If $A = \{a, b, c\}$, then $2^A = \{\emptyset, \{a\}, \{b\}, \{c\}, \{a, b\}, \{b, c\}, \{c, a\}, \{a, b, c\}\}$.

Theorem 1.2.1
For any finite set $A, \#(2^A) = 2^{\#(A)}$.

Proof Let $A = \{a_1, a_2, \ldots, a_n\}$. Every element of 2^A is a subset of A and can be constructed by opting to include or not include each element a_1, $\ldots$, a_n, in turn. There are 2^n different lists of such options and so there are 2^n different subsets of A.

Operations on Sets and Venn Diagrams

Definition 1.2.5
If A, B are two sets, then the set of elements contained in either A or B (or both) is $A \cup B$, the *union* of A and B.

The set of elements contained in both A and B is $A \cap B$, the *intersection* of A and B. If $A \cap B = \emptyset$, A and B are said to be *disjoint*.

The set of elements contained in A but not in B is $A \backslash B$, the *difference* of A and B.

Thus

$$A \cup B = \{x | x \in A \vee x \in B\}$$

$$A \cap B = \{x | x \in A \wedge x \in B\}$$

and

$$A \backslash B = \{x | x \in A \wedge \sim(x \in B)\}$$

Example 1.2.5

If $A = \{1, 2, 3, 4\}$, $B = \{3, 4, 5\}$ and $C = \{5, 6\}$, then

$$A \cup B = \{1, 2, 3, 4, 5\}, \quad B \cup C = \{3, 4, 5, 6\}, \quad C \cup A = \{1, 2, 3, 4, 5, 6\}$$

$$A \cap B = \{3, 4\}, \quad B \cap C = \{5\}, \quad C \cap A = \emptyset$$

$$A \backslash B = \{1, 2\}, \quad B \backslash C = \{3, 4\}, \quad C \backslash A = \{5, 6\}$$

$$B \backslash A = \{5\}, \quad C \backslash B = \{6\}, \quad A \backslash C = \{1, 2, 3, 4\}$$

A useful way of representing sets pictorially is by a *Venn diagram.* Each set is represented by a region such as the interior of a circle. If two sets have elements in common, then the regions overlap. If one set is a subset of the other, then one region is wholly contained within the other. The universe is represented by a surrounding rectangle in which every region representing a set must lie. Note that the area of the region does not relate to the cardinality of the set it represents.

The Venn diagrams in figure 1.1 illustrate commonly arising situations.

The operations $\cup$, $\cap$ and $\backslash$ can be conveniently represented in Venn diagrams by shading the appropriate region (figure 1.2).

Definition 1.2.6

The *complement* of a set A is A', where

$$A' = \{x | \sim(x \in A)\}$$

The *symmetric difference* of two sets A, B, is $A \Delta B$, where

$$A \Delta B = (A \backslash B) \cup (B \backslash A)$$

The Venn diagrams in figure 1.3 illustrate these operations.

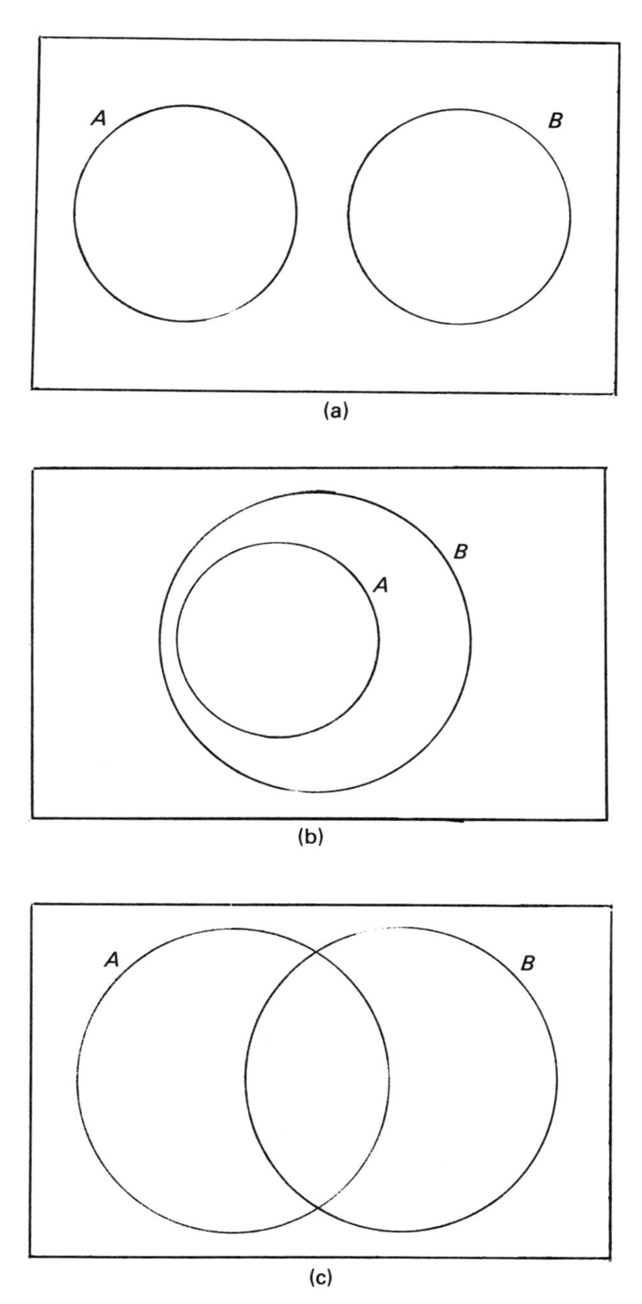

Figure 1.1 Venn diagrams: (a) A and B are disjoint; (b) A is a proper subset of B; (c) A and B are incomparable, and have a non-empty intersection

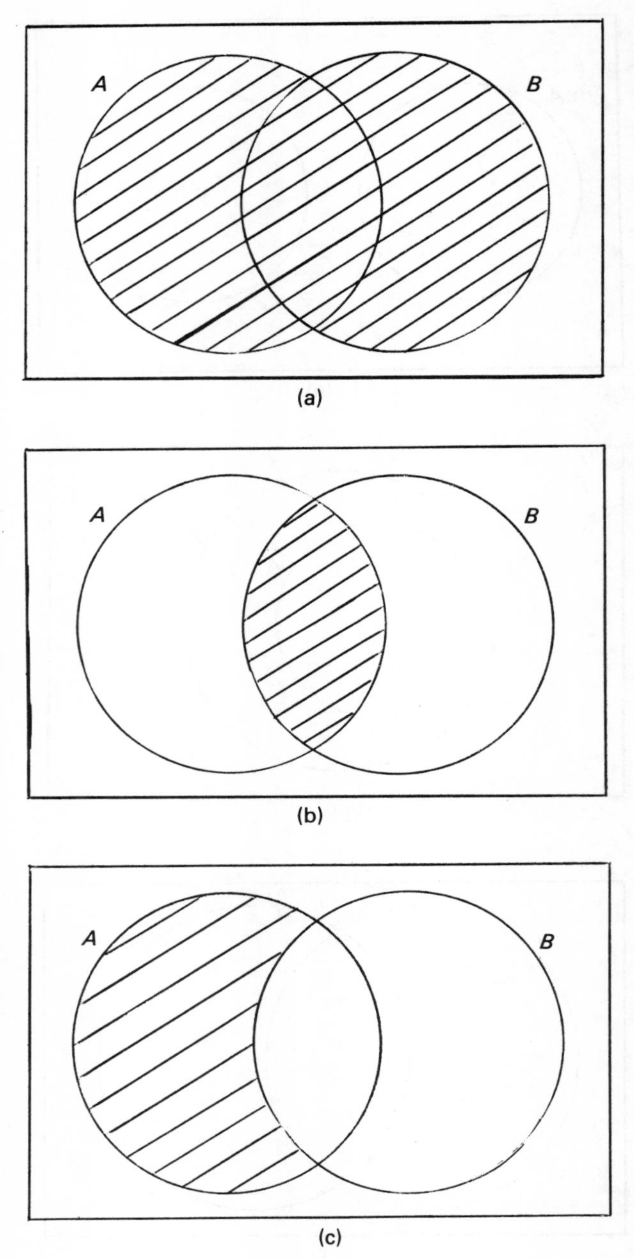

Figure 1.2 Venn diagrams: (a) $A \cup B$; (b) $A \cap \mathrm{B}$; (c) $A \backslash B$

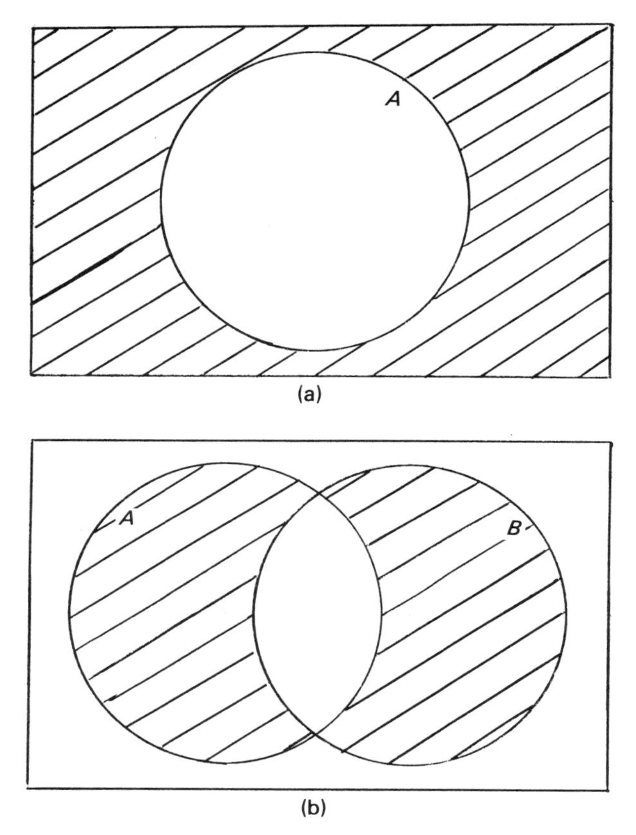

Figure 1.3 Venn diagrams: (a) A'; (b) $A\Delta B$

Example 1.2.6

In a group of students

10 study mathematics, 8 study computing, 9 study economics

7 study computing together with mathematics and/or economics

5 do no mathematics but study computing and/or economics

4 study none of mathematics, computing or economics.

Venn diagrams can be used to determine the total number of students in the group and to show that the number of students studying mathematics and economics is equal to the number of students studying mathematics but no economics.

Let M denote the set of students studying mathematics, C the set studying computing and E the set studying economics. A Venn diagram is drawn to illustrate the relationship between these sets. The universe is the

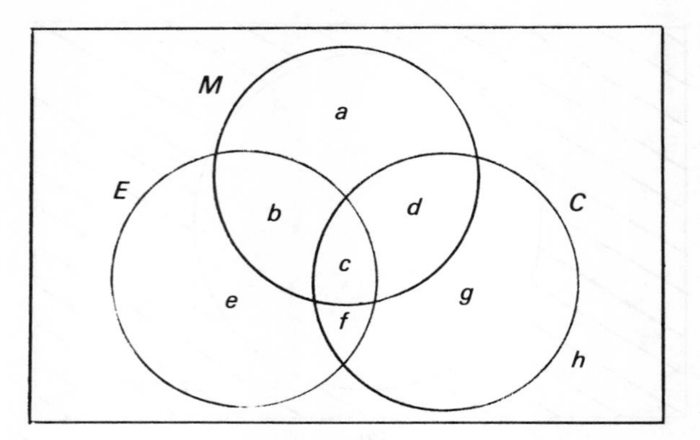

Figure 1.4 Venn diagram: example 1.2.6

group of students and the diagram divides the universe into eight disjoint regions, as in Figure 1.4. Each region represents a set and the integer variable associated with each region is used to determine the number of elements in the set.

The information given implies

$$a+b+c+d=10$$
$$c+d+f+g=8$$
$$b+c+e+f=9$$
$$c+d+f=7$$
$$e+f+g=5$$
$$h=4$$

Then the number of students in the group is

$$a+b+c+d+e+f+g+h=10+5+4=19$$

Now $g=(c+d+f+g)-(c+d+f)=1$

so $e+f=(e+f+g)-g=4$

Thus, the number of students doing both mathematics and economics is

$$b+c=(b+c+e+f)-(e+f)=9-4=5$$

and the number doing mathematics but no economics is

$$a+d=(a+b+c+d)-(b+c)=10-5=5$$

Laws of Set Theory

Expressions involving sets can be easily manipulated provided a few simple rules are remembered. Most of these are obvious once stated, but some require proof.

Theorem 1.2.2
For any sets A, B and C

(i) $A \cup B = B \cup A$; (ii) $A \cap B = B \cap A$;
(iii) $A \cup A = A$; (iv) $A \cap A = A$;
(v) $A \cup (B \cup C) = (A \cup B) \cup C$; (vi) $A \cap (B \cap C) = (A \cap B) \cap C$;
(vii) $A \cup (B \cap C) = (A \cup B) \cap (A \cup C)$; (viii) $A \cap (B \cup C) = (A \cap B) \cup (A \cap C)$;
(ix) $A \cup A' = \mathscr{U}$; (x) $A \cap A' = \emptyset$;
(xi) $(A \cap B)' = A' \cup B'$; (xii) $(A \cup B)' = A' \cap B'$.

Proof Parts (i) to (vi), (ix) and (x) are obvious. The proofs of (vii) and (xi) are given but the proofs of (viii) and (xii) are left as an exercise. All four of these proofs require results from section 1.1.

(vii) In this case

$$\begin{aligned} A \cup (B \cap C) &= \{x \mid x \in A\} \cup \{x \mid x \in B \wedge x \in C\} \\ &= \{x \mid (x \in A) \vee (x \in B \wedge x \in C)\} \\ &= \{x \mid (x \in A \vee x \in B) \wedge (x \in A \vee x \in C)\} \\ &= \{x \mid x \in A \vee x \in B\} \cap \{x \mid x \in A \vee x \in C\} \\ &= (A \cup B) \cap (A \cup C) \end{aligned}$$

(xi) In this case

$$\begin{aligned} (A \cap B)' &= \{x \mid x \in A \wedge x \in B\}' \\ &= \{x \mid \sim (x \in A \wedge x \in B)\} \\ &= \{x \mid \sim (x \in A) \vee \sim (x \in B)\} \\ &= \{x \mid \sim (x \in A)\} \cup \{x \mid \sim (x \in B)\} \\ &= A' \cup B' \end{aligned}$$

Parentheses in expressions involving sets and the operators $'$, $\cup$ and $\cap$ can be removed/restored according to rules similar to those for propositional calculus.

(a) The operators

$$' \quad \cap \quad \cup$$

are ranked from left to right, with $'$ having the highest priority.

(b) If there is only one operator in an expression, parentheses can be omitted.
(c) Outer parentheses can be omitted.

Example 1.2.7

(1) $A \cup B \cap C = (A \cup (B \cap C))$

(2) $A \cap B' \cup C = (A \cap (B')) \cup C$

(3) $A \cup (B \cup C) = (A \cup B) \cup C = A \cup B \cup C$

Exercise 1.2

1. If the universe is taken as the set of positive integers less than 20, then list the elements of the following sets

(a) $\{x | x+2 \text{ is less than } 10\}$;

(b) $\{x | x \text{ is prime}\}$;

(c) $\{x | x = x^2\}$;

(d) $\{x | 2x = 1\}$.

2. If $\mathscr{U} = \{0, 2, 4, 6, 8, 10\}$, $A = \{0, 4, 8\}$, $B = \{2, 4, 6, 8\}$, and $C = \{0, 6, 10\}$, determine

(a) A';

(b) $(A \cup B) \cap C'$;

(c) $A' \cap B' \cap C'$;

(d) $A | B'$;

(e) $A \Delta B'$.

3. Prove that, in general, $A \backslash B \neq B \backslash A$. If $A \backslash B = B \backslash A$, what can be said about A and B?

4. Place the following in numerically ascending order: $\#(A)$, $\#(A \cup B)$, $\#(A \cap B)$, $\#(\emptyset)$, $\#(\mathscr{U})$.

5. Prove, using results from propositional logic, that:

(a) $A \cap (B \cup C) = (A \cap B) \cup (A \cap C)$;

(b) $(A \cup B)' = A' \cap B'$;

(c) $A \Delta B = (A \cup B) \setminus (A \cap B)$.

6. In a mixed class of 33 children, boys do not play hockey and girls do not play football, but all play at least one of the games hockey, football and tennis. There are 10 boys who play football only, 2 boys who play tennis only, 8 girls who play both tennis and hockey and 9 girls who play hockey but not tennis.

How many pupils play tennis?

What can be deduced about

(a) the number of boys who play both football and tennis;
(b) the number of girls in the class?

7. Lewis Carroll is said to have made the following statements:

A dishonourable man is never perfect
An honourable man never tells lies
A man is not perfect unless he is always tactful
Every tactful man tells an occasional lie

Express each of these statements in set theoretic terms.

What can be deduced about the set of all perfect men?

8. Simplify

(a) $(A \cup B) \cap (C \cup A) \cap ((B \cup A) \cap C) \cup ((A \cup B)' \cap C)$;

(b) $(B \cap C') \cup (B \cap A) \cap C \cup (B \cap C)$;

(c) $(A \cup B)' \cup (A' \cap B) \cup (A \cap B')$.

1.3 NUMBERS

Number Systems

From a very early age, everyone is familiar with the sequence of numbers 1, 2, 3, 4, 5, 6, ... These numbers are called the *natural numbers* and they form the basis for all other number systems. The extension of natural numbers to include zero is usually met as a result of subtraction, which is also the motivation for the introduction of negative numbers. The resulting numbers, ... −3, −2, −1, 0, 1, 2, 3, ... are called the *integers*. Just as subtraction is used to motivate negative numbers, so division motivates the use of fractions or *rational numbers*.

Definition 1.3.1.

$N = \{1, 2, 3, \ldots\}$ is the set of *natural numbers*.

$Z = \{\ldots -3, -2, -1, 0, 1, 2, 3, \ldots\}$ is the set of *integers*.

Z^+ denotes the set of non-negative integers and Z^- the set of non-positive integers. Thus

$$Z^+ = N \cup \{0\}$$

$$Z^- = Z \backslash N$$

$Q = \{p/q \mid p \in Z,\ q \in N\}$ is the set of *rational numbers*.

If $x \in Z$, $x/1 \in Q$ and since $x/1 = x$, it follows that $Z \subset Q$. However, Q also contains elements not in Z such as $1/2$, $-1/3$, $2/3$. One may think that all numbers can be expressed as rational numbers, but this is not the case, as shown in theorem 1.3.1. Numbers which are not rational numbers form the set of *irrational numbers*, Q', and it is shown later that this set has infinite cardinality.

Theorem 1.3.1
$\sqrt{2}$ is not a rational number. This may be expressed concisely as

$$\sqrt{2} \notin Q$$

Proof The result is shown by assuming $\sqrt{2} \in Q$ and then deducing a contradiction. Let $\sqrt{2} \in Q$ then $\sqrt{2} = p/q$, where $p \in Z^+$ and $q \in N$. Without loss of generality, it can be assumed that p and q have no common divisor since, if they have, such a common divisor can be simply cancelled out.

Now, $p = \sqrt{2}q$, so $p^2 = 2q^2$. This means that p^2 is an even number and so p is an even number. Then, p^2 must be divisible by four, but since $p^2 = 2q^2$ it follows that q^2 is an even number. Thus q is an even number so p and q are both divisible by 2. This contradicts the assumption that p and q have no common divisor.

Definition 1.3.2
The set of *real numbers*, $\mathbb{R}$, consists of all the rational and irrational numbers. So, $\mathbb{R} = Q \cup Q'$.

Every real number can be represented as a point on the *real line* and every point on that line represents a real number. Notionally, the real line extends infinitely to the left and to the right (figure 1.5).

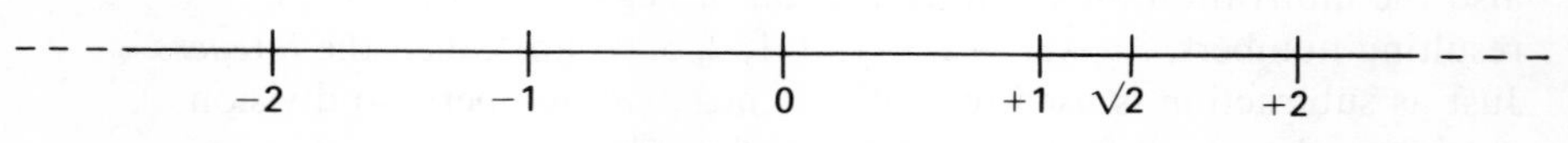

Figure 1.5 The real line

When handling real numbers, it is common to use a decimal notation. Thus, 1/2 is written as 0.5, $-14/5$ as -2.8, etc. All irrational and many rational numbers require an infinite string of digits if they are to be correctly expressed in such a notation, and thus it is customary to use only finite-length approximations. Thus, for example, 1/3 is said to be equal to 0.3333 correct to four decimal places.

Real numbers are commonly written in *scientific notation*. This is made up of a *mantissa*, consisting of a real number with just one non-zero digit before the decimal point, multiplied by 10 raised to the appropriate power, called the *exponent*. This enables all real numbers to be conveniently expressed even if they are very small or very large. If a real number is represented in scientific notation with a total of p digits in the mantissa, then the number is said to be given to p *significant digits*.

Example 1.3.1
In this example, the mantissa is correct to four significant digits.

Number	*Scientific notation*
1.414	1.414×10^0
$\sqrt{2}$	1.414×10^0
0.0032	3.200×10^{-3}
132.835	1.328×10^2
1/3	3.333×10^{-1}
c, the speed of light	2.998×10^8 m/s
e, the electronic charge	1.602×10^{-21} coulombs

Bases

Natural numbers are conventionally written to base 10. Thus, for example, 213 represents the number $2 \times 10^2 + 1 \times 10^1 + 3 \times 10^0$. The reason for using base 10 is historical, and arithmetic can be done using other bases. For example, if the base were 8, 213 would represent $2 \times 8^2 + 1 \times 8^1 + 3 \times 8^0$, that is, the equivalent of 139 base 10.

When handling numbers with a non-standard base, the base is usually written as a subscript of the number.

Example 1.3.2
Three examples of numbers written to a non-standard base are

$$3172_8 = 3 \times 8^3 + 1 \times 8^2 + 7 \times 8^1 + 2 \times 8^0 = 1658_{10}$$

$$2103_4 = 2 \times 4^3 + 1 \times 4^2 + 3 \times 4^0 \qquad = 147_{10}$$
$$1010_2 = 1 \times 2^3 + 1 \times 2^1 \qquad = 10_{10}$$

Definition 1.3.3
A *binary number* is a number written to base 2. Each digit occurring in a binary number is either 1 or 0 and such digits are known as *binary digits* or *bits*.

An *octal number* is a number written to base 8.

Although binary numbers are ideally suited for handling by digital computers, they are rather cumbersome for handling by people. For example, $1658_{10} = 11001111010_2$, that is, four decimal digits are replaced by eleven binary digits. However, since $2^3 = 8$, any octal digit may be represented by three bits, for example, $1_8 = 001_2$ and $7_8 = 111_2$, etc. Thus, any binary number may be simply converted into its shorter octal equivalent, and any octal number into its binary equivalent.

Algorithm 1.3.1

(a) *To Convert a Binary Number to an Equivalent Octal Number*

```
while  more than two bits remain in the binary number do
          compute the octal digit equivalent to the rightmost 3 bits of the
          binary number;
       delete these 3 bits from the binary number;
       place the octal digit in the leftmost position of the octal number
endwhile
if any bits remain then
       compute the octal digit equivalent to the remaining bits;
       place the octal digit in the leftmost position in the octal number;
endif;
output the octal number.
```

(b) *To Convert an Octal Number to an Equivalent Binary Number*

```
while  more digits remain in the octal number do
       obtain the 3-bit binary equivalent to the rightmost octal digit;
       delete this octal digit from the octal number;
       place the pattern of 3 bits in the leftmost position in the binary
       number
endwhile;
output the binary number.
```

Example 1.3.3

(a) 10010011100110_2

$= 2\ 2\ 3\ 4\ 6_8$

(b) $1\ 7\ 2\ 6\ 4_8$

$= 1111010110100_2$

Definition 1.2.4
If a number is written to a base greater than 10 then new symbols have to be used. For example, the *hexadecimal* method of writing numbers uses a base of 16. A, B, C, D, E, F can be used to represent the decimal numbers 10, 11, 12, 13, 14, 15, respectively.

Example 1.3.4

$$B79D_{16} = 11 \times 16^3 + 7 \times 16^2 + 9 \times 16^1 + 13 \times 16^0$$

To facilitate easy evaluation on a calculator, this expression can be rewritten in a *nested form* as $((11 \times 16 + 7) \times 16 + 9) \times 16 + 13$ which is equal to 47005. (See algorithm 3.3.3.)

Logarithms

Definition 1.3.5
The *logarithm* of $b(>0)$ to base a, $\log_a b$, is the number, x, to which a must be raised to obtain b, that is, the solution of $a^x = b$.

Example 1.3.5

$$\log_2 8 = 3, \quad \text{since} \quad 2^3 = 8$$

$$\log_2 1 = 0, \quad \text{since} \quad 2^0 = 1$$

$$\log_2 \tfrac{1}{4} = -2, \quad \text{since} \quad 2^{-2} = \tfrac{1}{4}$$

The use of logarithms enables two numbers, p and q, to be simply multiplied, since $\log_a(pq) = \log_a(p) + \log_a(q)$. The difficult task of multiplication is achieved by computing logarithms, adding the results and computing the antilogarithm of the number obtained. This process is summarised in figure 1.6.

Since $\log_a 1/q = -\log_a q$, if follows that $\log_a p/q = \log_a p - \log_a q$. Thus division can be achieved by computing logarithms, subtracting the results and computing the antilogarithm of the number obtained.

Tables of logarithms and antilogarithms to base 10 and to base e are commonly available, where e denotes a very important mathematical constant; see definition 3.3.9. Like $\sqrt{2}$ it is an irrational number but,

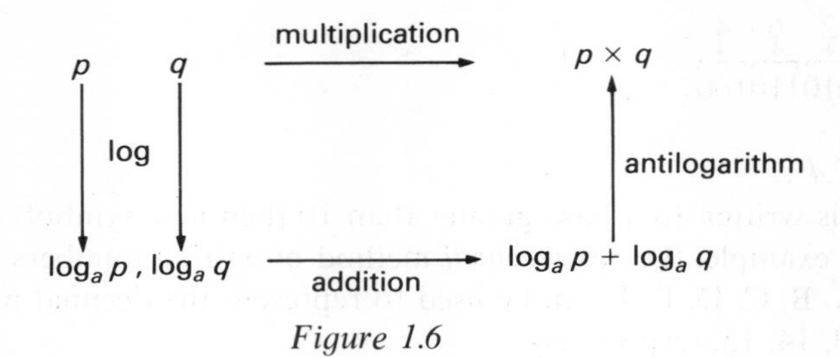

Figure 1.6

unlike $\sqrt{2}$, it does not satisfy an equation of the form $a_0 + a_1x + a_2x^2 + \ldots + a_nx^n = 0$ for any $n \in N$. (The equation of this form which $\sqrt{2}$ satisfies is $x^2 - 2 = 0$.) Logarithms to base e are sometimes called *Napierian logarithms* or *natural logarithms.*

When using the tables to base 10, it is important to be able to express the number in scientific notation.

Example 1.3.6

Using logarithms to base 10, evaluate 79.3×0.00123.

Now

$$79.3 = 7.93 \times 10^1$$

Hence

$$\log_{10}(79.3) = \log_{10}(7.93) + 1$$
$$= 0.8993 + 1$$

and

$$0.00123 = 1.23 \times 10^{-3}$$

Hence

$$\log_{10}(0.00123) = 0.0899 - 3$$

Therefore the solution is

$$\text{antilog } (1.8993 + 0.0899 - 3) = \text{antilog } (0.9892 - 2)$$
$$= 9.754 \times 10^{-2}$$
$$= 0.09754$$

The calculations in the above example are done using tables of four-figure logarithms. Six- and eight-figure tables are also published. Nowadays, of course, logarithms to base 10 and to base e are available as standard functions on most scientific calculators, so that the need for books of tables of such logarithms is not as great as it used to be. Nevertheless it is still important for the student fully to understand the manipulation of logarithms.

Representation of Numbers in a Computer

The main memory of a computer is divided into a number of units of equal size, called *words*. Each word comprises n bits, where n is usually 16, 24, 32 or 48. Provided a positive integer is not too large, it can thus be simply represented in its binary form in a single word of computer memory. The problem then is how to represent negative integers. There are a number of ways of doing this, in each of which the leftmost bit in each word is set aside as a *sign bit*. The most obvious method is the signed-magnitude representation.

Definition 1.3.6
In the *signed-magnitude* representation, the first bit of the word is used to denote the sign (0 for $+$ and 1 for $-$) and the remaining $n-1$ bits are used to represent the magnitude of the number in binary form.

Any number so stored must therefore have magnitude less than $2^{n-1}-1$.

Example 1.3.7

In a machine with 16-bit words, using a signed-magnitude representation

$+27$ is represented as 0000000000011011

-20 is represented as 1000000000010100

The largest number that can be stored is $+111111111111111_2 = 2^{15}-1$ and the smallest is $-111111111111111_2 = 1-2^{15}$.

While the signed-magnitude representation was used in several early computers, modern computers usually use either the one's-complement representation or the two's-complement representation.

Definition 1.3.7
If x is a binary number then the *complement* of x is denoted by $\bar{x}$ and is obtained by replacing each 0 in x by 1 and each 1 by 0. In the *one's-complement* representation, positive integers are represented as in the signed-magnitude representation, with a zero bit in the leftmost position,

but a negative integer is represented by the complement of the corresponding positive binary number. In the *two's-complement* representation, positive integers are represented as before, but a negative integer is represented by adding one to its one's-complement representation.

Thus, if x is a positive binary integer, the magnitude of $-x$ is represented by $\bar{x}$ in the one's-complement representation and by $\bar{x}+1$ in the two's-complement representation.

For both of these representations, the largest integer which can be represented is $2^{n-1}-1$. The smallest integer which can be represented is $1-2^{n-1}$ using one's-complement and -2^{n-1} using two's-complement.

By using a complement representation, the operation of subtraction is simplified on a computer.

Example 1.3.8

In a machine with 16-bit words, $+27$ is represented as

0000000000011011

in both one's-complement and two's-complement representations. Using one's-complement representation. -20 is represented as

1111111111101011

and, using two's-complement representation, -20 is represented as

1111111111101100

Now,

$27-20=7,$

that is

$$\begin{array}{r} 0000000000011011 \\ -0000000000010100 \\ \hline 0000000000000111 \end{array}$$

Using one's-complement, this subtraction is replaced by the addition

$$\begin{array}{r} 0000000000011011 \\ +1111111111101011 \\ \hline 10000000000000110 \end{array}$$

and the final carry is added to the rightmost bit.

Using two's-complement, the subtraction is replaced by the addition

$$\begin{array}{r} 0000000000011011 \\ +\,1111111111101100 \\ \hline 10000000000000111 \end{array}$$

and this time the final carry is simply deleted.

A full explanation of the use of one's-complement and two's-complement representations is given by Stone (1972).

Methods of representing real numbers in a computer vary widely between different manufacturers. However, they are all based on first representing the number in a *floating-point format*. The scientific notation met in example 1.3.1 is an example of such a format. If both the mantissa and the exponent are represented as signed binary numbers and the leading digit of the mantissa is non-zero, then this representation is known as a *normalised binary floating-point representation*. The exact assumed position of the binary point may vary from machine to machine, as may methods for representing the mantissa and the exponent.

Example 1.3.9

Consider the case where a real number is stored as two 16-bit words. The first word represents the mantissa and the second the exponent, both stored in two's-complement. The binary point of the mantissa is assumed to be between the first and second bits of the part of the word representing the magnitude of the mantissa

Mantissa		*Exponent*	
sign	magnitude	sign	magnitude
0	100001001100001	1	111111111000111

These two words represent the binary number $1.00001001100001_2 \times 2^{-57}$.

Clearly, whatever representation is used for real numbers in the computer, only a finite set of them can be represented. In the above example, the largest exponent that can be stored is $2^{15}-1$. Thus all the real numbers represented using this technique must lie in the range $\pm 2^m$ where $m=2^{15}$. Furthermore, the finite set of real numbers that can be represented using any particular technique corresponds to a set of non-equidistant points on the real line. Hence, in general, any given real number within the permitted range is only represented by a floating-point number close to it. The accuracy to which real numbers may be represented is determined by the number of bits used for the mantissa.

Inequalities

Definition 1.3.8
If $a,b \in \mathbb{R}$, then

$a<b$ iff a is less than b;

$a \leqslant b$ iff a is less than or equal to b;

$a>b$ iff a is greater than b;

$a \geqslant b$ iff a is greater than or equal to b.

Clearly, $a<b$ iff $b>a$ and $a<b$ implies that a lies to the left of b on the real line. The results of the following theorem are also obvious and the reader is advised to check them through. If $a<b$ and $b<c$, this is sometimes written $a<b<c$.

Theorem 1.3.2
If a, b, $c \in \mathbb{R}$ then

(i) $a<b \wedge b<c \supset a<c$;

(ii) $a<b \supset a+c<b+c$;

(iii) $a<b \wedge c>0 \supset ca<cb$;

(iv) $a<b \wedge c<0 \supset ca>cb$.

Manipulating inequalities requires some care, the most common error arising from multiplication of an inequality by a negative quantity; the inequality must then be reversed according to theorem 1.3.2.iv.

Example 1.3.10

Simplify $2x-3<4x+7$.
In this case

$$2x-3<4x+7 \supset 2x<4x+10 \quad \text{(by theorem 1.3.2.ii)}$$
$$\supset -2x<10 \quad \text{(by theorem 1.3.2.ii)}$$
$$\supset x>-5 \quad \text{(by theorem 1.3.2.iv)}$$

Definition 1.3.9
If $x \in \mathbb{R}$, the *absolute value* of x, denoted by $|x|$, is obtained from x by disregarding the sign. Thus, for example, $|2|=2$ and $|-2|=2$. Alternatively, $|x|$ can be defined by

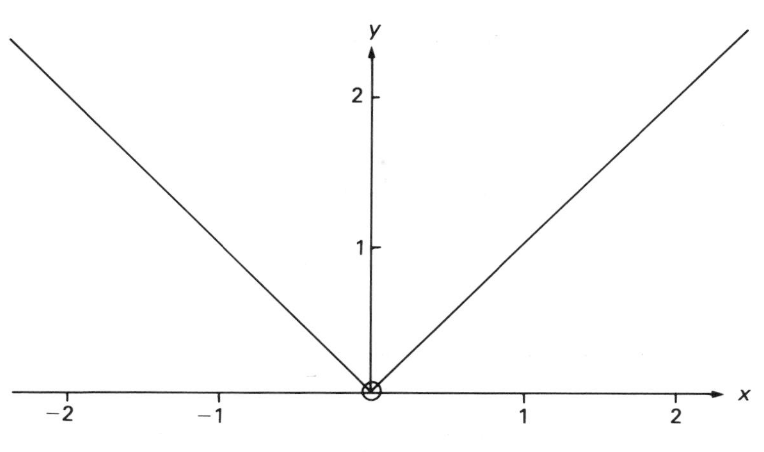

Figure 1.7

$$|x| = \begin{cases} x \text{ if } x \geqslant 0 \\ -x \text{ if } x < 0 \end{cases}$$

The graph of $y = |x|$ is given in Figure 1.7.

Definition 1.3.10
If $a, b \in \mathbb{R}$, then

$$[a, b] = \{x | x \in \mathbb{R} \text{ and } a \leqslant x \leqslant b\}$$

$$[a, b) = \{x | x \in \mathbb{R} \text{ and } a \leqslant x < b\}$$

$$(a, b] = \{x | x \in \mathbb{R} \text{ and } a < x \leqslant b\}$$

$$(a, b) = \{x | x \in \mathbb{R} \text{ and } a < x < b\}$$

The subsets are intervals in the real line and can be represented diagrammatically as in figure 1.8. To emphasise that a, b are included in the interval, the points at a and b are shaded. If an end point is not included in the interval, this is made clear by using an unshaded circle. If both end points are included in an interval the interval is said to be *closed* and if neither is included, the interval is said to be *open*.

Example 1.3.11

Express $|x-5| \leqslant 12$ in interval notation.

The relation $|x-5| \leqslant 12$ means that the distance of x from 5 (ignoring direction) must not exceed 12.

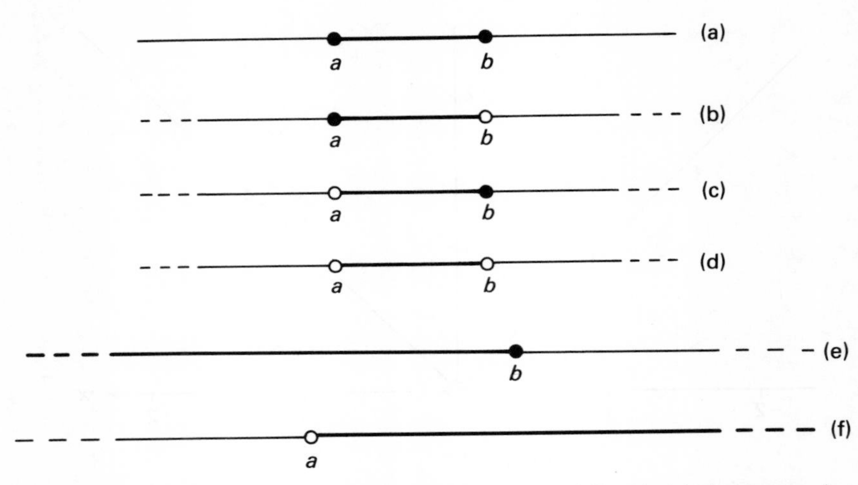

Figure 1.8 (a) The interval $[a, b]$; (b) the interval $[a, b)$; (c) the interval $(a, b]$; (d) the interval (a, b); (e) the interval $(-\infty, b]$; (f) the interval (a, ∞)

At one extreme x satisfies $x-5=12$, that is, $x=17$; at the other, $x-5=-12$, that is, $x=-7$. Thus, $|x-5|\leqslant 12$ implies $x\in[-7, 17]$. Figure 1.9 illustrates the result.

Definition 1.3.11
If A is a subset of the reals, $\mathbb{R}$, then

(a) $u\in\mathbb{R}$ is an *upper bound* of A iff $\forall a\in A, a\leqslant u$;

(b) $l\in\mathbb{R}$ is a *lower bound* of A iff $\forall a\in A, l\leqslant a$;

(c) $u_l\in\mathbb{R}$ is the *least upper bound* of A iff u_l is an upper bound of A and $u_l\leqslant u$ for every upper bound, u, of A;

(d) $l_g\in\mathbb{R}$ is the *greatest lower bound* of A iff l_g is a lower bound of A and $l\leqslant l_g$ for every lower bound, l, of A.

Note that $[a, b]$, (a, b), $(a, b]$ or $[a, b)$ all have a greatest lower bound (glb) of a and a least upper bound (lub) of b.

Countability

A set is *countable* if its elements can be listed such that there is a first element, a second element, a third element, etc. Clearly, any finite set is

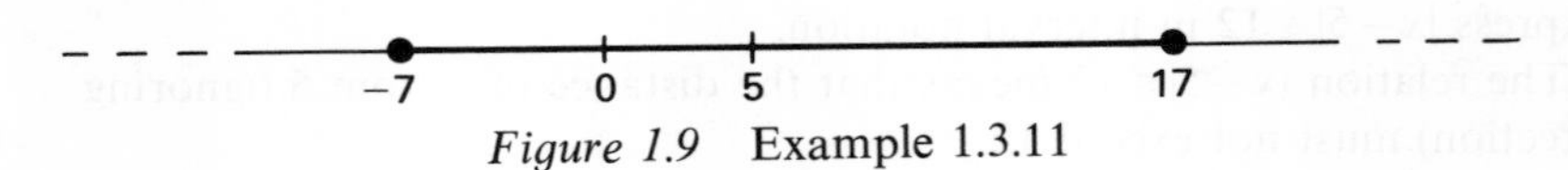

Figure 1.9 Example 1.3.11

countable, since the elements can be listed one after the other and, once done, that can be taken as the ordering. The problem arises when infinite sets are considered. The set, N, is countable since the elements can be listed 1, 2, 3, 4, ...Hence, it is valid to talk about the ith element in N. A surprising result is that not all sets are countable, for example, $\mathbb{R}$ is not countable and hence one cannot talk about 'the ith real number'.

Theorem 1.3.3
Q is countable.

Proof The elements of Q can be written in a table

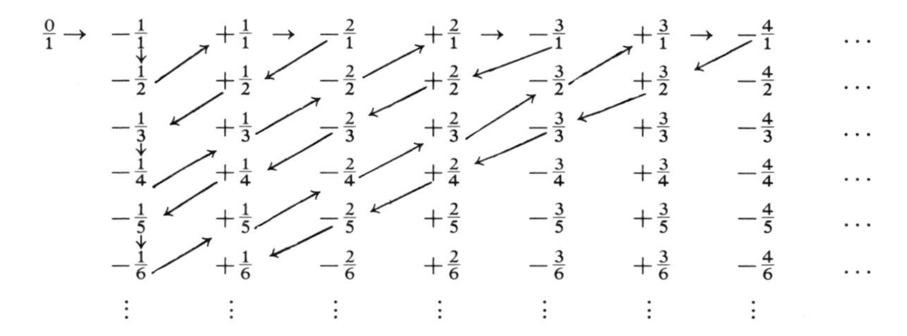

These elements can be ordered as indicated by the arrows. Every rational number will eventually occur in this list and hence the set is countable.

Theorem 1.3.4
$\mathbb{R}$ is uncountable.

Proof The proof relies on a technique known as Cantor's diagonalisation. Consider the real numbers contained in $[0, 1)$. If $\mathbb{R}$ were countable then this subset would be countable. Hence they could all be listed

$$\begin{aligned}
r_1 &= 0.a_{11}a_{12}a_{13}a_{14}a_{15}a_{16}\ldots \\
r_2 &= 0.a_{21}a_{22}a_{23}a_{24}a_{25}a_{26}\ldots \\
r_3 &= 0.a_{31}a_{32}a_{33}a_{34}a_{35}a_{36}\ldots \\
r_4 &= 0.a_{41}a_{42}a_{43}a_{44}a_{45}a_{46}\ldots \\
r_5 &= 0.a_{51}a_{52}a_{53}a_{54}a_{55}a_{56}\ldots \\
r_6 &= 0.a_{61}a_{62}a_{63}a_{64}a_{65}a_{66}\ldots \\
&\vdots
\end{aligned}$$

where a_{ij} is the jth digit following the decimal point in the ith real number listed.

Now, consider the real number constructed from the first digit of r_1, the second digit of r_2, etc., as indicated by the arrows. Let this real number be r. Thus $r=0.a_{11}a_{22}a_{33}a_{44}a_{55}a_{66}\ldots$ From r, construct a new real, $\hat{r}\in[0, 1)$, where $\hat{r}=0.b_{11}b_{22}b_{33}b_{44}b_{55}b_{66}\ldots$ and

$$b_{ii}=a_{ii}+1 \text{ if } a_{ii}<9$$
$$=0 \qquad \text{if } a_{ii}=9$$

Here $\hat{r}$ differs from each r_i in the list in at least the ith digit. Hence $\hat{r}$ does not occur in the list. Since the list was meant to contain all the real numbers in $[0, 1)$, this leads to a contradiction. Hence IR cannot be countable.

Mathematical Induction

A common task in computing science and mathematics is to prove that a result concerning an integer n holds for all possible positive values of n. For example, prove that 8^n-3^n is divisible by 5 for $n=1, 2, 3, \ldots$. This one result describes a countably infinite number of results, that is, 8^1-3^1 is divisible by 5, 8^2-3^2 is divisible by 5, 8^3-3^3 is divisible by 5, etc. It is clearly impossible to prove the result by considering each of the infinite number of cases, and thus a completely different, and very elegant approach, known as mathematical induction, is used.

Say there are a countable number of theorems, $\{T_n|n\in N\}$. If (1) T_1 is true and (2) it can be shown that T_k is true implies T_{k+1} is true, for all $k\in N$, then it follows that T_n is true for all $n\in N$ by *mathematical induction.*

The proof that this method is valid is shown by assuming the method is not valid and deducing a contradiction. Thus, assume $\{T_n|n\in N\}$ is a set of theorems which are not all true but such that (1) T_1 is true and (2) T_k is true implies T_{k+1} is true, for all $k\in N$. Since not all theorems are true, there must exist a least $j>0$ such that T_j is false. If $j=1$ this contradicts the fact that T_1 is true. Thus $j>1$. Now, T_{j-1} must be true but if T_{j-1} is true then T_j is true by (2), resulting in the required contradiction.

It is important that a proof by mathematical induction be correctly presented. A suitable layout for a proof by induction that T_n is true for all $n\in N$ is described as follows.

The proof is by mathematical induction	— The first line
The result holds for $n=1$ because...	Proof that T_1 is true
Assume that the result holds for $n=k$, that is . . .	Assume that T_k is true. This is sometimes called the *induction hypothesis*
Then . . . and so the result holds for $n=k+1$.	Proof that T_{k+1} is true then follows
Hence the result holds for all $n \in N$ by induction	The last line

Example 1.3.12

Prove that $8^n - 3^n$ is divisible by 5 for all $n \in N$.

The proof is by mathematical induction. The result holds for $n=1$ because

$$8^1 - 3^1 = 5$$

and this is clearly a multiple of 5.

Assume that the result holds for $n=k$, that is, that $8^k - 3^k$ is divisible by 5. Then

$$\begin{aligned} 8^{k+1} - 3^{k+1} &= 8 \times 8^k - 3 \times 3^k \\ &= 3(8^k - 3^k) + 5 \times 8^k \end{aligned}$$

Now, $8^k - 3^k$ is divisible by 5 by the induction hypothesis. Therefore $3(8^k - 3^k)$ is divisible by 5.

Clearly, 5×8^k is also divisible by 5, so $3(8^k - 3^k) + 5 \times 8^k = 8^{k+1} - 3^{k+1}$ is divisible by 5. Thus the result holds for $n = \mathrm{k}+1$.

Hence the result holds for all $n \in N$, by induction.

In section 3.2, a detailed study of *series* is presented. A finite series can be expressed as the sum, $a_1 + a_2 + \ldots + a_n$, of a sequence of terms a_1, a_2, ... a_n. For example, $1+2+\ldots+n$ is a series, as is $1^2+2^2+\ldots+n^2$. A useful notation when studying such series is the *sigma notation*, whereby

$a_1 + a_2 + \ldots + a_n$ is represented by

$$\sum_{i=1}^{n} a_i.$$

The variable i ranges from 1 to n, generating $a_1, a_2, \ldots, a_n$, which are summed to obtain the series.

Example 1.3.13

(1) If

$$S_n = \sum_{i=1}^{n} i = 1 + 2 + \ldots + n$$

then $S_1 = 1$, $S_2 = 1 + 2 = 3$, $S_3 = 1 + 2 + 3 = 6$, etc. The question arises: given n, what is S_n? It would be ideal if some simple formula could be found to enable one to compute the answer quickly. For this simple series, such a formula does indeed exist since

$$S_n = \frac{n}{2}(n+1)$$

for all $n \in N$.

The proof is by mathematical induction. The result holds for $n=1$ since

$$S_1 = \tfrac{1}{2}(1+1) = 1$$

Assume the result holds for $n=k$, that is

$$S_k = \frac{k}{2}(k+1)$$

Then

$$\begin{aligned} S_{k+1} &= S_k + (k+1) \\ &= \frac{k}{2}(k+1) + (k+1) \qquad \text{(by the induction hypothesis)} \\ &= \frac{(k+1)}{2}(k+2) \end{aligned}$$

Thus the result holds for $n=k+1$.

Hence the result holds for all $n \in N$, by induction.

(2)

$$\sum_{i=1}^{n} i^2 = 1^2 + 2^2 + \ldots + n^2$$

The reader is asked to provide an inductive proof that this series is equal to $n(n+1)(2n+1)/6$.

(3) In

$$\sum_{i=1}^{n} v_i w_i,$$

as i ranges from 1 through to n, the terms $v_1w_1, v_2w_2, \ldots, v_nw_n$ are generated. Thus

$$\sum_{i=1}^{n} v_i w_i = v_1 w_1 + v_2 w_2 + \ldots + v_n w_n$$

(4) In sigma notation, it is not essential to use i to indicate the variable nor is it necessary to always count from one through to n. Thus

$$\sum_{i=1}^{n} a_i = \sum_{j=1}^{n} a_j = a_1 + a_2 + \ldots + a_n$$

$$\sum_{i=1}^{10} a_i = a_1 + a_2 + \ldots + a_{10}$$

and

$$\sum_{j=3}^{10} a_j = a_3 + a_4 + \ldots + a_{10}$$

Exercise 1.3

1. Prove that $\log_2 3$ is an irrational number.

2. Compute the decimal equivalents of each of the following

(a) 3157_8;
(b) 1011101110_2;
(c) $12A_{16}$.

3. Construct an algorithm for transforming a binary number to a hexadecimal number and vice versa.

4. Use logarithms to base 10 to evaluate

(a) 7134×0.00917;
(b) $1.6298 \div 0.4718$

5. Using the representation of real numbers described in example 1.3.9, what is the accuracy that can be achieved?

6. Simplify

(a) $x-12<2x+3$;
(b) $18>3|x|$;
(c) $3\leqslant|x|+2x$.

7. Express the results of exercise 1.3.6 in interval notation.

8. If $A=[-15, 8]$, $B=(6, 14]$, $C=[-2, 1)$, express the following sets as intervals or unions of intervals

(a) $(A\cup B)'$;
(b) $(B\backslash C)'\cap A$;
(c) $(A\backslash B)\Delta C$.

9. Show that $|x+y|\leqslant|x|+|y|$ and $|xy|=|x|\cdot|y|$.

10. Prove that Z is countably infinite.

11. Prove that Q' is uncountably infinite.

12. Prove that (a) 5 divides 7^n-2^n, for all $n\in N$; and (b) 7 divides $3^{2n+1}+4^{2n+1}$, for all $n\in N$.

13. Prove by induction that

$$S_n=\sum_{i=1}^{n} i^2$$

is equal to $n(n+1)(2n+1)/6$ for all $n\in N$.

14. Let

$$C_n=\sum_{i=1}^{n} i^3$$

Evaluate C_1, C_2, C_3 and C_4. Hence postulate an expression in n for C_n. Prove your answer correct by using an induction argument. (Hint: see if you can see a connection between C_n and $\sum_{i=1}^{n} i$.)

1.4 COMPLEX NUMBERS

Polynomial Equations

Definition 1.4.1
A *polynomial of degree n in x* is an expression of the form $a_0+a_1x+a_2x^2 + \ldots + a_nx^n$, where $a_0, a_1, \ldots, a_n \in \mathbb{R}$ and $a_n \neq 0$. Equations of the form

$$P_n(x)=0$$

where $P_n(x)$ is a polynomial of degree n in x, commonly arise in mathematics, and there are various ways of finding values of x to satisfy such equations. Such values are called *roots* of the equation $P_n(x)=0$.

A polynomial equation of degree two is known as a *quadratic* equation. The roots x_1, x_2 of a quadratic equation

$$ax^2+bx+c=0$$

are given by the well-known formulae

$$x_1=\frac{-b+\sqrt{(b^2-4ac)}}{2a}, \quad x_2=\frac{-b-\sqrt{(b^2-4ac)}}{2a}$$

There are also formulae for the roots of *cubic* [degree 3] and *quartic* [degree 4] equations but it has been shown that, in general, no such formulae exist for the roots of higher-degree polynomial equations. It is then necessary to use numerical techniques.

Example 1.4.1

Consider the three quadratics

$$q_1(x) \equiv x^2-4x+3$$

$$q_2(x) \equiv x^2-4x+4$$

$$q_3(x) \equiv x^2-4x+5$$

Sketches of these graphs (figure 1.10) show that $q_1(x)=0$ has two distinct solutions, $q_2(x)=0$ has only one and $q_3(x)=0$ appears to have none.

If the roots of these three quadratics are calculated using the above formulae, $q_1(x)=0$ is found to have roots equal to 1 and 3, as expected; $q_2(x)=0$ is found to have two roots, both equal to two; this indicates that $q_2(x)$ just touches the x axis at $x=2$; $q_3(x)=0$ is found to have roots of $2+\sqrt{(-1)}$ and $2-\sqrt{(-1)}$, which are not real numbers since $\sqrt{(-1)}$ is not a real number.

Mathematicians have found it useful to allow the number $\sqrt{(-1)}$,

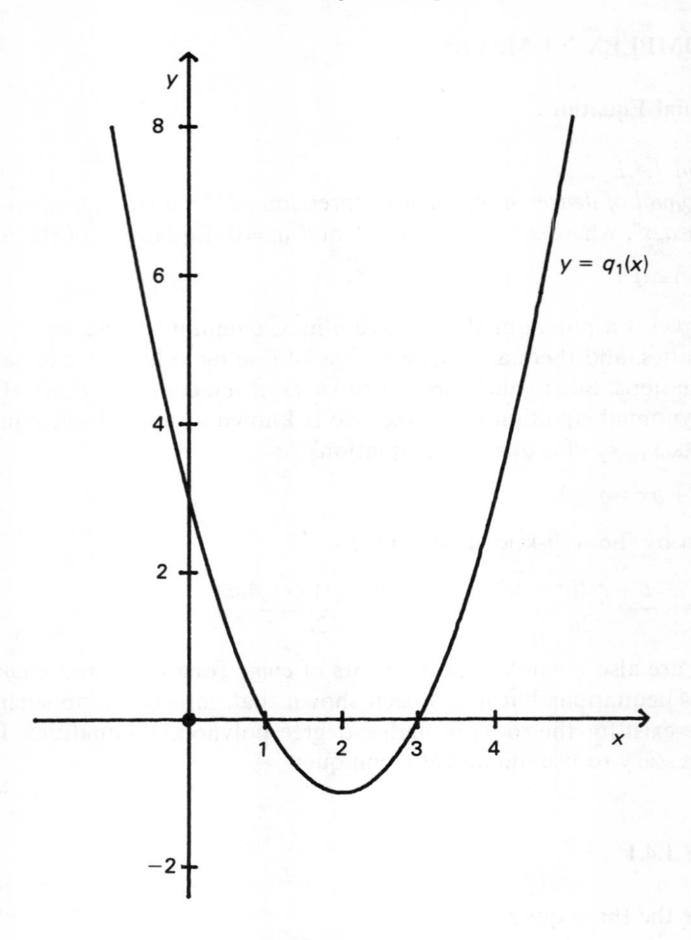

Figure 1.10a

denoting it by i. Hence, $i^2 = -1$. Since $\sqrt{(-1)}$ is clearly not a real number, it is called an *imaginary number*.

Definition 1.4.2
An expression of the form $a + ib$, where $a, b \in \mathbb{R}$, is called a *complex number*. The complex number consists of a *real part*, the real number, a, and an *imaginary part*, ib.

Thus, $2 + \sqrt{(-1)}$ is the complex number $2 + i$ and $2 - \sqrt{(-1)}$ the complex number $2 + (-1)i = 2 - i$. Thus the quadratic equation $q_3(x) = 0$ does have two roots, but they are both complex. If complex numbers are allowed, every quadratic equation has two roots and, in fact, every polynomial equation of degree n has n roots.

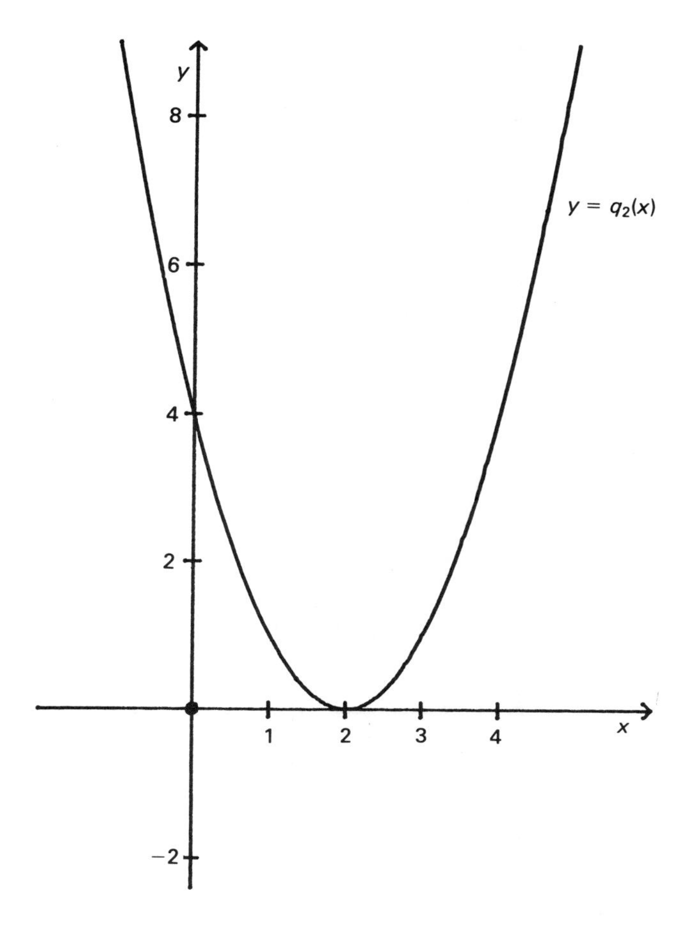

Figure 1.10b

Example 1.4.2

Find the three roots of the cubic equation $x^3-8=0$.

One root is clearly 2 since $2^3-8=0$.

Since $(x-2)$ is a factor of x^3-8, x^3-8 can be divided by $x-2$, giving $x^3-8=(x-2)(x^2+2x+4)$.

Now $x^2+2x+4=0$ has no real roots, since $b^2-4ac=-12<0$. However, using the expressions $[-b+\sqrt{(b^2-4ac)}]/2a$ and $[-b-\sqrt{(b^2-4ac)}]/2a$ the roots are seen to be $-1+i\sqrt{3}$ and $-1-i\sqrt{3}$.

Thus the three roots of $x^3-8=0$ are 2, $-1+i\sqrt{3}$ and $-1-i\sqrt{3}$.

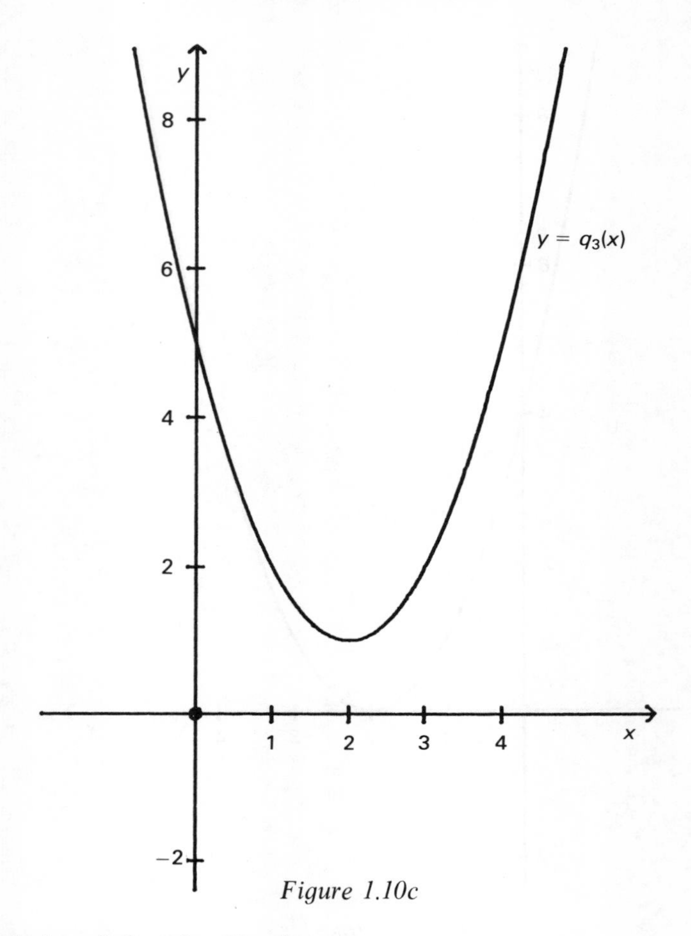

Figure 1.10c

Manipulation of Complex Numbers

Definition 1.4.3
The *set of all complex numbers* is denoted by $\mathscr{C}$. Thus, $\mathscr{C}=\{a+ib|a, b\in\mathbb{R}\}$.

From this definition, it is apparent that $\mathbb{R} \subsetneqq \mathscr{C}$.

If $\alpha\in \mathscr{C}$ is a complex number, then the real part of α is denoted by $\mathrm{Re}(\alpha)$ and the imaginary part by $\mathrm{Im}(\alpha)$.

If $\alpha=a+ib$ and $\beta=c+id$ are two complex numbers, then

$$\alpha+\beta=(a+c)+\mathrm{i}(b+d)$$

$$\alpha-\beta=(a-c)+\mathrm{i}(b-d)$$

$$\alpha\times\beta=(a+ib)\times(c+id)$$

$$=(ac-bd)+\mathrm{i}(ad+bc)$$

Example 1.4.3

If $\alpha=2+3i$ and $\beta=4-5i$, then

$$\alpha+\beta=6-2i$$

$$\alpha-\beta=-2+8i$$

$$\alpha\times\beta=23+2i$$

Definition 1.4.4
If $\alpha=a+ib$, then $\bar{\alpha}$, the *complex conjugate* of α, is equal to $a-ib$.

Theorem 1.4.1
For any complex numbers, α and β

$$\overline{\alpha+\beta}=\bar{\alpha}+\bar{\beta}$$

and

$$\overline{\alpha\times\beta}=\bar{\alpha}\times\bar{\beta}$$

Proof Let $\alpha=a+ib$ and $\beta=c+id$, then $\alpha+\beta=(a+c)+i(b+d)$ and so $\overline{\alpha+\beta}=(a+c)-i(b+d)$, but $\bar{\alpha}+\bar{\beta}=(a-ib)+(c-id)=(a+c)-i(b+d)$. Hence $\overline{\alpha+\beta}=\bar{\alpha}+\bar{\beta}$.

The result for multiplication is proved similarly and the proof is left as an exercise for the reader.

Definition 1.4.5
If $\alpha=a+ib$, then the *absolute value* or *modulus* of α, $|\alpha|$, is a real number defined by

$$|\alpha|=\sqrt{(a^2+b^2)}=\sqrt{\{[\mathrm{Re}(\alpha)]^2+[\mathrm{Im}(\alpha)]^2\}}$$

The square of this value, a^2+b^2, is equal to $\alpha\times\bar{\alpha}$.

Clearly, $|\alpha|=0$ if and only if both a and b are equal to zero and, for all other complex numbers, $|\alpha|>0$.

Theorem 1.4.2
For any complex numbers, α and β

$$|\alpha\times\beta|=|\alpha|\times|\beta|$$

Proof

$$\begin{aligned}|\alpha\times\beta|^2&=(\alpha\times\beta)\times(\overline{\alpha\times\beta})=\alpha\times\beta\times\bar{\alpha}\times\bar{\beta}\\&=\alpha\times\bar{\alpha}\times\beta\times\bar{\beta}=|\alpha|^2\times|\beta|^2\end{aligned}$$

Thus

$$|\alpha \times \beta| = |\alpha| \times |\beta|$$

Example 1.4.4

Definition 1.4.5 leads to the following

$$|\mathrm{i}| = |-\mathrm{i}| = 1$$

$$|2+\mathrm{i}| = |2-\mathrm{i}| = \sqrt{5}$$

$$|\cos\theta + \mathrm{i}\sin\theta| = \sqrt{(\cos^2\theta + \sin^2\theta)} = 1$$

If one complex number, α, is divided by another, β, the expression can be simplified by multiplying the numerator and the denominator by the complex conjugate of β

$$\frac{\alpha}{\beta} = \frac{\alpha \times \bar{\beta}}{\beta \times \bar{\beta}} = \frac{\alpha \times \bar{\beta}}{|\beta|^2}$$

Example 1.4.5

Express $(1+2\mathrm{i})/(2+3\mathrm{i})$ in the form $a+ib$, $a, b \in \mathbb{R}$.

$$\frac{1+2\mathrm{i}}{2+3\mathrm{i}} = \frac{(1+2\mathrm{i}) \times (2-3\mathrm{i})}{(2+3\mathrm{i}) \times (2-3\mathrm{i})} = \frac{8+\mathrm{i}}{13} = \frac{8}{13} + \frac{\mathrm{i}}{13}$$

Theorem 1.4.3
If $f(x) = a_0 + a_1x + \ldots + a_nx^n$ is a polynomial with real coefficients, then $f(\bar{\alpha}) = \overline{f(\alpha)}$ for any complex number α.

Proof Consider $f(\alpha) = a_0 + a_1\alpha + \ldots + a_n\alpha^n$.

Since $\overline{\alpha+\beta} = \bar{\alpha} + \bar{\beta}$, it follows that $\overline{f(\alpha)} = \bar{a}_0 + \overline{a_1\alpha} + \ldots + \overline{a_n\alpha^n}$.

Similarly, by repeated application of the rule $\overline{\alpha \times \beta} = \bar{\alpha} \times \bar{\beta}$, $\overline{f(\alpha)} = \bar{a}_0 + \bar{a}_1\bar{\alpha} \ldots + \bar{a}_n(\bar{\alpha})^n$. Each $a_i \in \mathbb{R}$ and so $\bar{a}_i = a_i$. Thus, $\overline{f(\alpha)} = a_0 + a_1\bar{\alpha} + \ldots + a_n(\bar{\alpha})^n = f(\bar{\alpha})$.

Now, suppose α is a root of the equation $f(x) = 0$, that is, $f(\alpha) = 0$. Then, $\overline{f(\alpha)} = \bar{0} = 0$. So, by theorem 1.4.3, $f(\bar{\alpha}) = 0$, that is, $\bar{\alpha}$ is also a root of the equation $f(x) = 0$. This proves the following important result.

Corollary 1.4.4
If $\alpha \in \mathscr{C}$ is a root of the polynomial equation $f(x) = 0$, then the complex conjugate of α is also a root of the equation.

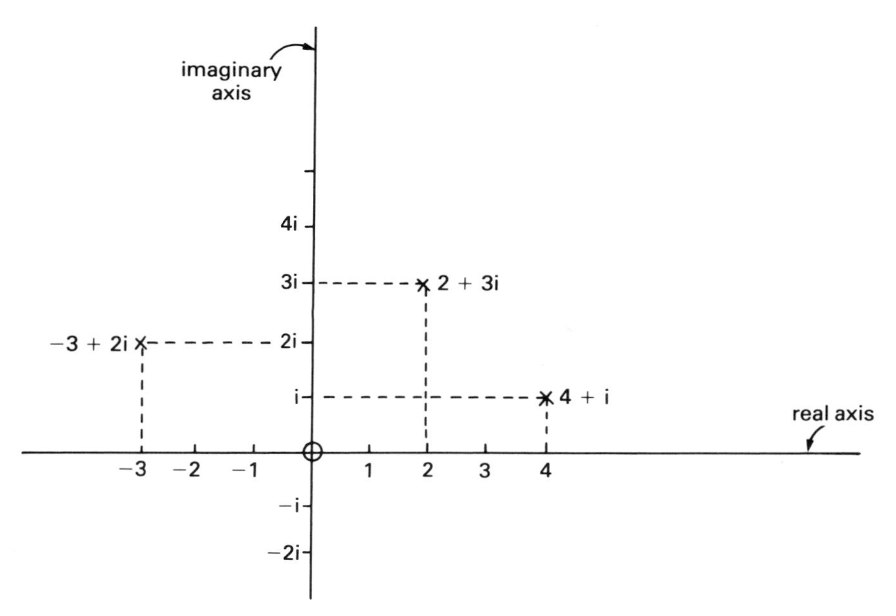

Figure 1.11 The Argand diagram

The Argand Diagram

The complex number $a+ib$ can be represented as an ordered pair (a, b) of real numbers. This means that a complex number can be represented as a point in a plane. The plane is known as the *Argand plane* or *Argand diagram* (figure 1.11).

If P is a point in the Argand diagram representing a complex number α, then the directed line (or vector) $\overrightarrow{\mathrm{OP}}$ can also be used to represent α. The summation of two complex numbers can be particularly well represented geometrically as illustrated in figure 1.12. If $\overrightarrow{\mathrm{OP}}$ represents $\alpha=a+ib$ and $\overrightarrow{\mathrm{OQ}}$ represents $\beta=c+id$ then $\overrightarrow{\mathrm{OR}}$ represents $\alpha+\beta=(a+c)+\mathrm{i}(b+d)$.

The point $\mathrm{P}=(a, b)$ representing the complex number $a+ib$ can also be defined using *polar coordinates*. A point in the plane is defined using polar coordinates by giving firstly the length, r, of $\overrightarrow{\mathrm{OP}}$ and secondly, θ, the inclination of $\overrightarrow{\mathrm{OP}}$ with respect to the real axis, as in figure 1.13.

If P represents the complex number $\alpha=a+ib$, then $r=|\alpha|$ and $\theta=\tan^{-1}(b/a)$. The angle θ is called the *argument* of α, written $\arg(\alpha)$, where $\arg(\alpha)$ is such that $0\leqslant\arg(\alpha)<2\pi$. (Here, and throughout the book, angles are measured in radians.)

Alternatively, if P is given in polar coordinates as (r, θ) then P represents the complex number $\mathrm{r}\cos\theta+\mathrm{ir}\sin\theta=r(\cos\theta+\mathrm{i}\sin\theta)$. The

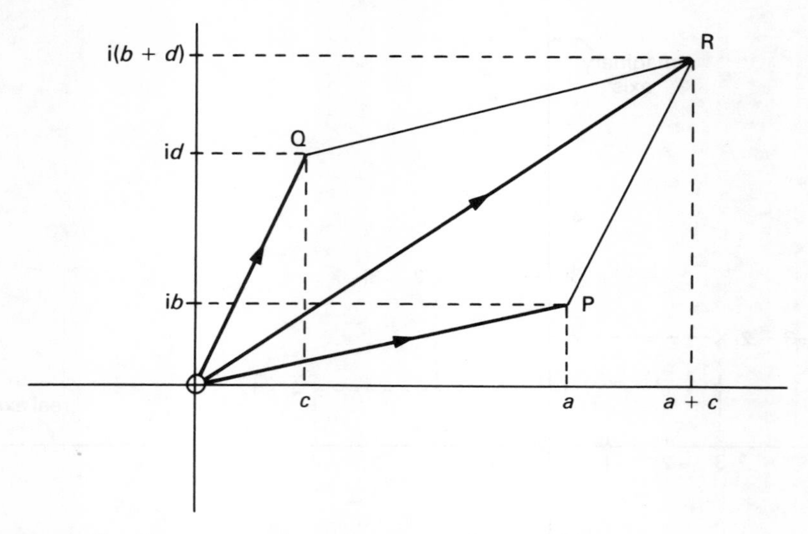

Figure 1.12 The summation of two complex numbers

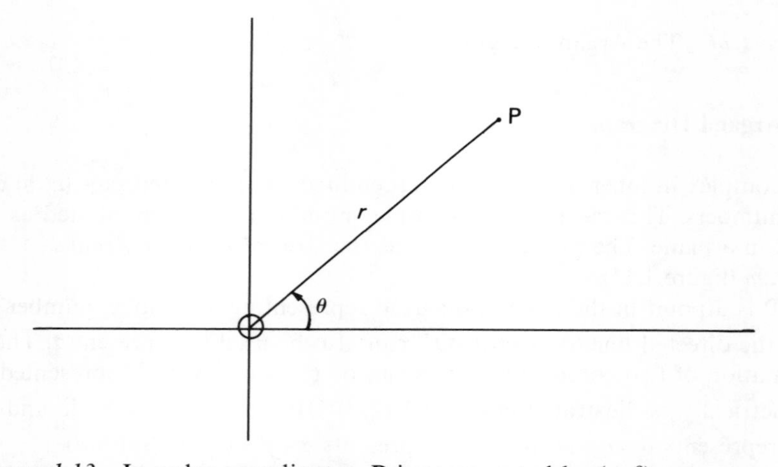

Figure 1.13 In polar coordinates, P is represented by (r, θ)

expression $\cos\theta + \mathrm{i}\sin\theta$ occurs so commonly in this context that it is often written simply as cis θ. Similarly, sic $\theta = \sin\theta + \mathrm{i}\cos\theta$.

Example 1.4.6

The complex number $\alpha = 3 + 4\mathrm{i}$ is represented by $\overrightarrow{\mathrm{OP}}$ in figure 1.14.

There is no need for a vector representing a complex number to be

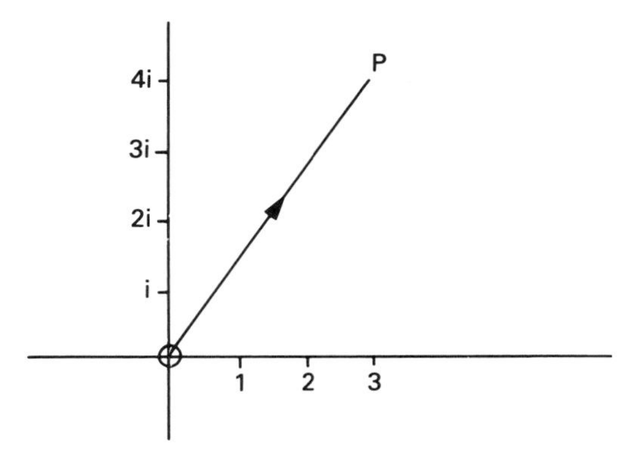

Figure 1.14 Example 1.4.6: $|\alpha| =$ length of $\overrightarrow{OP} = 5$; $\arg(\alpha) = \tan^{-1} 4/3$; in polar coordinates, P is $(5, \tan^{-1} 4/3)$

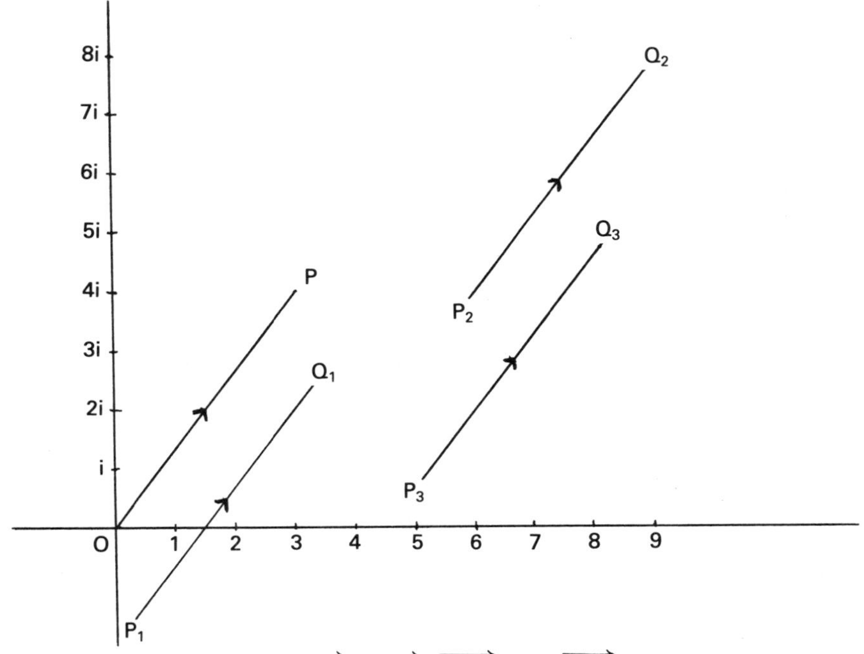

Figure 1.15 The vectors $\overrightarrow{OP}$, $\overrightarrow{P_1Q_1}$, $\overrightarrow{P_2Q_2}$ and $\overrightarrow{P_3Q_3}$ all represent the same complex number, $3+4i$

anchored at the origin. An arbitrary vector $\overrightarrow{PQ}$ can be regarded as representing the complex number α, where $|\alpha|$ is the length of $\overrightarrow{PQ}$ and $\arg(\alpha)$ is the inclination of $\overrightarrow{PQ}$ (figure 1.15).

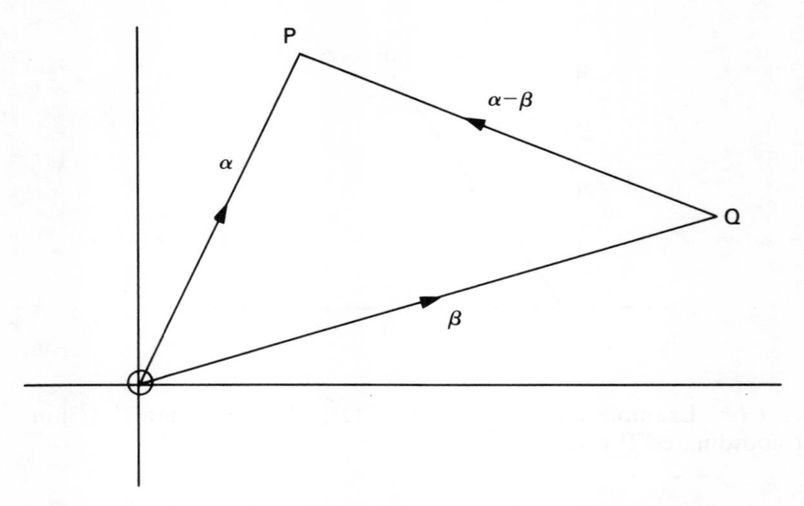

Figure 1.16 The subtraction of two complex numbers

The subtraction of two complex numbers can now also be simply represented geometrically, as in figure 1.16. If $\overrightarrow{OP}$ represents $\alpha = a + ib$ and $\overrightarrow{OQ}$ represents $\beta = c + id$ then $\overrightarrow{QP}$ represents $\alpha - \beta = (a-c) + i(b-d)$.

More generally, it can be easily shown that if $\overrightarrow{P_0P_1}$ represents a complex number α_1, $\overrightarrow{P_1P_2}$ represents α_2, $\overrightarrow{P_2P_3}$ represents $\alpha_3, \ldots$, $\overrightarrow{P_{n-1}P_n}$ represents α_n, then $\overrightarrow{P_0P_n}$ represents $\alpha_1 + \alpha_2 + \ldots + \alpha_n$ (figure 1.17).

In particular, if $\overrightarrow{P_0P_1}$ represents α and $\overrightarrow{P_1P_2}$ represents β then $\overrightarrow{P_0P_2}$ represents $\alpha + \beta$ (figure 1.18). An immediate result from this observation is as follows.

Theorem 1.4.5
For any complex numbers, α and β, $|\alpha + \beta| \leqslant |\alpha| + |\beta|$.

If $\alpha = r_1 \text{ cis } \theta_1$ and $\beta = r_2 \text{ cis } \theta_2$, then

$$\begin{aligned}\alpha \times \beta &= r_1 r_2 \text{ cis } \theta_1 \text{ cis } \theta_2 \\ &= r_1 r_2 (\cos \theta_1 + i \sin \theta_1) \times (\cos \theta_2 + i \sin \theta_2) \\ &= r_1 r_2 [(\cos \theta_1 \cos \theta_2 - \sin \theta_1 \sin \theta_2) + i(\sin \theta_1 \cos \theta_2 + \cos \theta_1 \sin \theta_2) \\ &= r_1 r_2 [\cos(\theta_1 + \theta_2) + i \sin(\theta_1 + \theta_2)]\end{aligned}$$

Thus, if the complex number α is represented by the point $P = (r_1, \theta_1)$, in polar coordinates, and β is represented by $Q = (r_2, \theta_2)$, then the product $\alpha \times \beta$ is represented by $R = (r_1 r_2, \theta_1 + \theta_2)$, as in figure 1.19.

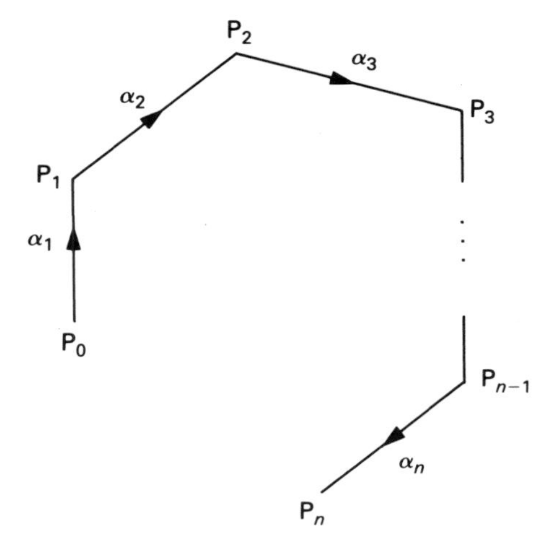

Figure 1.17

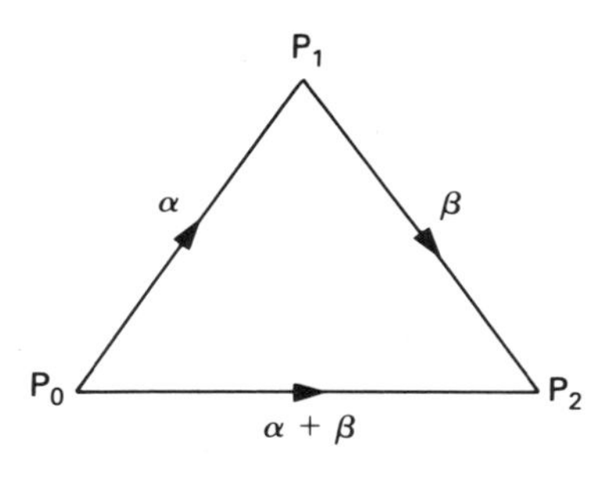

Figure 1.18

Example 1.4.7

The product $(3+2i)\times i$ is represented in figure 1.20.

The length of $\overrightarrow{OR}$ is equal to the length of $\overrightarrow{OP}$ since the length of $\overrightarrow{OQ}$ is 1. The inclination of $\overrightarrow{OR}$ is $\pi/2$ greater than that of $\overrightarrow{OP}$ since the inclination of $\overrightarrow{OQ}$ is $\pi/2$.

Euler's Formula

In chapter 3 it is shown that e^x, sin x and cos x can be written as infinite series

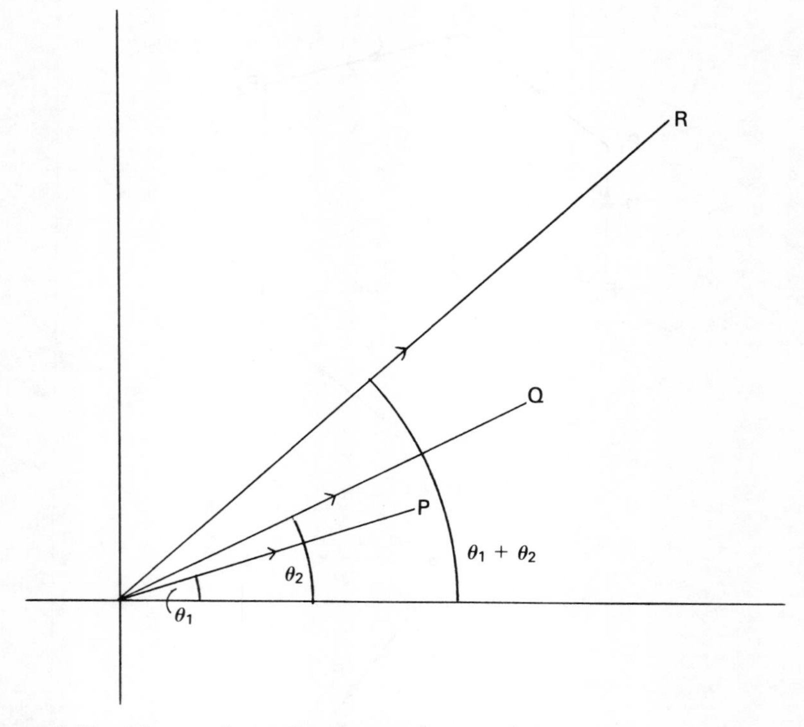

Figure 1.19 The product of two complex numbers

$$e^x = 1 + \frac{x}{1!} + \frac{x^2}{2!} + \frac{x^3}{3!} + \ldots + \frac{x^n}{n!} + \ldots$$

$$\cos x = 1 - \frac{x^2}{2!} + \frac{x^4}{4!} - \frac{x^6}{6!} + \ldots + \frac{(-1)^n x^{2n}}{(2n)!} + \ldots$$

$$\sin x = x - \frac{x^3}{3!} + \frac{x^5}{5!} - \frac{x^7}{7!} + \ldots + \frac{(-1)^n x^{2n+1}}{(2n+1)!} + \ldots$$

The infinite series for e^x also holds if x is a complex number. In particular, for $x = i\theta$

$$e^{i\theta} = 1 + i\theta + \frac{(i\theta)^2}{2!} + \frac{(i\theta)^3}{3!} + \ldots + \frac{(i\theta)^n}{n!} + \ldots$$

$$= \left(1 - \frac{\theta^2}{2!} + \frac{\theta^4}{4!} - \ldots\right) + i\left(\theta - \frac{\theta^3}{3!} + \frac{\theta^5}{5!} - \ldots\right)$$

$$= \cos\theta + i\sin\theta$$

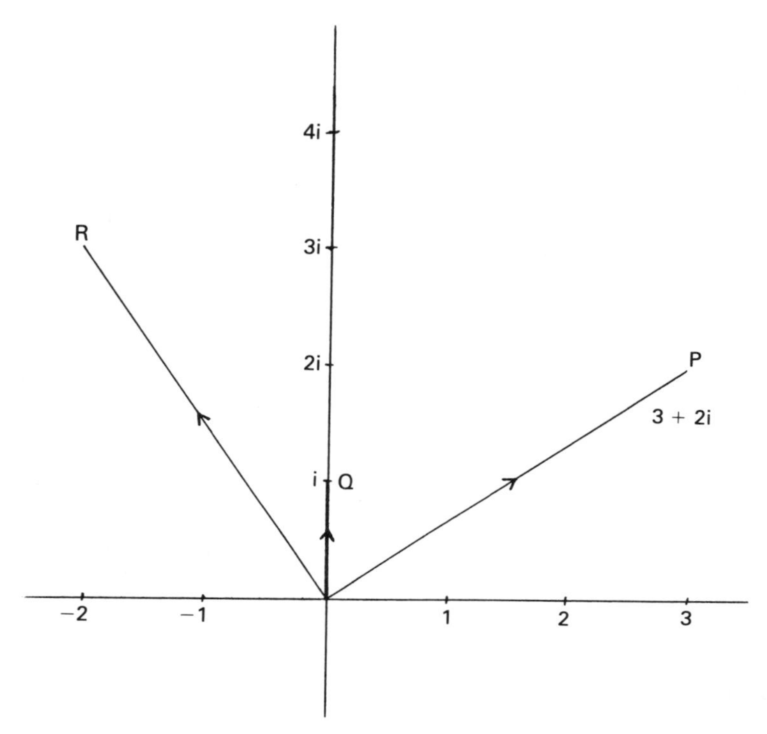

Figure 1.20 Example 1.4.7: let P represent $3+2i$ and Q represent i; then R represents $(3+2i)\times i=-2+3i$

Definition 1.4.6
The result $e^{i\theta}=\text{cis }\theta$ is known as *Euler's formula.*

Theorem 1.4.5
This theorem, attributed to De Moivre, states that $(\text{cis }\theta)^n=\text{cis }n\theta$.

Proof By Euler's formula, $(\text{cis }\theta)^n=(e^{i\theta})^n=e^{ni\theta}=\text{cis }n\theta$.

Example 1.4.7

Use De Moivre's theorem to show that $\cos 3\theta=\cos^3\theta-3\cos\theta\sin^2\theta$ and $\sin 3\theta= 3\cos^2\theta\sin\theta-\sin^3\theta$.

By De Moivre's theorem

$$\begin{aligned}\text{cis }3\theta&=(\text{cis }\theta)^3=(\cos\theta+i\sin\theta)^3\\&=\cos^3\theta+3\cos^2\theta\, i\sin\theta+3\cos\theta\,(i\sin\theta)^2-i\sin^3\theta\\&=\cos^3\theta-3\cos\theta\sin^2\theta+i(3\cos^2\theta\sin\theta-\sin^3\theta)\end{aligned}$$

Therefore

$$\cos 3\theta = \text{Re}(\text{cis } 3\theta) = \cos^3 \theta - 3 \cos \theta \sin^2 \theta$$

and

$$\sin 3\theta = \text{Im}(\text{cis } 3\theta) = 3 \cos^2 \theta \sin \theta - \sin^3 \theta$$

If α is a complex number represented in polar coordinates by the point $P = (r, \theta)$, then $\alpha = r \text{ cis } \theta = r\, e^{i\theta}$. Thus, any complex number, α, can be written as $|\alpha| e^{i \arg(\alpha)}$.

Definition 1.4.7
If a complex number is a solution of the equation $x^n = 1$, it is called an *nth root of unity*. From the earlier discussion, there are n such roots, one of which is 1.

Let α be such a complex number; then $\alpha^n = 1$. By theorem 1.4.2, $|\alpha|^n = |1| = 1$. Thus $|\alpha| = 1$ and α can be written $e^{i\theta}$. Since $\alpha^n = 1$, $e^{in\theta} = 1$, and so $n\theta = 2\pi k$, and $\theta = 2\pi k/n$, $k = 0, 1, \ldots, n-1$. In the Argand diagram, the n roots of unity are thus all equally spaced on the circumference of a circle of unit radius.

Example 1.4.9

The eight solutions of $x^8 = 1$ are

$$e^0 = 1, \; e^{i\pi/4} = \frac{1}{\sqrt{2}} + \frac{1}{\sqrt{2}} i$$

$$e^{i\pi/2} = i, \; e^{i3\pi/4} = -\frac{1}{\sqrt{2}} + \frac{1}{\sqrt{2}} i$$

$$e^{i\pi} = -1, \; e^{i5\pi/4} = -\frac{1}{\sqrt{2}} - \frac{1}{\sqrt{2}} i$$

$$e^{i6\pi/4} = -i, \; e^{i7\pi/4} = \frac{1}{\sqrt{2}} - \frac{1}{\sqrt{2}} i$$

These are illustrated in figure 1.21.

Exercise 1.4

1. Express the following in the form $a + ib$, $a, b \in \mathbb{R}$

(a) $(1+2i)(3+i)$;

(b) $(3+2i)^2/(5-7i)$;

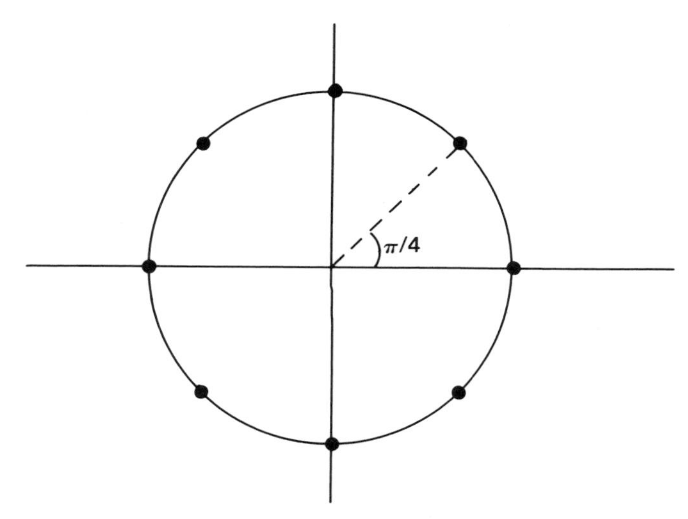

Figure 1.21 Example 1.4.9: the eight eighth roots of unity

(c) $(4+\mathrm{i})/(1-\mathrm{i})^2$;

(d) $(2+2\mathrm{i})^{1/3}$

2. If $\alpha=3-7\mathrm{i}$ and $\beta=6+2\mathrm{i}$, evaluate

(a) $\alpha\times\bar{\beta}$;

(b) $|\alpha|+|\beta|$;

(c) $\bar{\alpha}^2+2\bar{\alpha}\bar{\beta}+\bar{\beta}^2$.

3. Complete the proof of theorem 1.4.1.

4. Find all the roots of the following equations and represent them on an Argand diagram

(a) $\mathrm{x}^2-4\mathrm{x}+9=0$;

(b) $\mathrm{x}^6-1=0$.

5. Use De Moivre's theorem to find expressions for sin 4θ and cos 4θ in terms of powers of sin θ and cos θ only.

6. Let α be a complex number such that $|\alpha-2|=|\alpha+2|$. Show that all such numbers α must lie on the same straight line in the Argand diagram. If α also satisfies $|\alpha-\mathrm{i}|=2$, show that there are only two possible values for α and express them in the form $x+\mathrm{i}y$.

1.5 FUNCTIONS

Definitions

Definition 1.5.1
A *function* f: $A \to B$ assigns to every element $a \in A$ a unique element $b \in B$. The unique element $b \in B$ assigned to a is called the *image* of a under f and is commonly denoted by $f(a)$. The *domain* of the function f is the set A and the *codomain* is the set B. The *image set* of f or the *range* of f is the set $f(A) = \{f(a) | a \in A\}$. The codomain does not have to be equal to the image set but must contain it.

Example 1.5.1

The function f: $\{1, 2, 3, 4\} \to \{a, b, c, d\}$, where $f(1) = a, f(2) = b, f(3) = b$, $f(4) = c$, can be represented as in figure 1.22. The image set of f is $\{a, b, c\}$, which is a proper subset of the codomain $\{a, b, c, d\}$.

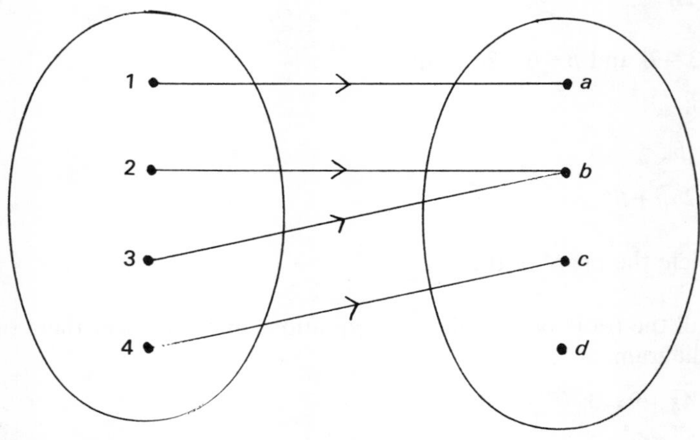

Figure 1.22 Example 1.5.1

Rather than define a function by listing the images associated with each element of the domain, it is usual to give a rule associating images with elements. This is clearly essential if the domain is infinite. The rule is usually of the form

$$\langle \text{variable} \rangle \mapsto \langle \text{some expression in the variable} \rangle$$

Example 1.5.2

The function sq which associates each integer with its square is defined by the rule

$$x \mapsto x^2$$

Of course, it is not essential to use x as the variable, so an alternative definition is

$$i \mapsto i^2$$

When defining a function, the domain must be explicitly stated and functions with different domains cannot be equal. If the codomain is not explicitly stated, it is taken to be equal to the image set.

Thus a full definition of sq is

$$\text{sq: } Z \to Z^+ \text{ is defined by the rule } x \mapsto x^2$$

Alternatively, a shorthand version is

$$\text{sq: } x \mapsto x^2 \ (x \in Z)$$

This is read as 'the function sq takes each x to x^2 where $x \in Z$'. This means that the domain is Z but the codomain is not explicitly stated. It is thus assumed to be $\{x^2 | x \in Z\}$, not Z^+ as before. It would be a pedantic mathematician who would argue that since the codomains differ in the two definitions the definitions cannot be taken as equivalent.

Definition 1.5.2
A function f: $A \to B$ is *one–one*, often written 1–1, if different elements in the domain have different images. An alternative way of saying this is to say: $f(a) = f(b) \supset a = b$.

A function f: $A \to B$ is *onto* if every $b \in B$ is the image of some $a \in A$, that is, the codomain is equal to the image set (figure 1.23).

Definition 1.5.3
A 1–1 function is known as an *injection*, an onto function is known as a *surjection* and a function that is both an injection and a surjection is known as a *bijection*.

Example 1.5.3

(1) The expression f: $x \mapsto \pm\sqrt{x}$ ($x \in \mathbb{R}^+$) is not a function since the image of x is only unique when $x = 0$. But g: $x \mapsto +\sqrt{x}$ ($x \in \mathbb{R}^+$) is a bijection $\mathbb{R}^+ \to \mathbb{R}^+$.

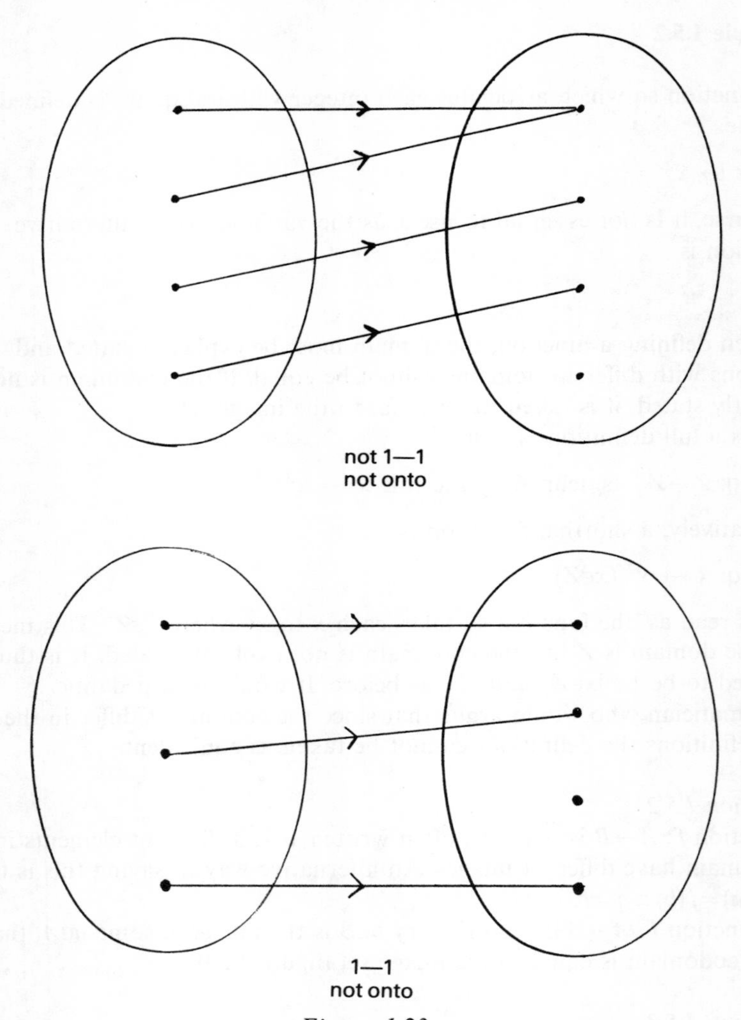

Figure 1.23

(2) $f\colon x \mapsto \begin{Bmatrix} x \text{ if } x \geqslant 0 \\ -x \text{ if } x < 0 \end{Bmatrix}$ $(x \in \mathbb{R})$

is a function $\mathbb{R}$ *onto* $\mathbb{R}^+$.

(3) $f\colon \mathbb{R} \to \mathbb{R}$ defined by $f\colon x \mapsto x+1$ is a bijection.

(4) If a set is countably infinite, then there exists a bijection $X \to N$, defined by associating the ith element of the set X with the integer i.

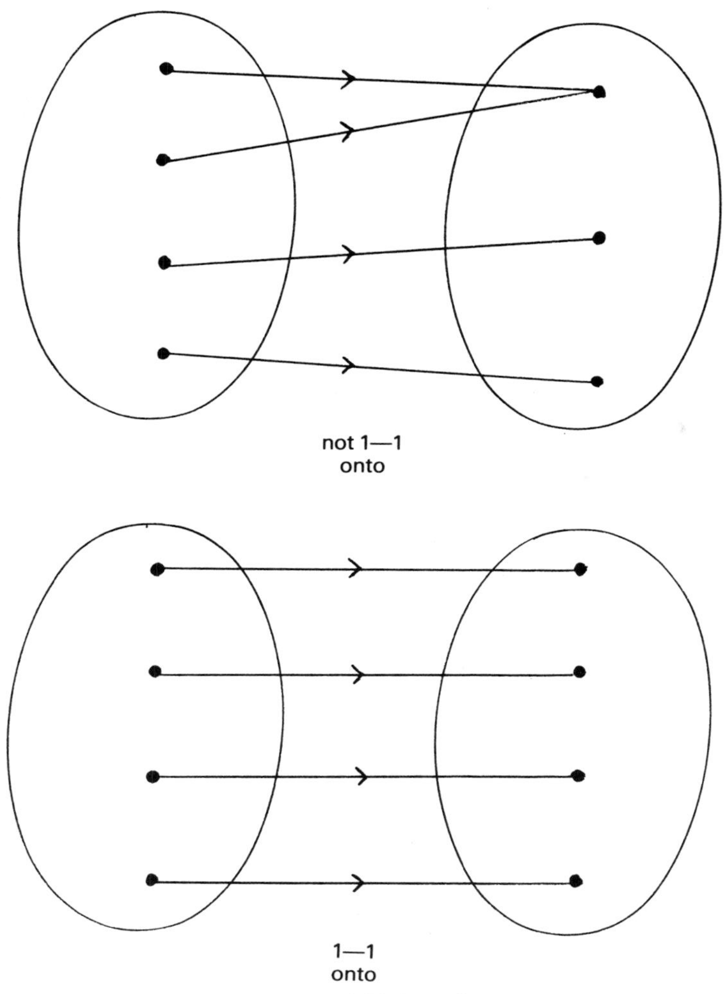

Figure 1.23 (*contd*)

(5) The factorial function, fact: $Z^+ \rightarrow Z^+$, defined by fact: $n \mapsto$ **if** $n=0$ **then** 1 **else** $n*(n-1)*(n-2)*\ldots*1$ **endif** is an injection but not a surjection. The value $n*(n-1)*(n-2)*\ldots*1$ is commonly denoted by $n!$ and, by convention, $0!=1$.

Definition 1.5.4
Let f: $A \rightarrow B$ and g: $C \rightarrow D$ be two functions. If the image set of f is a subset of the domain of g, then the *composition*, $g_0 f$, of the two functions can be formed where $g_0 f$: $x \mapsto g(f(x))$ $(x \in A)$. The image set of $g_0 f$ is $g(f(A))$, which is a subset of $g(C)$ (figure 1.24).

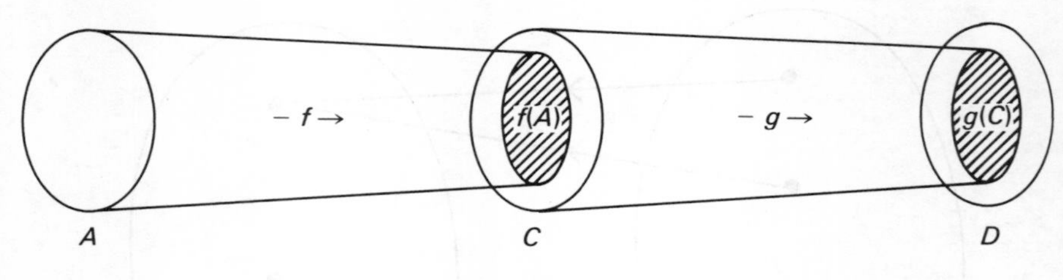

Figure 1.24

Example 1.5.4

If f: $x \mapsto x^2$ ($x \in \mathbb{R}$) and g: $x \mapsto x+1$ ($x \in \mathbb{R}$), then $f_0 g$: $x \mapsto (x+1)^2$ ($x \in \mathbb{R}$) and $g_0 f$: $x \mapsto x^2+1$ ($x \in \mathbb{R}$).

Definition 1.5.5
For any set A, the *identity function* on A, i_A, is defined by i_A: $x \mapsto x$ ($x \in A$).
If f: $A \to B$ and f^{-1}: $B \to A$ are such that $f^{-1}{}_0 f$ is the identity function on A and $f_0 f^{-1}$ is the identity function on B, then f^{-1} is the *inverse function* of f.
For a function to have an inverse it must be a bijection.

Example 1.5.5

(1) If f: $x \mapsto x^2$ ($x \in \mathbb{R}$), then f has no inverse since it is not a bijection. But g: $x \mapsto x^2$ ($x \in \mathbb{R}^+$) is a bijection and has inverse g^{-1}: $x \mapsto +\sqrt{x}$ ($x \in \mathbb{R}^+$).

(2) If f: $x \mapsto x+1$ ($x \in \mathbb{R}$), then f is a bijection and its inverse is f^{-1}: $x \mapsto x-1$ ($x \in \mathbb{R}$). Both $f_0 f^{-1}$ and $f^{-1}{}_0 f$ are equal to the identity function on $\mathbb{R}$.

Functions over the Reals

If f: $X \to \mathbb{R}$ and $X \subset \mathbb{R}$, then f is said to be a *function over the reals* or, simply, a *real function*. Such functions are particularly common and every computer scientist must be familiar with the classic examples.

Definition 1.5.6
If f: $X \to \mathbb{R}$ and $X \subset \mathbb{R}$ then the *graph* of f is the set of points $\{(x, y) | y = f(x), x \in X\}$ plotted in the cartesian plane.
The values of x where $f(x) = 0$ are known as the *roots* of the equation $f(x) = 0$ or, sometimes, simply as the roots of f.

It is important to have a good appreciation of real functions and their graphs and so, in the following examples, sketches of the graphs of some commonly occurring functions are given. The functions are all defined in the form $y=f(x)$.

Example 1.5.6

(1) A sketch of $y=ax+b$ is shown in figure 1.25.

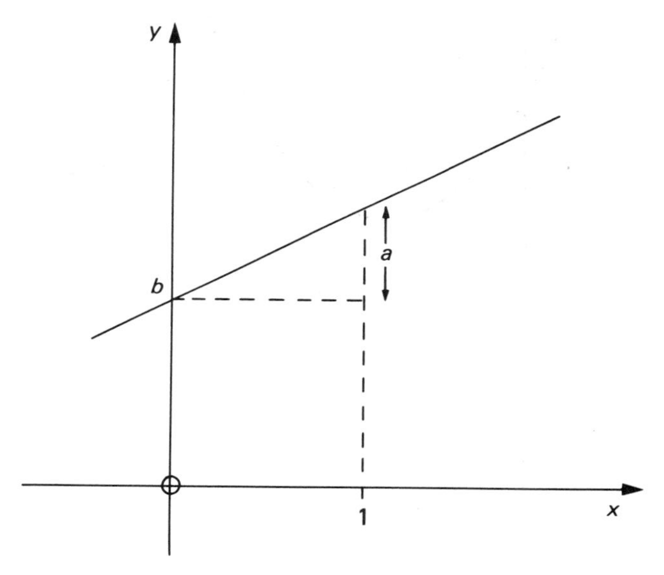

Figure 1.25 Example 1.5.6(1) $y=ax+b$

(2) Sketches of $y=ax^2+bx+c$, $a>0$, assuming two positive real roots, x_1, x_2, that is, assuming that $y=a(x-x_1)(x-x_2)$, $a>0$, and then assuming no real roots are shown in figure 1.26a and b, respectively.

(3) Sketches of $y=a(x-x_1)(x-x_2)(x-x_3)$, assuming $x_1<0$, a, x_2, $x_3>0$, and then assuming $a<0$ and just one real root, $x_1>0$, are shown in figure 1.27a and b, respectively.

(4) Figure 1.28 is a sketch of $y=1/(x-k)$, $k>0$.

(5) A sketch of $y=\sin x$ is shown in figure 1.29.

(6) The graph of $y=\cos x$ is sketched in figure 1.30.

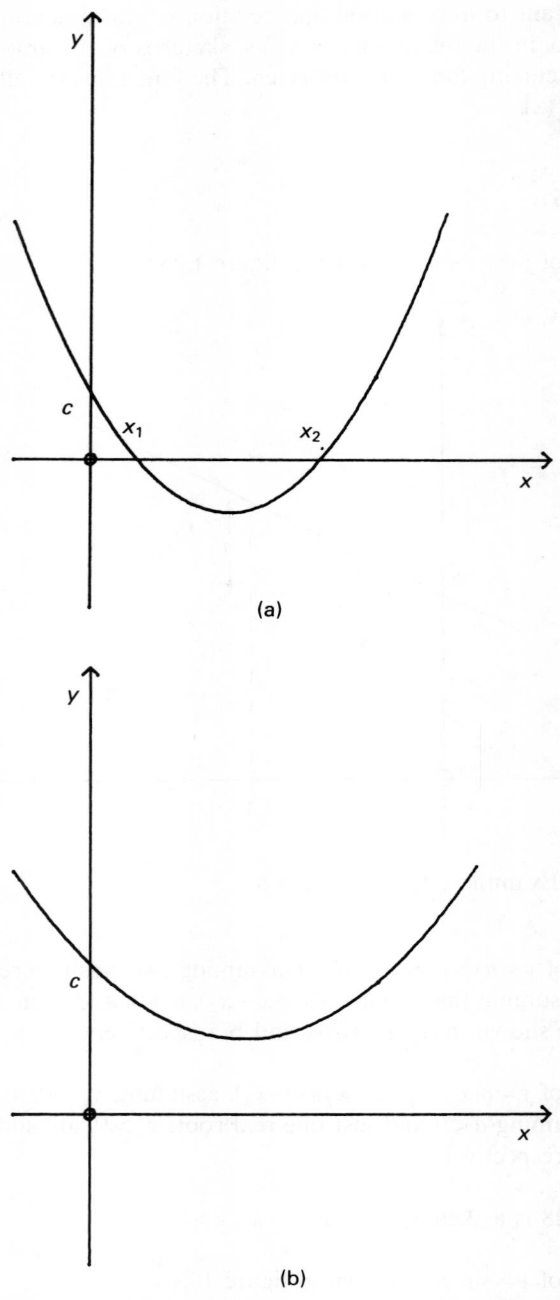

Figure 1.26 Example 1.5.6(2) $y = ax^2 + bx + c$, $a > 0$, with two positive real roots (a) and no real roots (b)

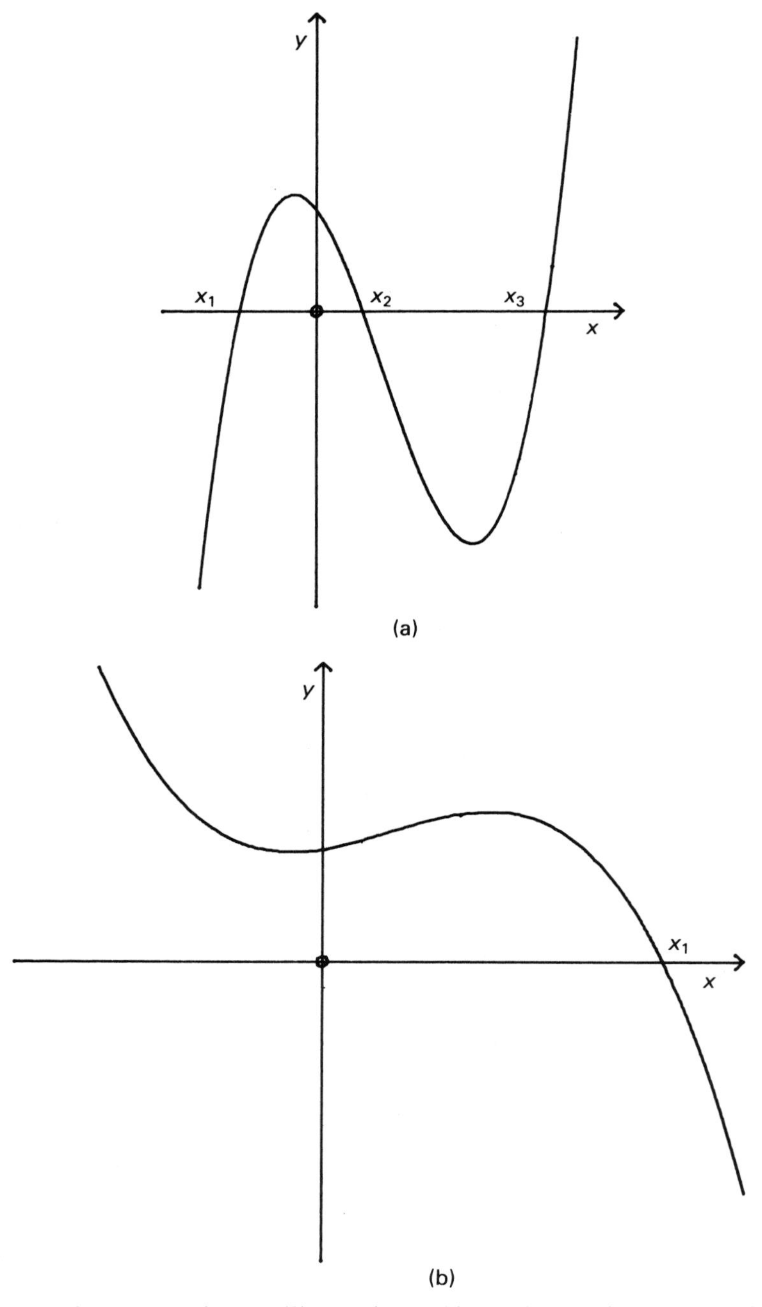

Figure 1.27 Example 1.5.6(3) $y=a(x-x_1)(x-x_2)(x-x_3)$ when $x_1<0$, a, x_2, $x_3>0$ (a) and when $a<0$ and with one real root (b)

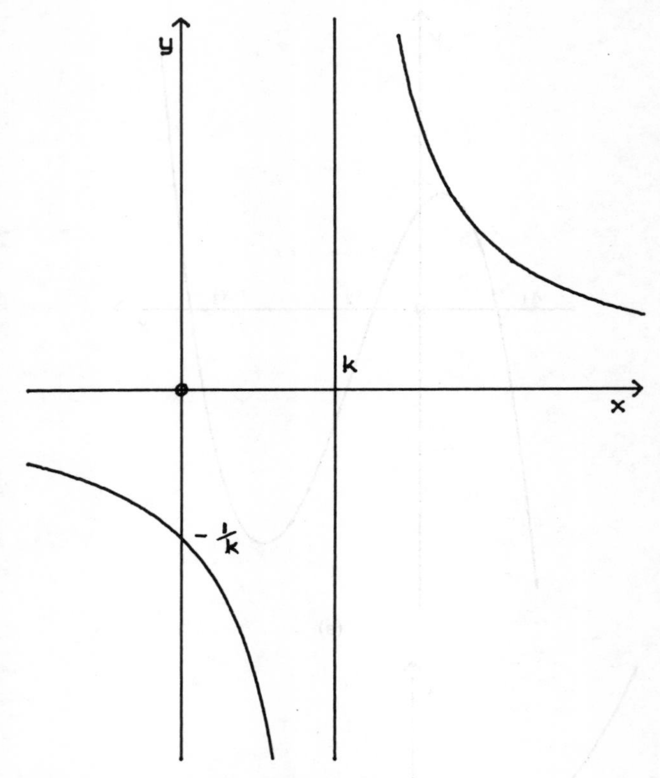

Figure 1.28 Example 1.5.6(4) $y=1/(x-k)$, $k>0$

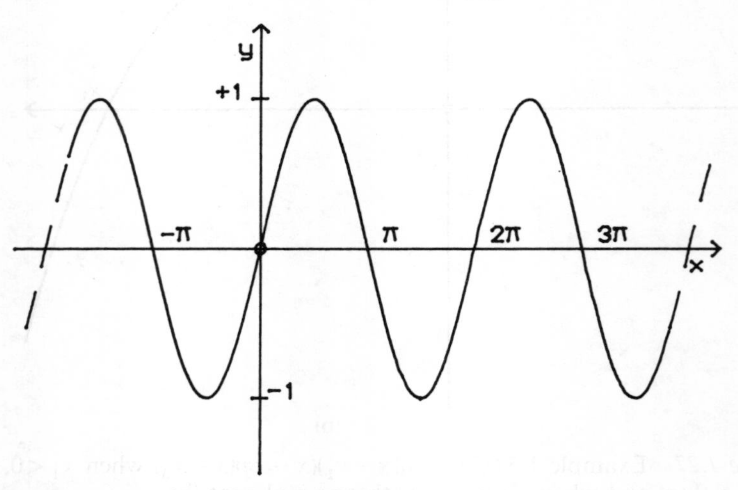

Figure 1.29 Example 1.5.6(5) $y=\sin x$

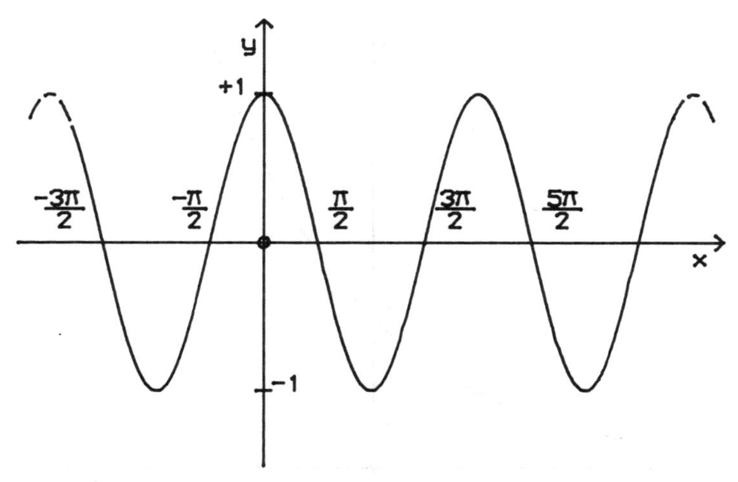

Figure 1.30 Example 1.5.6(6) $y = \cos x$

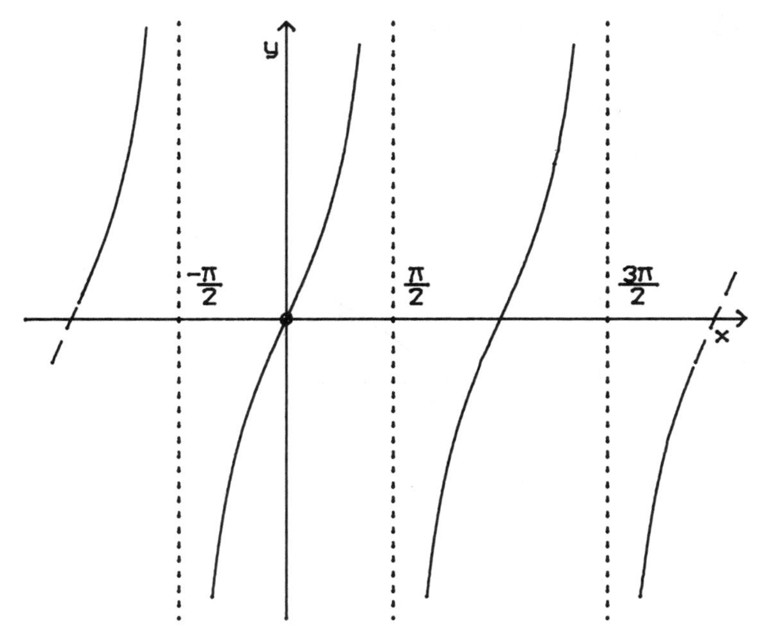

Figure 1.31 Example 1.5.6(7) $y = \tan x$

(7) Figure 1.31 sketches the graph of $y = \tan x$.

(8) The graph of $y = e^x$ is sketched in figure 1.32.

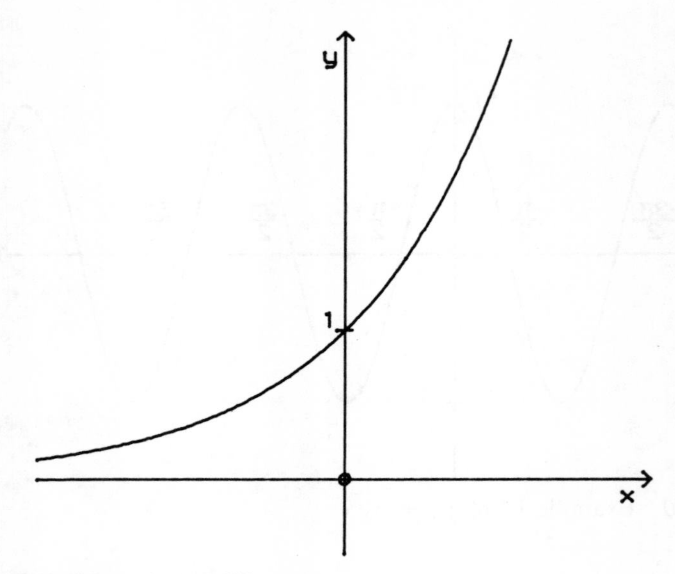

Figure 1.32 Example 1.5.6(8) $y=e^x$

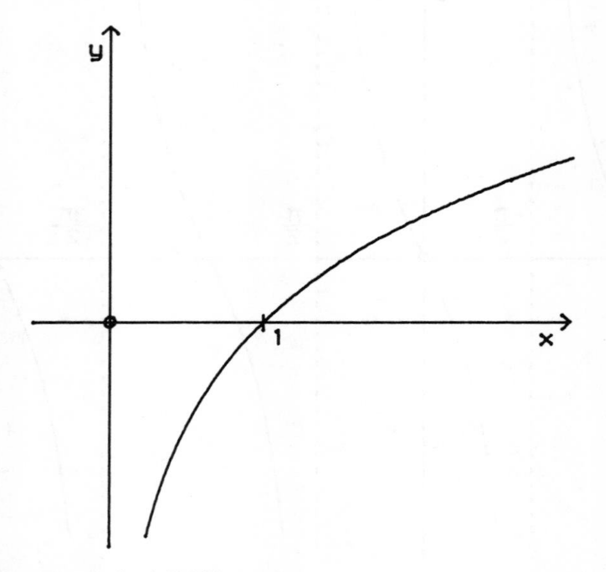

Figure 1.33 Example 1.5.6(9) $y=\log_e x$

(9) A sketch of $y=\log_e x$ is shown in figure 1.33.

(10) Finally, figure 1.34 sketches the graph of $y=\lceil x \rceil$, that is, the least integer greater than or equal to x.

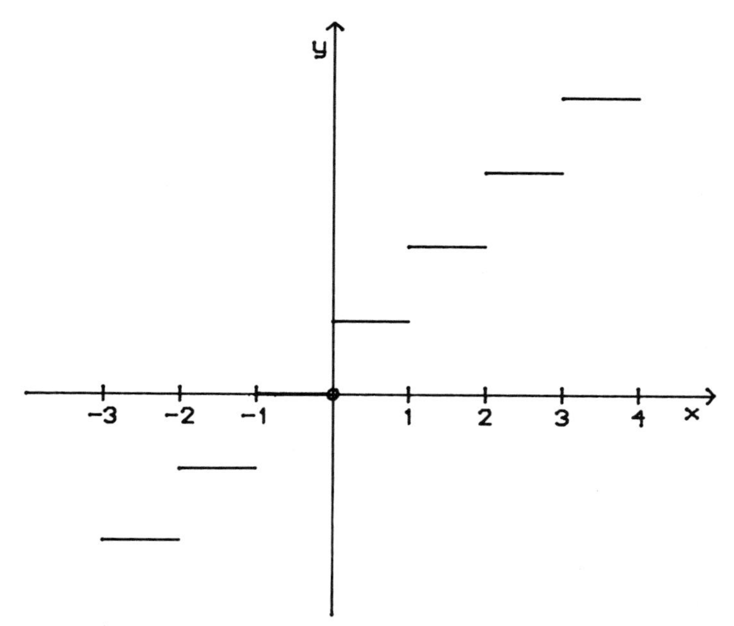

Figure 1.34 Example 1.5.6(10) $y=\lceil x \rceil$

If the graph of $y=f(x)$ has been drawn, then a horizontal shift to the right by α will result in the graph of $y=f(x-\alpha)$. A vertical shift upwards by β of the graph $y=f(x)$ will result in the graph of $y-\beta=f(x)$ that is, $y=f(x)+\beta$.

Example 1.5.7

(1) The graph of $f(x)=x^2-3x+2$ is shown in figure 1.35.

(2) For comparison, figure 1.36 sketches the graph of $f(x-2)=(x-2)^2-3(x-2)+2=x^2-7x+12$.

(3) Finally, the graph of $f(x)+3=x^2-3x+5$ is shown in figure 1.37.

If f is the bijection, $\mathbb{R}\to\mathbb{R}$, then for every value of $y\in\mathbb{R}$ there is precisely one value $x\in\mathbb{R}$ such that $y=f(x)$. In such a case the inverse of f exists and its graph is a reflection of f through the line $y=x$, as illustrated in figure 1.38.

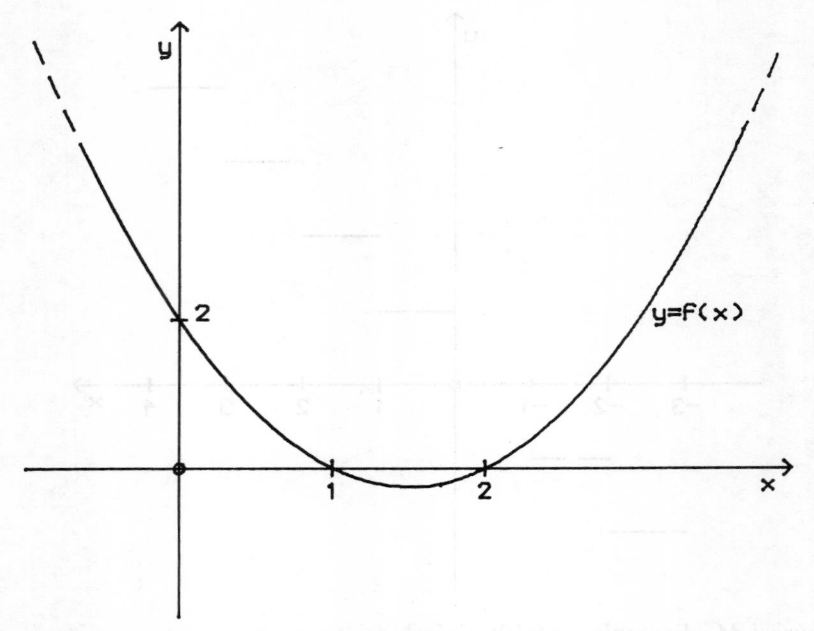

Figure 1.35 Example 1.5.7(1) $f(x)=x^2-3x+2$

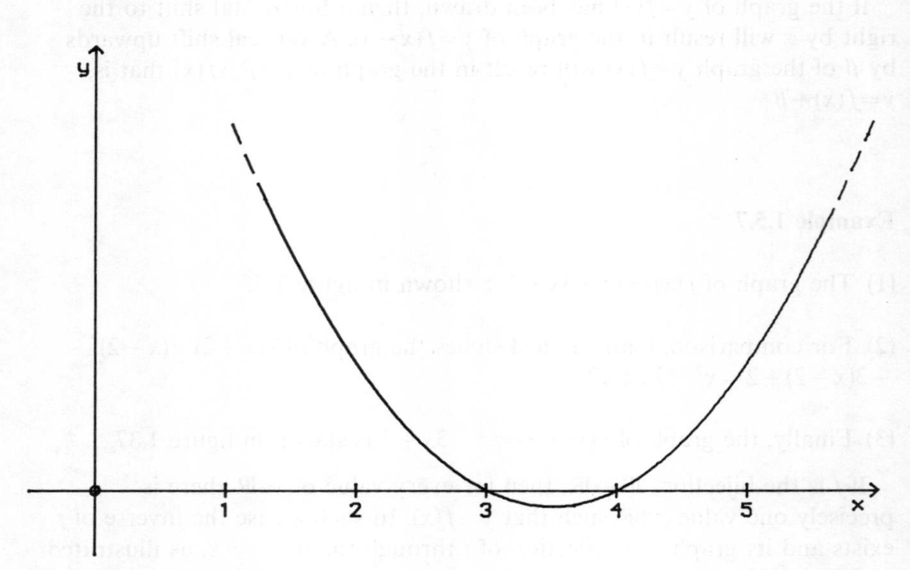

Figure 1.36 Example 1.5.7(2) $f(x-2)=x^2-7x+12$

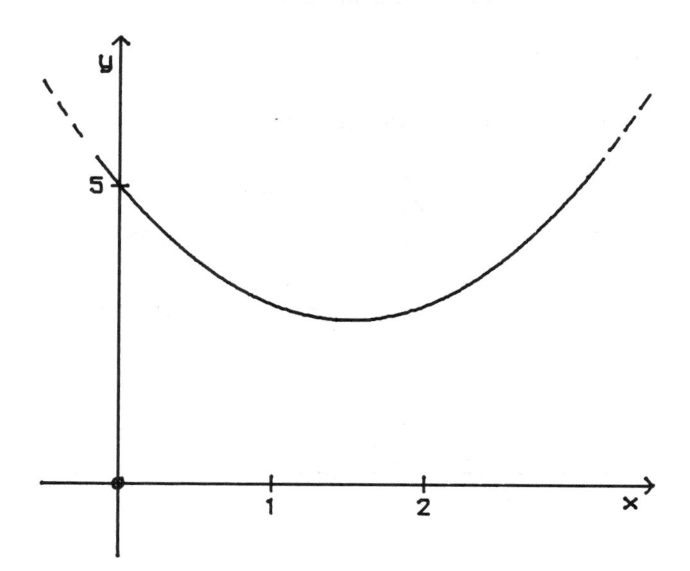

Figure 1.37 Example 1.5.7(3) $f(x)+3=x^2-3x+5$

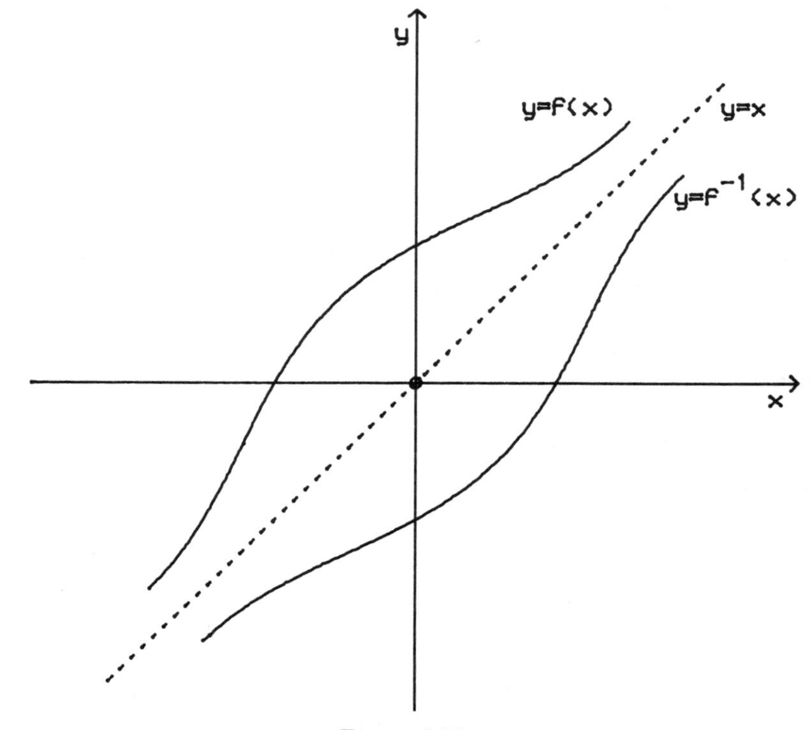

Figure 1.38

Example 1.5.8

If $f(x) = x + 1$, then $f^{-1}(x) = x - 1$ (figure 1.39).

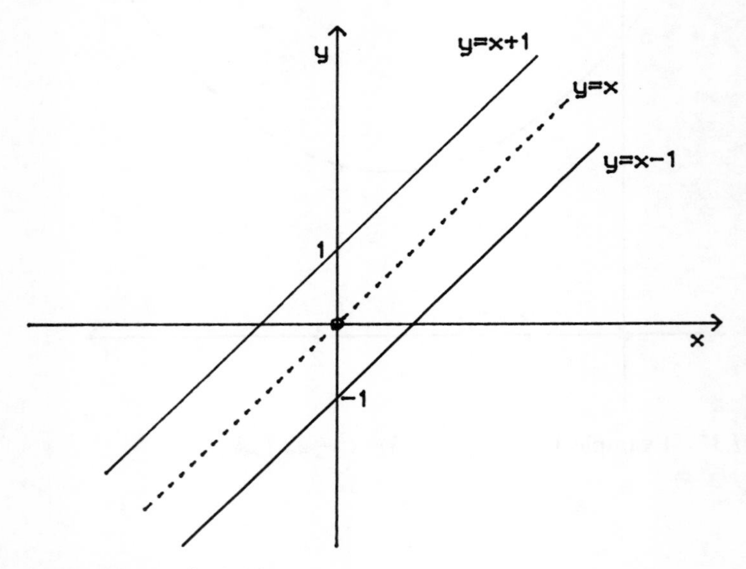

Figure 1.39 Example 1.5.8

Conics

Conics are an important class of curves, most of which do *not* represent functions, since for some of their values there is more than one y value.

Example 1.5.9

(1) The circle with radius r and centre (a, b)

$$(x-a)^2 + (y-b)^2 = r^2$$

is *not* a function (figure 1.40).

(2) The rectangular hyperbola, $xy = c^2$, is a function $\mathbb{R}\backslash\{0\} \to \mathbb{R}$ (figure 1.41).

(3) The parabola, $y^2 = 4ax$, for which every point on the curve is such that its distance from the *focus* $(a, 0)$ is equal to its distance from the *directrix* $x = -a$, is *not* a function (figure 1.42).

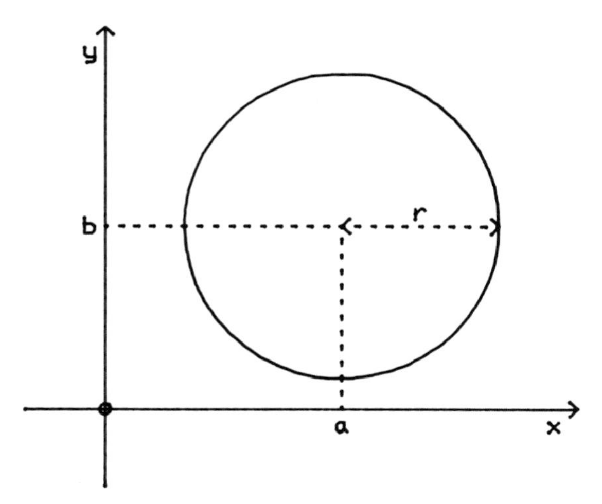

Figure 1.40 Example 1.5.9(1) $(x-a)^2+(y-b)^2=r^2$ is *not* a function

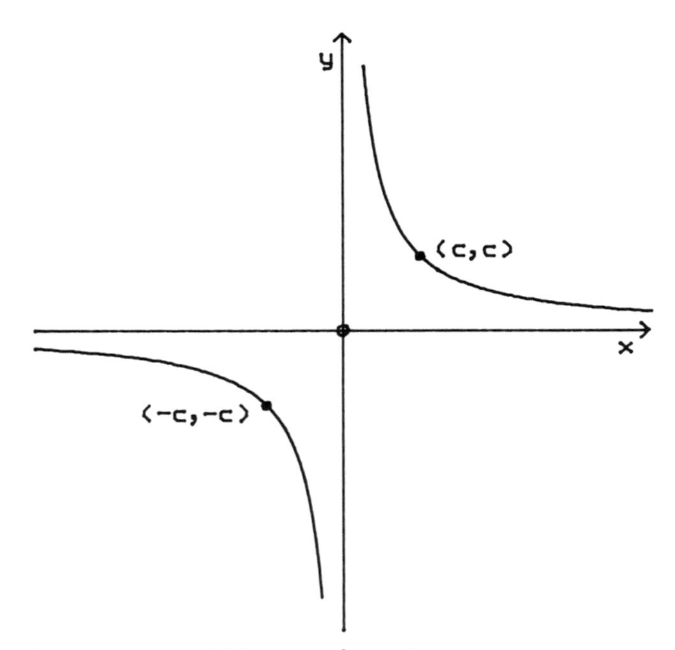

Figure 1.41 Example 1.5.9(2) $xy=c^2$ *is* a function

(4) The ellipse, $x^2/a^2+y^2/b^2=1$, where $b=ae$ and the *eccentricity* e is a constant satisfying $0<e<1$, is *not* a function (figure 1.43). For the ellipse, every point on the curve is such that its distance from the focus $(ae, 0)$ is equal to e times its distance from the directrix $x=a/e$ (PS$=e$PN). Also, for

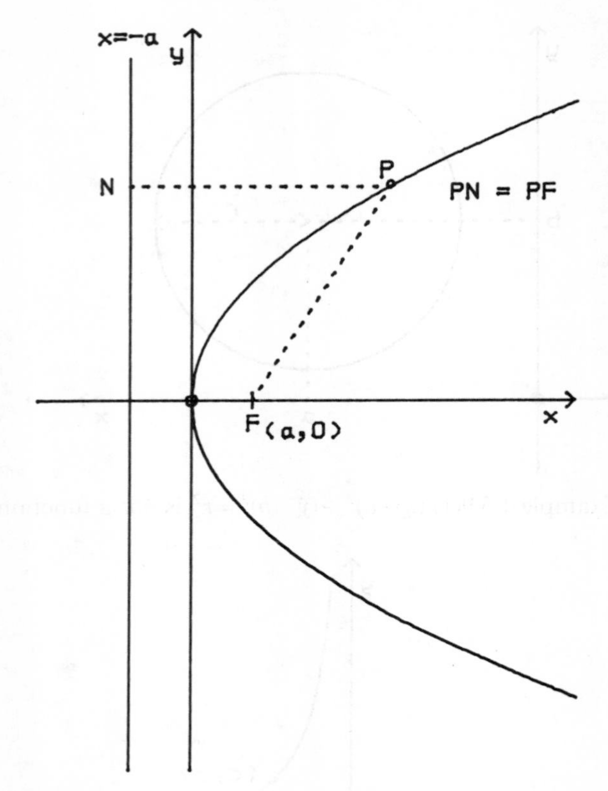

Figure 1.42 Example 1.5.9(3) $y^2 = 4ax$ is *not* a function

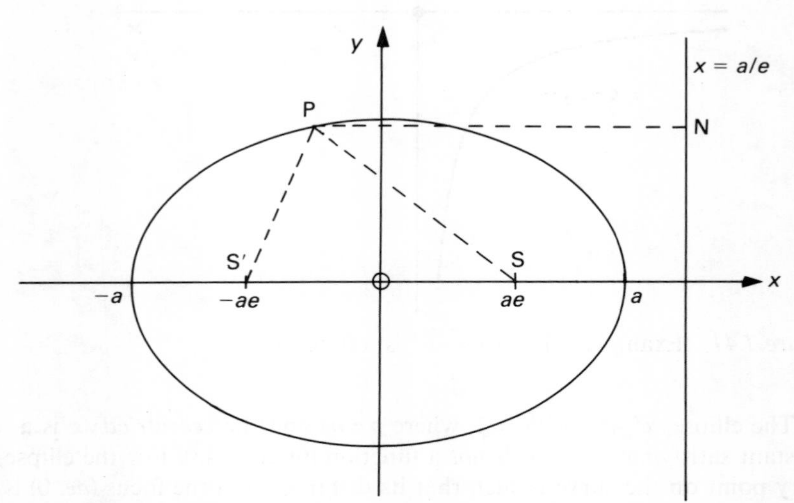

Figure 1.43 Example 1.5.9(4) $x^2/a^2 + y^2/b^2 = 1$ is *not* a function

every point on the curve, the sum of its distances from $(ae, 0)$ and from $(-ae, 0)$ is constant ($PS + PS' = 2a$).

(5) The general hyperbola, $x^2/a^2 - y^2/b^2 = 1$, where $b^2 = a^2(e^2 - 1)$ and the eccentricity $e > 1$ is *not* a function, in general (figure 1.44). As with the ellipse, every point on the curve is such that its distance from the focus is e times its distance from the directrix ($PS = ePN$).

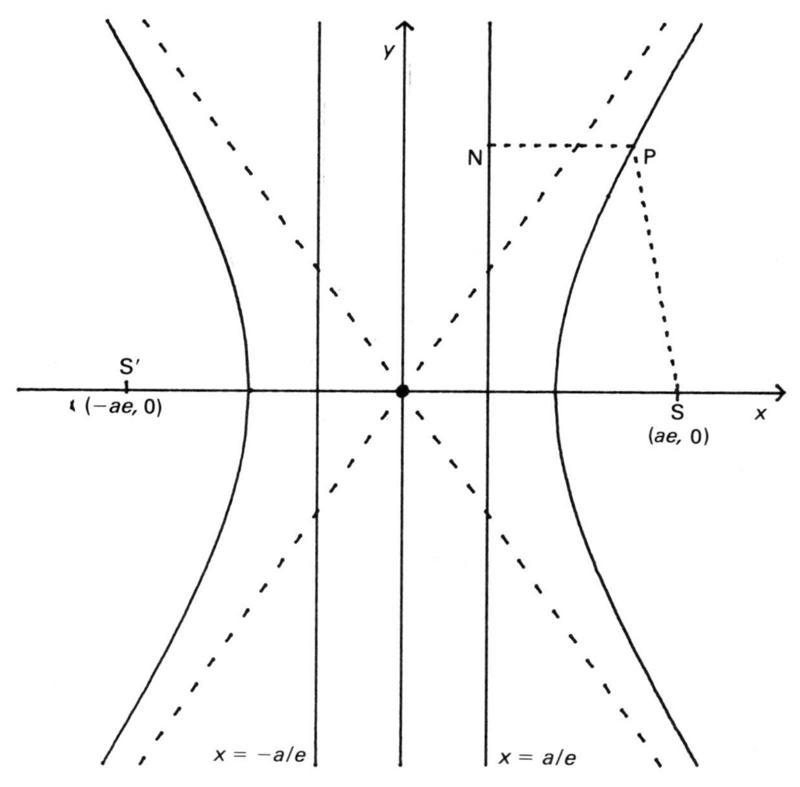

Figure 1.44 Example 1.5.9(5) $x^2/a^2 - y^2/b^2 = 1$ is *not* a function

Arithmetic Operations on Real Functions

Consider two real functions, f: $X \to \mathbb{R}$ and g: $X \to \mathbb{R}$ with the same domain $X \in \mathbb{R}$. For any $x \in X$, the real numbers $f(x)$ and $g(x)$ both exist, and thus $f(x) + g(x)$, $f(x) - g(x)$ and $f(x) \times g(x)$ are all defined. However, $f(x)/g(x)$ only exists when $g(x) \neq 0$. These observations lead to the following definitions of arithmetic operations on real functions.

Definition 1.5.7
Let f: $X \to \mathbb{R}$ and g: $X \to \mathbb{R}$ be two real functions. Then the real functions $f+g, f-g$ and $f \times g$ are defined as follows

$$f+g\colon x \mapsto f(x)+g(x) \quad (x \in X)$$

$$f-g\colon x \mapsto f(x)-g(x) \quad (x \in X)$$

$$f \times g\colon x \mapsto f(x) \times g(x) \quad (x \in X)$$

and if X' denotes the set $X \setminus \{x | g(x)=0\}$ then f/g: $X' \mapsto \mathbb{R}$ is defined by

$$f/g\colon x \mapsto f(x)/g(x) \quad (x \in X')$$

Example 1.5.10
If f: $x \mapsto x^2+3$, g: $x \mapsto e^x$ and h: $x \mapsto \sin x$ are real functions, each having domain $\mathbb{R}$, then

$$f+g\colon x \mapsto x^2+3+e^x \quad (x \in \mathbb{R})$$

$$f-g\colon x \mapsto x^2+3-e^x \quad (x \in \mathbb{R})$$

$$f \times g\colon x \mapsto (x^2+3)e^x \quad (x \in \mathbb{R})$$

$$f/g\colon \ x \mapsto (x^2+3)e^{-x} \quad (x \in \mathbb{R})$$

However, f/h is not defined for all $x \in \mathbb{R}$ since $\sin x = 0$ if $x = n\pi$, $n \in Z$. Thus

$$f/h\colon \ x \mapsto (x^2+3)/\sin x \quad (x \in \mathbb{R} \setminus \{n\pi | n \in Z\})$$

Functions of More Than One Argument

Definition 1.5.8
An *ordered pair* (a, b) consists of two elements a and b, where a is designated to be the first element and b is designated to be the second element.

An *ordered n-tuple* $(a_1, \ldots, a_n)$ consists of n (>1) elements, $a_1, \ldots, a_n$, where a_1 is the first element, a_2 is the second, etc.

Definition 1.5.9
If A, B are sets, then their *cartesian product* $A \times B$ is defined by

$$A \times B = \{(a, b) | a \in A, b \in B\}$$

Similarly if $A_1, \ldots, A_n$ are sets, then

$$A_1 \times A_2 \times \ldots \times A_n$$
$$= \{(a_1, a_2, \ldots, a_n) | a_1 \in A_1, a_2 \in A_2, \ldots, a_n \in A_n\}$$

By convention, A^2 is used to denote $A \times A$, A^3 for $A \times A \times A$, etc.

Example 1.5.11

If $A=\{1, 2, 3\}$ and $B=\{1, 4\}$, then

$$A\times B=\{(1, 1), (1, 4), (2, 1), (2, 4), (3, 1), (3, 4)\}$$

$$A^2=\{(1, 1), (1, 2), (1, 3), (2, 1), (2, 2), (2, 3), (3, 1), (3, 2), (3, 3)\}$$

$$B^3=\{(1, 1, 1), (1, 1, 4), (1, 4, 1), (1, 4, 4), (4, 1, 1), (4, 1, 4), (4, 4, 1), (4, 4, 4)\}$$

The use of cartesian products enables functions of more than one variable to be handled using the techniques that have been developed earlier in this section. For example, the function of two real variables, x and y, defined by

$$f(x,y)=x^2+y-3$$

is formally defined as

$$f\colon \mathbb{R}\times\mathbb{R}\to\mathbb{R} \text{ where } (x, y)\mapsto x^2+y-3.$$

The domain of f is $\mathbb{R}\times\mathbb{R}$ and the codomain is $\mathbb{R}$.

Example 1.5.12

(1) $\mathbb{R}^2$ denotes the cartesian plane and $f\colon \mathbb{R}^2\to\mathbb{R}$ defined by $f\colon (x, y)\mapsto x^2+y^2$ associates with any point in the plane the square of the distance of the point from the origin.

(2) $\mathbb{R}^3$ denotes three-dimensional real space and $f\colon \mathbb{R}^3\to\mathbb{R}$ defined by $f\colon (x, y, z)\mapsto x^2+y^2+z^2$ associates with any point in $\mathbb{R}^3$ the square of the distance of the point from the origin.

Exercise 1.5

1. If $f\colon x\mapsto 2x^3-3$ ($x\in\mathbb{R}$), what is: (a) $f(0)$; (b) $f(1)$; (c) the domain of f; (d) the range of f?

2. The sum of two positive reals is 1. If x represents the first of these numbers, express as a function of x: (a) their product; (b) the sum of their squares; and (c) their quotient.

3. Say which of the following functions are onto and which are 1–1.

(a) $f\colon \mathbb{R}\to\mathbb{R}$ where $f\colon x\mapsto x^2-1$;

(b) $f\colon \mathbb{R}\to Z$ where $f\colon x\mapsto \lceil x\rceil$;

(c) $f\colon Z\to Z$ where $f\colon x\mapsto x^3$;

(d) $f: \mathbb{R} \to \mathbb{R}$ where $f: x \mapsto x^3$.

4. Show that if $f: A \to B$ and $g: B \to C$ are bijections then the function $(g_0 f)^{-1}$ is equal to $f^{-1}{}_0 g^{-1}$. Hence, or otherwise, compute the inverse of $h: x \mapsto 2x^3+3$ $(x \in \mathbb{R})$.

5. Show that for any function $f: A \to B$, $i_{B\,0}f = f = f_0 i_A$.

6. Prove that if $f: A \to B$ and $g: B \to A$ satisfy $g_0 f = i_A$, then f is 1–1 and g is onto.

7. Determine the number of different functions from $\{1, 2\}$ into $\{1, 2, 3\}$. How many of these are injections and how many are surjections?

8. Prove by induction on n, that if $g: Z^+ \to Z^+$ is defined *recursively* by $g: n \mapsto$ **if** $n=0$ **then** 1 **else** $n * g(n-1$ **endif** then $g(n)=n!$ for all integer $\mathbf{n} \geqslant 0$. Note that the recursive definition of functions is very important in computer science.

9. Sketch the graphs of the following real functions.

(a) $y=3x^2-2x-5$;

(b) $y=x^3-x+2$;

(c) $y=(x+2)^3-3$;

(d) $y=(x+2)/(x-3)$;

(e) $y=\lceil \cos x \rceil$;

(f) $y=\lfloor x \rfloor$, where $\lfloor x \rfloor$ denotes the greatest integer less than x;

(g) $$y=\begin{cases} x^2-2 \text{ if } x \leqslant 0, \\ x-2 \text{ if } x>0. \end{cases}$$

10. Give the equation of a circle whose diameter has end points $(2, -3)$ and $(-1, 4)$.

11. Find the eccentricity, focus and directrix of the ellipse defined by $x^2/32+y^2/30=1$.

12. Sketch $9x^2-4y^2=36$.

13. If $A=\{a, b\}$, $B=\{c, d\}$, list the elements of $A \times B$ and of $B \times A$.

14. Show that if both A and B are countably infinite then so is $A \times B$.

2 Linear Algebra

2.1 VECTORS

Vectors and matrices are particularly important structures in both mathematics and computing. Linear algebra, which encompasses the study of both of these structures, is thus a subject of particular importance to the computer scientist.

An essential difference between a vector and a finite set is that the elements of a vector are ordered, that is, there is a first element, a second element and so on. Throughout this introductory section all the vectors will be real vectors, that is, they will consist of real numbers. The theory developed can also be used to handle vectors of rational numbers or even vectors of complex numbers. However for a more generalised approach to the subject, the reader must wait until section 5.4. At that stage, it will become apparent why the term *scalar* is used as a synonym for a real number throughout this chapter.

Definition 2.1.1
A (*real*) *vector*, $\boldsymbol{v}$, is a list of scalars (real numbers) comprising $v_1, v_2, \ldots, v_n$, ordered in a column

$$\boldsymbol{v} = \begin{bmatrix} v_1 \\ v_2 \\ \vdots \\ v_n \end{bmatrix}$$

Bold symbols will be used to distinguish vectors from scalars, as above.

The elements $v_1, v_2, \ldots, v_n$ of a vector $\boldsymbol{v}$ are usually called the *components* or *elements* of $\boldsymbol{v}$. Each component, v_i, of a vector $\boldsymbol{v}$ has an associated integer, i, which denotes the position of the component in the vector. This integer is called the *index* of the component in the vector.

The number of components in a vector is the *dimension* (or *order*) of the vector. The symbol $\mathbb{R}^n$ is used to denote all the real vectors of dimension n.

Two vectors, $\boldsymbol{v}$ and $\boldsymbol{w}$, are equal if and only if they have the same dimension and the same corresponding components, that is, if

$$v = \begin{bmatrix} v_1 \\ v_2 \\ \vdots \\ v_n \end{bmatrix} \text{ and } w = \begin{bmatrix} w_1 \\ w_2 \\ \vdots \\ w_m \end{bmatrix}$$

then $\boldsymbol{v} = \boldsymbol{w}$ iff $m = n$ and $v_i = w_i$ for $i = 1, \ldots, n$.

Example 2.1.1

(1) If

$$\boldsymbol{v} = \begin{bmatrix} 3 \\ 4 \\ -1 \end{bmatrix}, \boldsymbol{w} = \begin{bmatrix} 1 \\ -1 \\ 1 \end{bmatrix}, \boldsymbol{x} = \begin{bmatrix} 3 \\ -1 \\ 4 \end{bmatrix}, \boldsymbol{y} = \begin{bmatrix} 3 \\ 4 \\ -1 \end{bmatrix}, \boldsymbol{z} = \begin{bmatrix} 1 \\ 2 \end{bmatrix}$$

then $\boldsymbol{v}$, $\boldsymbol{w}$, $\boldsymbol{x}$, $\boldsymbol{y}$ all have dimension 3 and are thus elements of $\mathbb{R}^3$, while $\boldsymbol{z}$ has dimension 2 and is an element of $\mathbb{R}^2$. The only two vectors that are equal are $\boldsymbol{v}$ and $\boldsymbol{y}$.

(2) The vector of dimension n having all zero components is denoted by $\mathbf{0}_n$ and is called the *null vector* of dimension n. Thus

$$\mathbf{0}_3 = \begin{bmatrix} 0 \\ 0 \\ 0 \end{bmatrix}$$

is the null vector of dimension 3. The subscript n may be dropped when the dimension of the space is clear from the context.

A real vector has been defined as a column of scalars but the essential ordering property of vectors could equally well be portrayed by writing the components $v_1, v_2, \ldots, v_n$ of a vector $\boldsymbol{v}$ as a row $[v_1 \; v_2 \; \ldots \; v_n]$. In common with most textbooks, this book uses both notations and, when it is necessary to distinguish between them, the terms *column vector* or *row vector* will be used, as appropriate.

Example 2.1.2

$\begin{bmatrix} 2 \\ 1 \\ 3 \end{bmatrix}$ is a column vector and $[2 \; 1 \; 3]$ is a row vector.

Vector Algebra

Definition 2.1.2
Let $\boldsymbol{v}$ and $\boldsymbol{w}$ be vectors of the same dimension. The *sum of* $\boldsymbol{v}$ *and* $\boldsymbol{w}$, denoted by $\boldsymbol{v}+\boldsymbol{w}$, is the vector obtained by adding the corresponding components of $\boldsymbol{v}$ and $\boldsymbol{w}$

$$\boldsymbol{v}+\boldsymbol{w}=\begin{bmatrix} v_1 \\ v_2 \\ \vdots \\ v_n \end{bmatrix}+\begin{bmatrix} w_1 \\ w_2 \\ \vdots \\ w_n \end{bmatrix}=\begin{bmatrix} v_1+w_1 \\ v_2+w_2 \\ \vdots \\ v_n+w_n \end{bmatrix}$$

Definition 2.1.3
The product of a scalar, λ, and a vector, $\boldsymbol{v}$, denoted by $\lambda\boldsymbol{v}$, is the vector obtained from $\boldsymbol{v}$ by multiplying each component by λ

$$\lambda\boldsymbol{v}=\lambda\begin{bmatrix} v_1 \\ v_2 \\ \vdots \\ v_n \end{bmatrix}=\begin{bmatrix} \lambda v_1 \\ \lambda v_2 \\ \vdots \\ \lambda v_n \end{bmatrix}$$

As a special case, the scalar multiplication of -1 and $\boldsymbol{v}$ is written $-\boldsymbol{v}$ and then $\boldsymbol{v}+(-1)\boldsymbol{w}$ is written as $\boldsymbol{v}-\boldsymbol{w}$.

Let $\boldsymbol{v}_1, \boldsymbol{v}_2, \ldots, \boldsymbol{v}_r$ be r n-dimensional vectors. A vector $\boldsymbol{v}$ is a *linear combination of* $\boldsymbol{v}_1, \boldsymbol{v}_2, \ldots, \boldsymbol{v}_r$ if there exist scalars $\mu_1, \mu_2, \ldots, \mu_r \in \mathbb{R}$ such that $\boldsymbol{v}=\mu_1\boldsymbol{v}_1+\mu_2\boldsymbol{v}_2+\ldots+\mu_r\boldsymbol{v}_r$.

Example 2.1.3

(1) If

$$\boldsymbol{v}=\begin{bmatrix} 3 \\ 4 \\ -1 \end{bmatrix}, \quad \boldsymbol{w}=\begin{bmatrix} 1 \\ -1 \\ -1 \end{bmatrix}, \quad \boldsymbol{x}=\begin{bmatrix} 3 \\ -1 \\ 4 \end{bmatrix}$$

then

$$\boldsymbol{v}+\boldsymbol{w}=\begin{bmatrix} 4 \\ 3 \\ -2 \end{bmatrix}, \quad \boldsymbol{w}+\boldsymbol{x}=\begin{bmatrix} 4 \\ -2 \\ 3 \end{bmatrix}, \quad 2\boldsymbol{v}=\begin{bmatrix} 6 \\ 8 \\ -2 \end{bmatrix}, \quad \boldsymbol{v}-\boldsymbol{w}=\begin{bmatrix} 2 \\ 5 \\ 0 \end{bmatrix}$$

$$\boldsymbol{w}+2\boldsymbol{x}=\begin{bmatrix} 7 \\ -3 \\ 7 \end{bmatrix}, \quad 3\boldsymbol{x}-2\boldsymbol{w}=\begin{bmatrix} 7 \\ -1 \\ 14 \end{bmatrix} \quad 2\boldsymbol{v}-\boldsymbol{w}+3\boldsymbol{x}=\begin{bmatrix} 14 \\ 6 \\ 11 \end{bmatrix}$$

(2) Any vector

$$\boldsymbol{v} = \begin{bmatrix} v_1 \\ v_2 \\ \vdots \\ v_n \end{bmatrix}$$

can be written as a linear combination of the *n unit vectors*

$$\boldsymbol{e}_1 = \begin{bmatrix} 1 \\ 0 \\ 0 \\ \vdots \\ 0 \end{bmatrix}, \boldsymbol{e}_2 = \begin{bmatrix} 0 \\ 1 \\ 0 \\ \vdots \\ 0 \end{bmatrix}, \ldots, \boldsymbol{e}_n = \begin{bmatrix} 0 \\ 0 \\ 0 \\ \vdots \\ 1 \end{bmatrix}$$

since $\boldsymbol{v} = v_1\boldsymbol{e}_1 + v_2\boldsymbol{e}_2 + \ldots + v_n\boldsymbol{e}_n$.

The main properties of vector algebra are summarised below in theorem 2.1.1, and the reader should convince himself that the statements in this theorem are correct. In section 5.4, the properties listed in this theorem are used to define the abstract mathematical structure called a *vector space.* $\mathbb{R}^n$ is an example of such a vector space.

Theorem 2.1.1
For any $\boldsymbol{v}, \boldsymbol{w}, \boldsymbol{x} \in \mathbb{R}^n$ and $\lambda_1, \lambda_2 \in \mathbb{R}$:

(i) $(\boldsymbol{v}+\boldsymbol{w})+\boldsymbol{x} = \boldsymbol{v}+(\boldsymbol{w}+\boldsymbol{x})$;

(ii) $\boldsymbol{v}+\boldsymbol{w} = \boldsymbol{w}+\boldsymbol{v}$;

(iii) $\boldsymbol{v}+\boldsymbol{0} = \boldsymbol{0}+\boldsymbol{v} = \boldsymbol{v}$;

(iv) $\boldsymbol{v}+(-\boldsymbol{v}) = (-\boldsymbol{v})+\boldsymbol{v} = \boldsymbol{0}$;

(v) $\lambda_1(\boldsymbol{v}+\boldsymbol{w}) = \lambda_1\boldsymbol{v}+\lambda_1\boldsymbol{w}$;

(vi) $(\lambda_1\lambda_2)\boldsymbol{v} = \lambda_1(\lambda_2\boldsymbol{v})$;

(vii) $1\boldsymbol{v} = \boldsymbol{v}$;

(viii) $0\boldsymbol{v} = \boldsymbol{0}$.

Linear Dependence

Definition 2.1.4
The r n-dimensional vectors $\boldsymbol{v}_1, \boldsymbol{v}_2, \ldots, \boldsymbol{v}_r$ are said to be *linearly dependent* if there exist scalars $\lambda_1, \lambda_2, \ldots, \lambda_r$, not all zero, such that
$\lambda_1\boldsymbol{v}_1 + \lambda_2\boldsymbol{v}_2 + \ldots + \lambda_r\boldsymbol{v}_r = \boldsymbol{0}$.

If r n-dimensional vectors are not linearly dependent, they are said to be *linearly independent.*

Example 2.1.4

(1)

$$\begin{bmatrix}1\\2\\3\end{bmatrix}, \begin{bmatrix}-1\\4\\0\end{bmatrix}, \begin{bmatrix}3\\0\\6\end{bmatrix}$$

are linearly dependent since

$$2\begin{bmatrix}1\\2\\3\end{bmatrix} -1\begin{bmatrix}-1\\4\\0\end{bmatrix} -1\begin{bmatrix}3\\0\\6\end{bmatrix} = \begin{bmatrix}2\\4\\6\end{bmatrix} + \begin{bmatrix}1\\-4\\0\end{bmatrix} + \begin{bmatrix}-3\\0\\-6\end{bmatrix} = \begin{bmatrix}0\\0\\0\end{bmatrix}$$

(2) In $\mathbb{R}^n$, the n unit vectors, $\boldsymbol{e}_1, \boldsymbol{e}_2, \ldots, \boldsymbol{e}_n$ are linearly independent.

Theorem 2.1.2
The vectors $\boldsymbol{v}_1, \boldsymbol{v}_2, \ldots, \boldsymbol{v}_r$ are linearly dependent iff there is at least one vector, $\boldsymbol{v}_i$, which can be written as a linear combination of the remaining vectors $\boldsymbol{v}_1, \boldsymbol{v}_2, \ldots, \boldsymbol{v}_{i-1}, \boldsymbol{v}_{i+1}, \ldots, \boldsymbol{v}_r$.

Proof Assume that the vectors $\boldsymbol{v}_1, \boldsymbol{v}_2, \ldots, \boldsymbol{v}_r$ are linearly dependent. Then there exist $\lambda_1, \lambda_2, \ldots, \lambda_r$, not all zero, such that $\lambda_1\boldsymbol{v}_1 + \lambda_2\boldsymbol{v}_2 + \ldots + \lambda_r\boldsymbol{v}_r = \boldsymbol{0}$. Let i be such that λ_i is the first non-zero value of $\lambda_1, \lambda_2, \ldots, \lambda_r$. Then

$$\lambda_i\boldsymbol{v}_i = -\lambda_1\boldsymbol{v}_1 - \lambda_2\boldsymbol{v}_2 - \ldots - \lambda_{i-1}\boldsymbol{v}_{i-1} - \lambda_{i+1}\boldsymbol{v}_{i+1} - \ldots - \lambda_r\boldsymbol{v}_r$$

and so

$$\boldsymbol{v}_i = -\frac{\lambda_1\boldsymbol{v}_1}{\lambda_i} - \frac{\lambda_2\boldsymbol{v}_2}{\lambda_i} - \ldots - \frac{\lambda_{i-1}\boldsymbol{v}_{i-1}}{\lambda_i} - \frac{\lambda_{i+1}\boldsymbol{v}_{i+1}}{\lambda_i} - \ldots - \frac{\lambda_r\boldsymbol{v}_r}{\lambda_i}$$

$$= 0\boldsymbol{v}_1 + 0\boldsymbol{v}_2 + \ldots + 0\boldsymbol{v}_{i-1} - \frac{\lambda_{i+1}\boldsymbol{v}_{i+1}}{\lambda_i} - \ldots - \frac{\lambda_r\boldsymbol{v}_r}{\lambda_i}$$

Thus $\boldsymbol{v}_i$ is a linear combination of the remaining vectors.

On the other hand, if $\boldsymbol{v}_i$ is a linear combination of $\boldsymbol{v}_1, \boldsymbol{v}_2, \ldots, \boldsymbol{v}_{i-1}, \boldsymbol{v}_{i+1}, \ldots, \boldsymbol{v}_r$, then there exist scalars $\mu_1, \mu_2, \ldots, \mu_{i-1}, \mu_{i+1}, \ldots, \mu_r \in \mathbb{R}$ such that

$$\boldsymbol{v}_i = \mu_1\boldsymbol{v}_1 + \mu_2\boldsymbol{v}_2 + \ldots + \mu_{i-1}\boldsymbol{v}_{i-1} + \mu_{i+1}\boldsymbol{v}_{i+1} + \ldots + \mu_r\boldsymbol{v}_r$$

Thus

$$\mu_1\boldsymbol{v}_1 + \mu_2\boldsymbol{v}_2 + \ldots + \mu_{i-1}\boldsymbol{v}_{i-1} - 1\boldsymbol{v}_i + \mu_{i+1}\boldsymbol{v}_{i+1} + \ldots + \mu_r\boldsymbol{v}_r = \boldsymbol{0}$$

Hence $\boldsymbol{v}_1, \boldsymbol{v}_2, \ldots, \boldsymbol{v}_r$ are linearly dependent.

In section 5.4, a formal proof is given to show that any collection of more than n vectors in $\mathbb{R}^n$ must be linearly dependent. In particular, if $\boldsymbol{v}_1$, $\boldsymbol{v}_2, \ldots, \boldsymbol{v}_n, \boldsymbol{v}_{n+1}$ are any $n+1$ vectors in $\mathbb{R}^n$, then each of them can be written as a linear combination of the remaining n vectors.

Example 2.1.5

The three vectors

$$\begin{bmatrix} 1 \\ -2 \end{bmatrix}, \begin{bmatrix} 1 \\ -4 \end{bmatrix}, \begin{bmatrix} 1 \\ 7 \end{bmatrix}$$

must be linearly dependent. There exist, therefore, scalars μ_1 and μ_2 such that

$$\begin{bmatrix} 1 \\ -2 \end{bmatrix} = \mu_1 \begin{bmatrix} -1 \\ 4 \end{bmatrix} + \mu_2 \begin{bmatrix} 1 \\ 7 \end{bmatrix}$$

that is

$$1 = -\mu_1 + \mu_2$$
$$-2 = 4\mu_1 + 7\mu_2$$

Solving these two simultaneous equations gives

$$\mu_1 = -\frac{9}{11} \text{ and } \mu_2 = \frac{2}{11}$$

Similarly

$$\begin{bmatrix} -1 \\ 4 \end{bmatrix} = -\frac{11}{9} \begin{bmatrix} 1 \\ -2 \end{bmatrix} + \frac{2}{9} \begin{bmatrix} 1 \\ 7 \end{bmatrix}$$

and

$$\begin{bmatrix} 1 \\ 7 \end{bmatrix} = \frac{11}{2} \begin{bmatrix} 1 \\ -2 \end{bmatrix} + \frac{9}{2} \begin{bmatrix} -1 \\ 4 \end{bmatrix}$$

Definition 2.1.5
A collection of vectors $\boldsymbol{v}_1, \boldsymbol{v}_2, \ldots, \boldsymbol{v}_r$ in $\mathbb{R}^n$ is called a *basis* of $\mathbb{R}^n$ if: (i) $\boldsymbol{v}_1$, $\boldsymbol{v}_2, \ldots, \boldsymbol{v}_r$ are linearly independent; and (ii) every vector $\boldsymbol{v} \in \mathbb{R}^n$ can be written as a linear combination of $\boldsymbol{v}_1, \boldsymbol{v}_2, \ldots, \boldsymbol{v}_r$.

By the remark above, in order to satisfy the linear independence criterion, $r \leqslant n$, and in order to satisfy the second criterion, $r \geqslant n$. Hence every basis of $\mathbb{R}^n$ consists of exactly n linearly independent vectors. A particularly important basis consists of the n unit vectors $\boldsymbol{e}_1, \boldsymbol{e}_2, \ldots, \boldsymbol{e}_n$.

Scalar Product

Definition 2.1.6
The *scalar product*, $\boldsymbol{v}\cdot\boldsymbol{w}$, of the row vector $\boldsymbol{v}$ with the column vector $\boldsymbol{w}$ of the same dimension, is defined by

$$\boldsymbol{v}\cdot\boldsymbol{w}=[v_1 \ v_2 \ \dots \ v_n]\begin{bmatrix} w_1 \\ w_2 \\ \vdots \\ w_n \end{bmatrix}=\sum_{i=1}^{n} v_i w_i$$

that is

$$\boldsymbol{v}\cdot\boldsymbol{w}= v_1 w_1+v_2 w_2+\dots+v_n w_n$$

The scalar product is thus a scalar quantity, hence its name (although it is also commonly called the *dot product* or the *inner product*).

Theorem 2.1.3
For any n-dimensional row vector $\boldsymbol{v}$ and for any n-dimensional column vectors $\boldsymbol{w}$ and $\boldsymbol{x}$

$$\boldsymbol{v}\cdot(\boldsymbol{w}+\boldsymbol{x})=\boldsymbol{v}\cdot\boldsymbol{w}+\boldsymbol{v}\cdot\boldsymbol{x}$$

Proof From the definition of the scalar product

$$\boldsymbol{v}\cdot(\boldsymbol{w}+\boldsymbol{x})=\sum_{i=1}^{n} v_i(w_i+x_i)=\sum_{i=1}^{n} v_i w_i+\sum_{i=1}^{n} v_i x_i=\boldsymbol{v}\cdot\boldsymbol{w}+\boldsymbol{v}\cdot\boldsymbol{x}$$

Example 2.1.6

Using the vectors of example 2.1.3(1)

$$\boldsymbol{v}\cdot\boldsymbol{w}=3\times 1+4\times(-1)+(-1)\times(-1)=0$$

$$\boldsymbol{v}\cdot\boldsymbol{x}=3\times 3+4\times(-1)+(-1)\times 4=1$$

$$\boldsymbol{w}\cdot\boldsymbol{x}=1\times 3+(-1)\times(-1)+(-1)\times 4=0$$

In the above example, there are two cases where the inner product of two vectors is zero. Such vectors are called *orthogonal.* It is important to realise that $\boldsymbol{v}\cdot\boldsymbol{w}=0$ does not necessarily imply that either $\boldsymbol{v}$ or $\boldsymbol{w}$ is a null vector.

Norms

Consider the two vectors

$$\boldsymbol{v} = \begin{bmatrix} 1 \\ 2 \\ 3 \end{bmatrix}, \boldsymbol{w} = \begin{bmatrix} 4 \\ 1 \\ 0 \end{bmatrix}$$

Which of these is the 'bigger'? One might argue that $\boldsymbol{v}$ is bigger than $\boldsymbol{w}$ since

$$\sum_{i=1}^{3} v_i > \sum_{i=1}^{3} w_i,$$

but on the other hand one could say that $\boldsymbol{w}$ is bigger than $\boldsymbol{v}$ since $\boldsymbol{w}$ has a larger component than $\boldsymbol{v}$. It all depends on what is meant by 'bigger'. To define the 'size' of a vector, it is necessary to find a measure that satisfies certain properties. Such a measure is called a norm.

Definition 2.1.7
The *norm* of a vector $\boldsymbol{v}$, written $||\boldsymbol{v}||$, is a scalar measure satisfying the following three axioms.

(N1) $||\boldsymbol{v}|| = 0$ if $\boldsymbol{v}$ is the null vector and otherwise $||\boldsymbol{v}|| > 0$;

(N2) $||\lambda\boldsymbol{v}|| = |\lambda| \, ||\boldsymbol{v}||$ for any $\lambda \in \mathbb{R}$, $\boldsymbol{v} \in \mathbb{R}^n$;

(N3) $||\boldsymbol{v} + \boldsymbol{w}|| \leqslant ||\boldsymbol{v}|| + ||\boldsymbol{w}||$, $\forall \boldsymbol{v}, \boldsymbol{w} \in \mathbb{R}^n$.

This last axiom is known as the *triangle inequality*.

A norm commonly used to define the 'size' of a vector is the *Euclidean norm.* If

$$\boldsymbol{v} = \begin{bmatrix} v_1 \\ v_2 \\ \vdots \\ v_n \end{bmatrix}$$

the Euclidean norm of $\boldsymbol{v}$ is denoted by $||\boldsymbol{v}||_2$, which is often abbreviated to $|\boldsymbol{v}|$, where

$$||\boldsymbol{v}||_2 = \left(\sum_{i=1}^{n} v_i^2 \right)^{1/2} = (\boldsymbol{v} \cdot \boldsymbol{v})^{1/2}$$

The Euclidean norm is sometimes called the l_2 *norm.*

Example 2.1.7

If

$$v = \begin{bmatrix} 3 \\ 4 \\ -1 \end{bmatrix}, \quad w = \begin{bmatrix} 1 \\ -1 \\ 1 \end{bmatrix}$$

then $|v| = \sqrt{26}$ and $|w| = \sqrt{3}$. In this case, $|v+w| = \sqrt{25} < \sqrt{26} + \sqrt{3} = |v| + |w|$, so the triangle inequality is satisfied.

It is a simple matter to check that the Euclidean norm satisfies axioms N1 and N2. The proof that the triangle inequality is satisfied is a little tricky, and is given below.

Theorem 2.1.4
For any real vectors v, w of the same dimension

(i) (*Cauchy–Schwarz inequality*) $|v \cdot w| \leqslant |v| \cdot |w|$,

(ii) (*Triangle inequality*) $|v+w| \leqslant |v| + |w|$.

Proof (i) Let v, w be of dimension n. If either v or w is null the result is immediate, so assume that neither is null.

Now $|v \cdot w| = |v_1 w_1 + v_2 w_2 + \ldots + v_n w_n| \leqslant |v_1 w_1| + |v_2 w_2| + \ldots + |v_n w_n| = \sum_{i=1}^{n} |v_i w_i|$, so if $\sum_{i=1}^{n} |v_i w_i| \leqslant |v| \cdot |w|$, the result follows.

For any real numbers $x, y \in \mathbb{R}$, $0 \leqslant (x-y)^2 = x^2 - 2xy + y^2$. Thus $2xy \leqslant x^2 + y^2$. Putting $x = |v_i|/|v|$ and $y = |w_i|/|w|$ results in

$$\frac{2|v_i| \cdot |w_i|}{|v| \cdot |w|} \leqslant \frac{|v_i|^2}{|v|^2} + \frac{|w_i|^2}{|w|^2}$$

This result holds for $i = 1, 2, \ldots, n$.

By definition, $|v|^2 = \sum_{i=1}^{n} v_i^2 = \sum_{i=1}^{n} |v_i|^2$ and $|w|^2 = \sum_{i=1}^{n} w_i^2 = \sum_{i=1}^{n} |w_i|^2$. Thus

$$\frac{2\sum_{i=1}^{n} |v_i| \cdot |w_i|}{|v| \cdot |w|} \leqslant \frac{\sum_{i=1}^{n} |v_i|^2}{|v|^2} + \frac{\sum_{i=1}^{n} |w_i|^2}{|w|^2} = \frac{|v|^2}{|v|^2} + \frac{|w|^2}{|w|^2} = 2$$

Hence, $\sum_{i=1}^{n} |v_i| \cdot |w_i| \leqslant |v| \cdot |w|$ and the result is proved.

(ii) If $|v+w| = 0$, the inequality clearly holds and so it can be assumed that $|v+w| \neq 0$.

Now, $|v_i + w_i| \leqslant |v_i| + |w_i|$ for any real numbers v_i, $w_i \in \mathbb{R}$. Hence

$$|\boldsymbol{v}+\boldsymbol{w}|^2 = \sum_{i=1}^{n} (v_i+w_i)^2 = \sum_{i=1}^{n} |v_i+w_i|^2$$

$$\leqslant \sum_{i=1}^{n} |v_i+w_i| \cdot (|v_i|+|w_i|)$$

$$= \sum_{i=1}^{n} |v_i+w_i| \cdot |v_i| + \sum_{i=1}^{n} |v_i+w_i| \cdot |w_i|$$

From the proof of the Cauchy–Schwarz inequality, $\sum_{i=1}^{n} |v_i| \cdot |w_i| \leqslant |\boldsymbol{v}| \cdot |\boldsymbol{w}|$ for any vectors $\boldsymbol{v}$, $\boldsymbol{w}$ of the same dimension. Since $\boldsymbol{v}+\boldsymbol{w}$ and $\boldsymbol{v}$ are of the same dimension, it follows that $\sum_{i=1}^{n} |v_i+w_i| \cdot |v_i| \leqslant |\boldsymbol{v}+\boldsymbol{w}| \cdot |\boldsymbol{v}|$. Similarly

$$\sum_{i=1}^{n} |v_i+w_i| \cdot |w_i| \leqslant |\boldsymbol{v}+\boldsymbol{w}| \cdot |\boldsymbol{w}|$$

Hence

$$|\boldsymbol{v}+\boldsymbol{w}|^2 \leqslant |\boldsymbol{v}+\boldsymbol{w}| \cdot |\boldsymbol{v}| + |\boldsymbol{v}+\boldsymbol{w}| \cdot |\boldsymbol{w}|$$

$$= |\boldsymbol{v}+\boldsymbol{w}| \cdot (|\boldsymbol{v}| + |\boldsymbol{w}|)$$

Dividing by $|\boldsymbol{v}+\boldsymbol{w}|$ yields the desired result.

Definition 2.1.8
As well as the Euclidean norm, two other norms are also commonly used: the l_1 *norm* and the l_∞ (*infinity*, *uniform* or *maximum*) *norm*. If

$$\boldsymbol{v} = \begin{bmatrix} v_1 \\ v_2 \\ \vdots \\ v_n \end{bmatrix}$$

the l_1 norm is defined by

$$\|\boldsymbol{v}\|_1 = \sum_{i=1}^{n} |v_i|$$

and the l_∞ norm by

$$\|\boldsymbol{v}\|_\infty = \max_i |v_i|$$

Example 2.1.8

If $\boldsymbol{v}=\begin{bmatrix}3\\4\\-1\end{bmatrix}$, $\|\boldsymbol{v}\|_1=3+4+1=8$ and $\|\boldsymbol{v}\|_\infty=4$.

It is a simple exercise to check that both the l_1 and the l_∞ norms satisfy the three axioms N1, N2 and N3. All the norms considered so far have been special cases of the norm defined by

$$\|\boldsymbol{v}\|_p=\left(\sum_{i=1}^{n}|v_i|^p\right)^{1/p}$$

where $p\geqslant 1$. The Euclidean norm is obtained by setting $p=2$, the l_1 norm by setting $p=1$ and the infinity norm by letting p tend to infinity.

Geometric Interpretation of Vectors

In section 1.4, complex numbers were represented both as directed lines in a plane, that is, two-dimensional space, and as ordered pairs of real numbers. Since two-dimensional vectors can be represented as ordered pairs of real numbers it follows that they can also be represented as directed lines in two-dimensional space. Similarly, three-dimensional vectors can be represented as directed lines in three-dimensional space, etc.

Example 2.1.9

(1) In figure 2.1a, all the directed lines represent the same vector

$$\begin{bmatrix}2\\1\end{bmatrix}$$

The first component is represented by the length of the projection of the line on the x axis and the second component by the length of the projection of the line on the y axis.

(2) In figure 2.1b, the directed line represents the vector

$$\begin{bmatrix}1\\2\\1\end{bmatrix}$$

The three components are represented, in order, by the length of the projection of the line on the x, y and z axes, respectively.

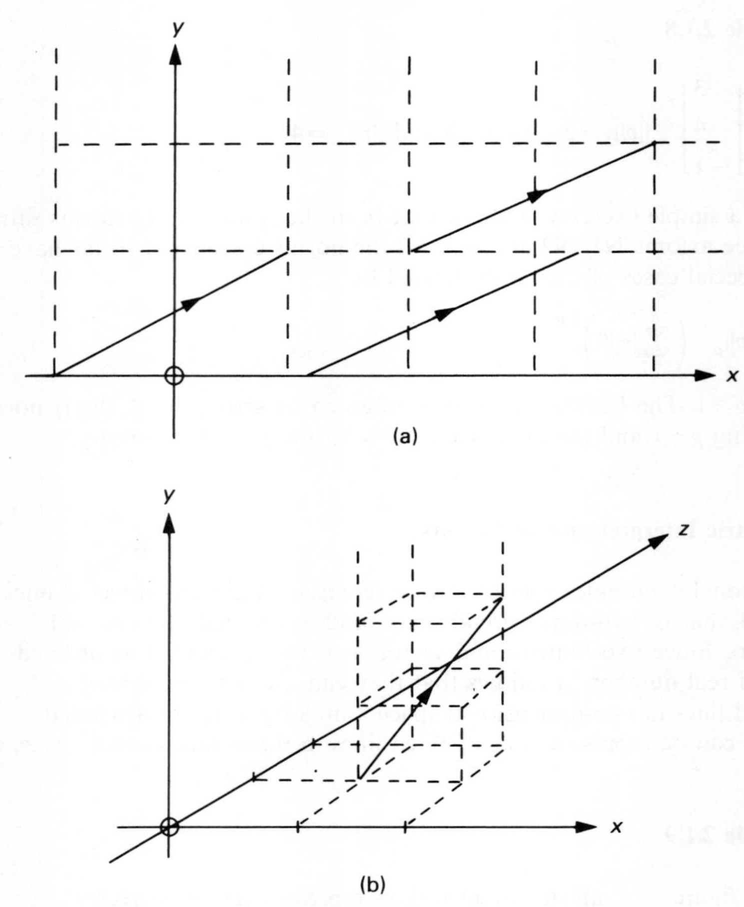

Figure 2.1 (a) Example 2.1.9(1); (b) example 2.1.9(2)

All the operations on vectors so far discussed have particularly simple geometric interpretations. This is not very surprising, since historically it was the study of the geometric object that motivated the definitions. Mathematicians have simply generalised concepts useful in two- and three-dimensional space to the abstract *n*-dimensional space.

Since a page in a book is a planar object, the geometric illustrations given in the rest of this section will be confined to the two-dimensional case. The addition of vectors is illustrated in figure 2.2, and scalar multiplication in figure 2.3. It should be clear that any two non-parallel lines in the plane will represent some basis of $\mathbb{R}^2$. In particular, lines of unit length parallel to the x and y axis, respectively, represent the basis consisting of the unit vectors $\boldsymbol{e}_1$ and $\boldsymbol{e}_2$.

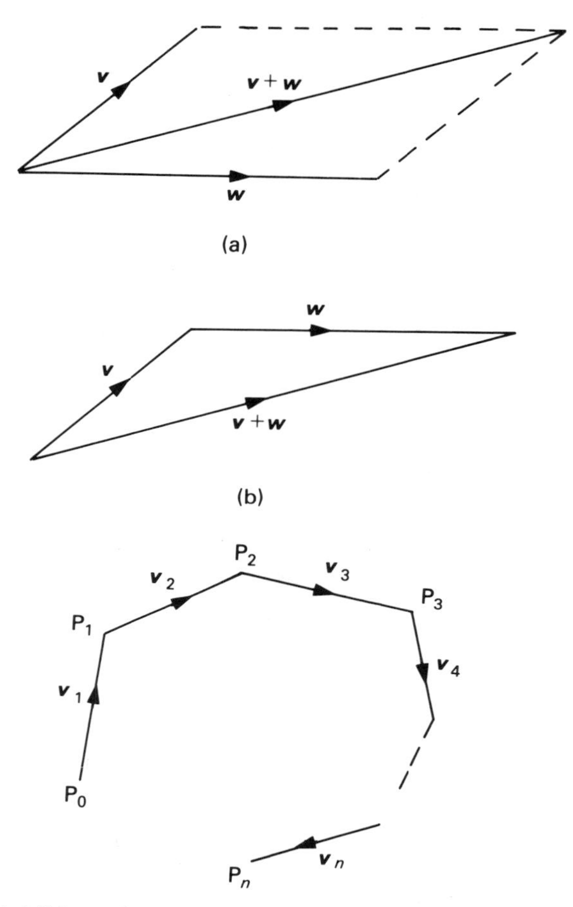

Figure 2.2 Addition of vectors: (a) construction of $\boldsymbol{v}+\boldsymbol{w}$ by parallelogram; (b) construction of $\boldsymbol{v}+\boldsymbol{w}$ by completing the triangle; (c) $\boldsymbol{v}_1 + \boldsymbol{v}_2 + \ldots + \boldsymbol{v}_n$ is represented by the directed line from P_0 to P_n

If

$$\boldsymbol{v} = \begin{bmatrix} v_1 \\ v_2 \end{bmatrix}$$

is represented by a directed line $\overrightarrow{P_1P_2}$ as in figure 2.4, then the length of the line $\overrightarrow{P_1P_2}$ is $(v_1^2 + v_2^2)^{1/2}$, that is, $|\boldsymbol{v}|$, the Euclidean norm of $\boldsymbol{v}$. The triangle inequality in two-dimensional space follows immediately from figure 2.2b since the shortest distance between any two points must be obtained from following the straight line connecting the points.

In figure 2.4, if θ denotes the angle of inclination of the line P_1P_2 then

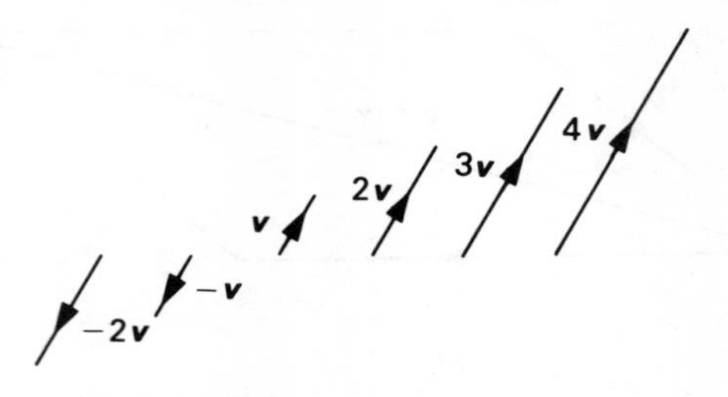

Figure 2.3 Scalar multiplication

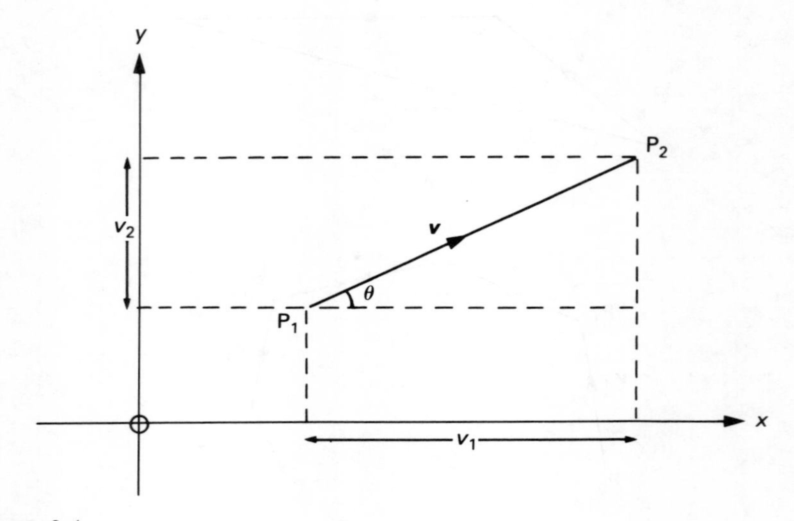

Figure 2.4

$\sin\theta = v_2/|\boldsymbol{v}|$, $\cos\theta = v_1/|\boldsymbol{v}|$ and $\tan\theta = v_2/v_1$. Now consider two vectors $\boldsymbol{v}$, $\boldsymbol{w}$ as in figure 2.5, and let α denote the angle between them. If θ denotes the angle of inclination of $\boldsymbol{v}$ and φ denotes the angle of inclination of $\boldsymbol{w}$ then $\alpha = \theta - \varphi$. Hence

$$\cos\alpha = \cos(\theta - \varphi) = \cos\theta\cos\varphi + \sin\theta\sin\varphi$$

$$= \frac{v_1}{|\boldsymbol{v}|}\cdot\frac{w_1}{|\boldsymbol{w}|} + \frac{v_2}{|\boldsymbol{v}|}\cdot\frac{w_2}{|\boldsymbol{w}|}$$

$$= \frac{\boldsymbol{v}\cdot\boldsymbol{w}}{|\boldsymbol{v}|\cdot|\boldsymbol{w}|}$$

This proves the following result.

Theorem 2.1.5
Let $\boldsymbol{v}$, $\boldsymbol{w}$ be two vectors in two-dimensional space, and let α be the angle between them; then $\boldsymbol{v}\cdot\boldsymbol{w} = |\boldsymbol{v}|\,|\boldsymbol{w}|\cos\alpha$.

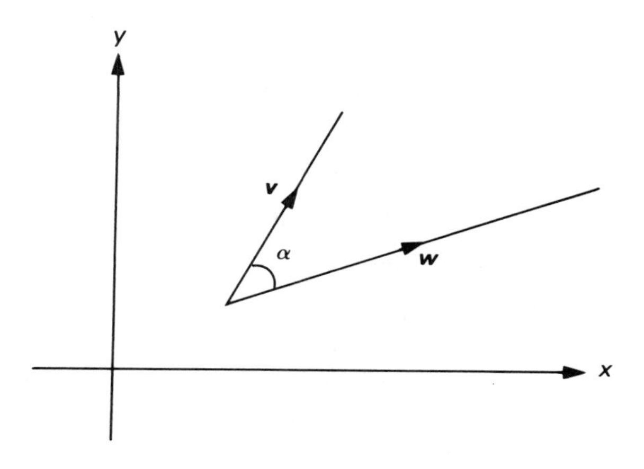

Figure 2.5

An immediate corollary of this theorem is the Cauchy–Schwarz inequality for two-dimensional vectors. Also, $\boldsymbol{v} \cdot \boldsymbol{w} = 0$ if and only if $\alpha = \pi/2$ or $3\pi/2$, provided neither $\boldsymbol{v}$ nor $\boldsymbol{w}$ is the null vector.

Having the above geometric representations of vectors, the tables can be turned and vectors can be used for geometric purposes, as in example 2.1.10.

Example 2.1.10

(1) Use vectors to prove that the diagonals of a square are perpendicular. Let ABCD be a square with diagonals AC and BD (figure 2.6). The task is to show that $AC \perp BD$.

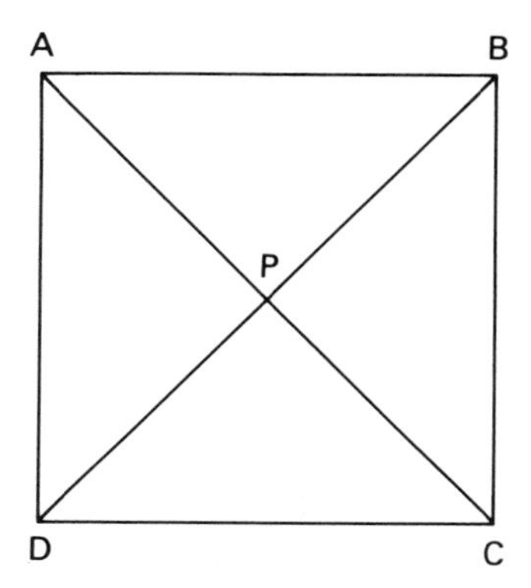

Figure 2.6 Example 2.1.10(1)

Denote $\overrightarrow{AB}$ by the vector $\boldsymbol{a}$ and $\overrightarrow{BC}$ by the vector $\boldsymbol{b}$. Since $AB \perp BC$ it follows that $\boldsymbol{a} \cdot \boldsymbol{b} = \boldsymbol{b} \cdot \boldsymbol{a} = 0$. Further, since ABCD is a square, it follows that $|\boldsymbol{a}| = |\boldsymbol{b}|$.

$\overrightarrow{AC}$ is denoted by $\boldsymbol{a}+\boldsymbol{b}$ and $\overrightarrow{BD}$ by $-\boldsymbol{a}+\boldsymbol{b}$. Since

$$(\boldsymbol{a}+\boldsymbol{b})\cdot(-\boldsymbol{a}+\boldsymbol{b}) = -\boldsymbol{a}\cdot\boldsymbol{a} - \boldsymbol{b}\cdot\boldsymbol{a} + \boldsymbol{a}\cdot\boldsymbol{b} + \boldsymbol{b}\cdot\boldsymbol{b}$$
$$= -|\boldsymbol{a}|^2 + |\boldsymbol{b}|^2 = 0$$

AC must be perpendicular to BD.

(2) Let ABC denote a triangle with $\hat{B} = \pi/2$, $AB = 1$ cm and $BC = 2$ cm. Let D divide AC in the ratio 1:2 (figure 2.7). Find the length of BD.

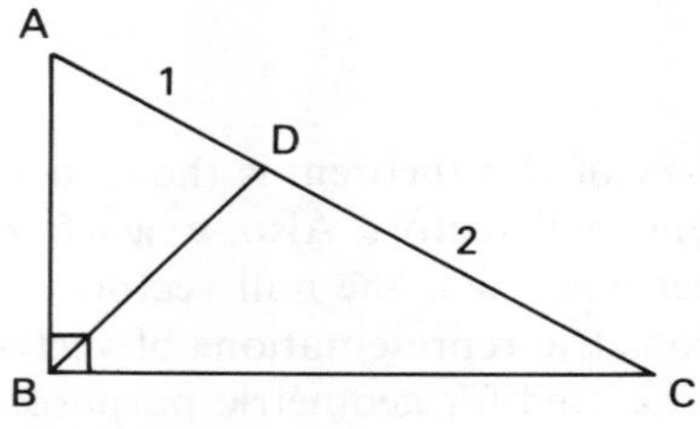

Figure 2.7 Example 2.1.10(2)

Denote $\overrightarrow{BA}$ by $\boldsymbol{a}$ and $\overrightarrow{BC}$ by $\boldsymbol{b}$. Then $\boldsymbol{a}\cdot\boldsymbol{b} = 0$, $|\boldsymbol{a}| = 1$ and $|\boldsymbol{b}| = 2$. Let $\overrightarrow{BD}$ be denoted by $\boldsymbol{c}$. The task is to find $|\boldsymbol{c}|$.

Now, from triangle ABC, $\overrightarrow{AC}$ can be denoted by $\boldsymbol{b}-\boldsymbol{a}$. Hence $\overrightarrow{AD}$ is $(\boldsymbol{b}-\boldsymbol{a})/3$ and $\overrightarrow{DC}$ is $2(\boldsymbol{b}-\boldsymbol{a})/3$. From triangle BAD, $\boldsymbol{c} = \boldsymbol{a} + (\boldsymbol{b}-\boldsymbol{a})/3$ $= 2\boldsymbol{a}/3 + \boldsymbol{b}/3$. Hence

$$|\boldsymbol{c}|^2 = \boldsymbol{c}\cdot\boldsymbol{c} = \frac{4}{9}\boldsymbol{a}\cdot\boldsymbol{a} + \frac{4}{9}\boldsymbol{a}\cdot\boldsymbol{b} + \frac{1}{9}\boldsymbol{b}\cdot\boldsymbol{b}$$

$$= \frac{4}{9}|\boldsymbol{a}|^2 + \frac{1}{9}|\boldsymbol{b}|^2 = \frac{4}{9} + \frac{4}{9} = \frac{8}{9}\ \text{cm}^2$$

Thus

$$|\boldsymbol{c}| = \frac{2}{3}\sqrt{2}\ \text{cm}$$

Exercise 2.1

1. If $\boldsymbol{x} = \begin{bmatrix} 4 \\ 0 \\ 6 \end{bmatrix}$, $\boldsymbol{y} = \begin{bmatrix} -1 \\ 2 \\ 3 \end{bmatrix}$ and $\boldsymbol{z} = \begin{bmatrix} 5 \\ 6 \\ 3 \end{bmatrix}$

evaluate: (a) $(\boldsymbol{x}+\boldsymbol{y})$, (b) $(2\boldsymbol{x}+3\boldsymbol{y})$, (c) $|\boldsymbol{z}|$. Are the three vectors linearly dependent?

2. If $\boldsymbol{a}=\begin{bmatrix}3\\7\end{bmatrix}$ and $\boldsymbol{b}=\begin{bmatrix}2\\3\end{bmatrix}$, find a row vector $\boldsymbol{x}$ such that $\boldsymbol{x}\cdot\boldsymbol{a}=7$ and $\boldsymbol{x}\cdot\boldsymbol{b}=4$.

3. Express

$$\begin{bmatrix}1\\4\\2\end{bmatrix}$$

as a linear combination of the three linearly independent vectors

$$\begin{bmatrix}3\\2\\1\end{bmatrix}, \quad \begin{bmatrix}1\\2\\3\end{bmatrix} \quad \text{and} \quad \begin{bmatrix}2\\1\\3\end{bmatrix}$$

4. Find a unit vector (that is one whose Euclidean norm is 1) that is orthogonal to $\begin{bmatrix}-3\\4\end{bmatrix}$. Represent both vectors as directed lines in the cartesian plane.

5. Draw the vectors $\overrightarrow{AB}=\begin{bmatrix}1\\2\end{bmatrix}$ and $\overrightarrow{BC}=\begin{bmatrix}3\\1\end{bmatrix}$ as directed lines in the cartesian plane.

(a) Verify $\overrightarrow{AC}=\overrightarrow{AB}+\overrightarrow{BC}$.

(b) Measure $|\overrightarrow{AC}|$.

(c) Find a vector of length one perpendicular to $\overrightarrow{AC}$.

(d) Find the angle between $\overrightarrow{AB}$ and $\overrightarrow{BC}$.

6. Let $\overrightarrow{AB}$ and $\overrightarrow{AC}$ be two two-dimensional vectors. Find an expression for $\overrightarrow{BC}$ in terms of $\overrightarrow{AB}$ and $\overrightarrow{AC}$. Hence, if P is a point on $\overrightarrow{BC}$ such that $m\overrightarrow{BP}=n\overrightarrow{PC}$, prove that $\overrightarrow{AP}=(n\overrightarrow{AC}+m\overrightarrow{AB})/(m+n)$.

7. What *geometrical* conditions must a basis for $\mathbb{R}^3$ satisfy?

8. Which of the following sets of vectors are linearly independent?

(a) $\begin{bmatrix}1\\2\\3\\4\end{bmatrix}, \begin{bmatrix}2\\-3\\2\\3\end{bmatrix}, \begin{bmatrix}-4\\13\\0\\-1\end{bmatrix}$

(b) $\begin{bmatrix}1\\2\\3\\4\end{bmatrix}, \begin{bmatrix}2\\3\\4\\1\end{bmatrix}, \begin{bmatrix}3\\4\\1\\2\end{bmatrix}$

(c) $\begin{bmatrix}1\\2\\1\\3\end{bmatrix}, \begin{bmatrix}2\\1\\3\\0\end{bmatrix}, \begin{bmatrix}-1\\2\\1\\3\end{bmatrix}, \begin{bmatrix}3\\1\\1\\-1\end{bmatrix}$

(d) $\begin{bmatrix}2\\1\\3\end{bmatrix}, \begin{bmatrix}1\\2\\3\end{bmatrix}, \begin{bmatrix}3\\3\\5\end{bmatrix}$

2.2 MATRICES

In everyday life, one is often presented with information in a tabular form, for example, mathematical tables, railway timetables and seminar schedules. Most tables can be divided into rows or columns of information, each of which have several entries. A matrix is a mathematical way of representing such a structure. To access an element of an n-dimensional vector requires just one index but to access an element of a matrix requires two, the first to determine in which row the element can be found and the second to determine in which column. As with vectors, the discussion in this section is restricted to matrices of real numbers but it should be understood that the definitions can equally well apply to matrices of integers or matrices of complex numbers, etc. A more generalised approach is given in section 5.4.

Definition 2.2.1
An $m \times n$*(real) matrix* is a rectangular array of real numbers arranged as m rows, each consisting of n scalars.

If A denotes such a matrix, then a_{ij} will be used to denote the jth scalar in the ith row (or, equivalently, the ith scalar in the jth column). Thus

$$A=\begin{bmatrix} a_{11} & a_{12} & \dots & a_{1n} \\ a_{21} & a_{22} & \dots & a_{2n} \\ \vdots & \vdots & & \vdots \\ a_{m1} & a_{m2} & \dots & a_{mn} \end{bmatrix}$$

An alternative notation for A is $[a_{ij}]$. This is particularly useful in proofs where writing out all the elements becomes a tedious exercise. Usually, a_{ij} is called the i,jth *element* of A.

Example 2.2.1

(1) $$A=\begin{bmatrix} 2 & 1 & 3 \\ -4 & 1 & 0 \\ 3 & -1 & 2 \end{bmatrix}$$

is a 3×3 matrix. $a_{11}=2$, $a_{12}=1$, $a_{13}=3$, $a_{21}=-4$, $a_{22}=1$, $a_{23}=0$, $a_{31}=3$, $a_{32}=-1$, $a_{33}=2$.

(2) $$B=\begin{bmatrix} -2 & 4 \\ 3 & 1 \\ 4 & 2 \end{bmatrix}$$

is a 3×2 matrix. b_{32} is equal to 2 but b_{23} is not defined since there is no third column.

(3) An n-dimensional row vector $[v_1\ v_2\ \dots\ v_n]$ can be thought of as a $1\times n$ matrix and, similarly, an n-dimensional column vector as an $n\times 1$ matrix.

Matrices occur so frequently in computer science that they can be considered to be one of the most important structures discussed in this book. Very large matrices often arise and it is particularly important that these are handled efficiently. Thus, the computer scientist must be aware of various special types of matrix and of techniques for their manipulation.

Definition 2.2.2

A *square* matrix has the same number of rows as columns.

If A is a square matrix of *order n*, that is with n rows and n columns, then the elements $a_{11}, a_{22}, \dots, a_{nn}$ are said to lie on the *leading diagonal* (see figure 2.8).

A *diagonal matrix* is a square matrix which has non-zero values only on the leading diagonal. Thus, A is a diagonal matrix if and only if $a_{ij}=0$ provided $i\neq j$.

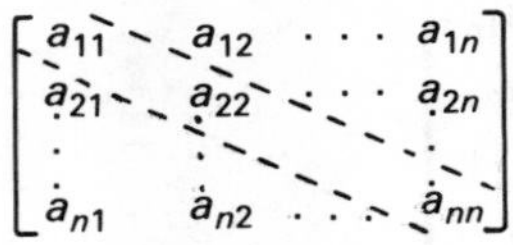

$$\begin{bmatrix} a_{11} & a_{12} & \cdots & a_{1n} \\ a_{21} & a_{22} & \cdots & a_{2n} \\ \vdots & \vdots & & \vdots \\ a_{n1} & a_{n2} & \cdots & a_{nn} \end{bmatrix}$$

Figure 2.8 The leading diagonal is contained within the broken lines

The $n \times n$ *unit matrix* is a diagonal matrix of n rows and n columns such that all numbers on the leading diagonal are equal to 1. This matrix is denoted by I_n but the subscript n may be dropped when the order of I_n is clear from the context.

Example 2.2.2

(1) $\begin{bmatrix} 3 & 2 & 1 \\ 4 & -2 & 3 \\ 5 & 6 & 1 \end{bmatrix}$

is a 3×3 square matrix but neither

$$\begin{bmatrix} 1 & 2 \\ 2 & 1 \\ 3 & -4 \end{bmatrix} \text{ nor } \begin{bmatrix} 3 & -5 & 2 \\ 2 & 3 & 4 \end{bmatrix}$$

is square since, in both cases, the number of rows is not equal to the number of columns.

(2) $\begin{bmatrix} 2 & 0 & 0 \\ 0 & 3 & 0 \\ 0 & 0 & 4 \end{bmatrix}$ is a 3×3 diagonal matrix.

(3) $I_2 = \begin{bmatrix} 1 & 0 \\ 0 & 1 \end{bmatrix}$, $I_3 = \begin{bmatrix} 1 & 0 & 0 \\ 0 & 1 & 0 \\ 0 & 0 & 1 \end{bmatrix}$

Rather than store all n^2 elements of a diagonal matrix in a computer, it is sufficient to store just the n non-zero elements on the leading diagonal. If A is such a matrix and if $\boldsymbol{b}$ is the stored vector containing the elements $[a_{11}\ a_{22}\ \ldots\ a_{nn}]$ then the following code will deliver a_{ij} given $1 \leqslant i, j \leqslant n$.

```
if i=j then b_i
else 0
endif
```

Since computer memory is essentially linear in structure, the problem of representing matrices in computer memory can be separated into two stages. The first is to map the elements of the matrix onto a vector and the second is to store the elements of this vector in adjacent locations in store. To map the elements of an $m \times n$ matrix onto a vector, there are two main techniques, the *row-major* technique and the *column-major* technique. These techniques are illustrated in figure 2.9; in the row-major technique, the elements are ordered row by row, and in the column-major technique, the elements are ordered column by column.

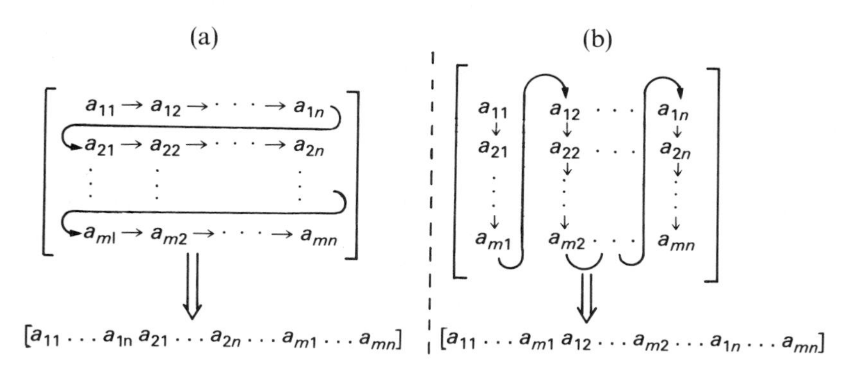

Figure 2.9 (a) Row-major technique: a_{ij} is the kth element of the vector, where $k=n\times(i-1)+j$; (b) column-major technique: a_{ij} is the kth element of the vector, where $k=m\times(j-1)+i$.

Using the row-major technique, a_{ij} is mapped onto the kth component of the vector, where $k=n\times(i-1)+j$. This formula is easily deduced, since each row of the array is mapped onto n consecutive components. Since there are $i-1$ rows to be mapped before the one in which a_{ij} appears, this uses $n\times(i-1)$ components. The scalar a_{ij} then occurs as the jth element in the next row mapped.

A similar argument can be used to show that if the column-major technique is used, a_{ij} is mapped onto the kth component of the vector, where $k=m\times(j-1)+i$.

Definition 2.2.3
If a matrix has a high proportion of zero entries then it is called a *sparse matrix*.

Rather than store all the entries of a sparse $m \times n$ matrix it may be more efficient to store the matrix as a sequence of triples (i, j, a_{ij}) for all the non-zero elements. Assuming that a real number requires two words of storage and an integer requires only one, storing the matrix as triples will save on storage provided $2s<mn$, where s is the number of non-zero elements.

Example 2.2.3

The sparse matrix

$$\begin{bmatrix} 4.1 & 0.0 & 0.0 & 0.0 & 2.3 \\ 0.0 & 1.5 & 0.0 & 0.0 & 0.0 \\ -1.0 & 0.0 & 0.0 & 0.0 & 0.0 \\ 2.1 & 0.0 & 0.0 & 1.3 & 0.0 \end{bmatrix}$$

can be stored as a sequence of triples ((1, 1, 4.1), (1, 5, 2.3), (2, 2, 1.5), (3, 1, −1.0), (4, 1, 2.1), (4, 4, 1.3)). This requires $6 \times 4 = 24$ storage locations, a saving of 16 over the 40 locations required to store the original matrix. In such a case, the saving is hardly worth while because of the necessary overheads involved in searching the triples to see if a particular a_{ij} value is non-zero.

However, applications in which $m \approx 500$ and $n \approx 1000$ with 99 per cent sparsity (that is 99 per cent zero entries) are not uncommon. In such a case, the above 'packing' technique is essential.

Some special cases of square matrices are worthy of special consideration since they arise so often in practical problems.

Definition 2.2.4

A *lower triangular matrix* is a square matrix, A, such that $a_{ij} = 0, \forall i < j$.

An *upper triangular matrix* is a square matrix, A, such that $a_{ij} = 0, \forall i > j$.

Example 2.2.4

(1) $\begin{bmatrix} 1 & 0 & 0 \\ 3 & 5 & 0 \\ 9 & 6 & 8 \end{bmatrix}$ is a 3×3 lower triangular matrix.

(2) $\begin{bmatrix} 1 & -2 & 3 & -2 \\ 0 & 5 & -6 & 3 \\ 0 & 0 & 4 & 7 \\ 0 & 0 & 0 & 9 \end{bmatrix}$ is a 4×4 upper triangular matrix.

An $n \times n$ triangular matrix can be represented as a vector of only $n(n+1)/2$ elements by mapping just the non-zero elements into the vector. This can be achieved either by listing the elements row by row or column by column.

Consider the lower triangular matrix, A, of figure 2.10 where the non-zero elements are ordered row by row. (The large circle in the right-hand

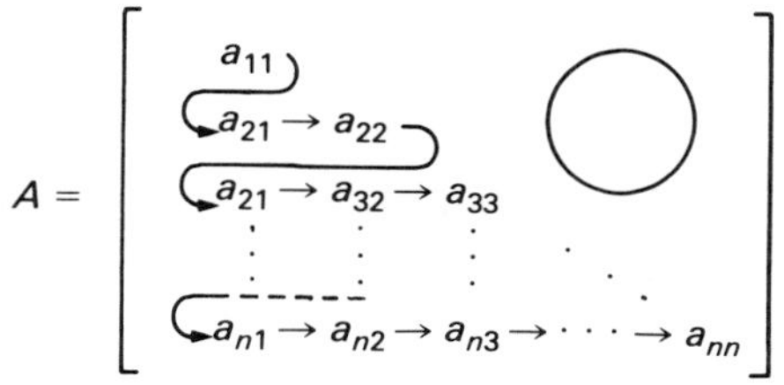

$$A = \begin{bmatrix} a_{11} & & & & \\ a_{21} \rightarrow a_{22} & & & & \\ a_{21} \rightarrow a_{32} \rightarrow a_{33} & & & & \\ \vdots & & & \ddots & \\ a_{n1} \rightarrow a_{n2} \rightarrow a_{n3} \rightarrow \cdots & & & & \rightarrow a_{nn} \end{bmatrix}$$

Figure 2.10

corner is used to denote that all elements above the diagonal are zero.) The elements of the matrix A are mapped onto a vector, $\boldsymbol{b}$, such that if a_{ij} is the kth component of $\boldsymbol{b}$, then $k=1+2+\ldots+(i-1)+j=i(i-1)/2+j$ (using the result proved in example 1.3.13). $\boldsymbol{b}$ thus contains $1+2+\ldots+n$ $=n(n+1)/2$ elements which, for large n, is a considerable saving on the n^2 elements of A. Assuming only $\boldsymbol{b}$ is stored, the code to recover a_{ij}, given $1 \leqslant i,j \leqslant n$ is

if $i<j$ **then** 0

else $k \leftarrow (i-1) * i/2+j;\ b_k$

endif

Just as there are techniques for efficient storage of matrices, so there are techniques for their efficient manipulation, and many of these rely on the important concept of a submatrix.

Definition 2.2.5
A *submatrix* of an $m \times n$ matrix, A, is a matrix, B, obtained from A by the deletion of any number of rows and/or columns.

Since a $1 \times n$ submatrix can be viewed as a row vector, an $m \times n$ matrix can be viewed as a column of m row vectors. Alternatively, since an $m \times 1$ submatrix can be viewed as a column vector, an $m \times n$ matrix can be viewed as a row of n column vectors.

If $\boldsymbol{a}_j$ denotes the column vector with components equal to the elements of the jth column of A and if $\hat{\boldsymbol{a}}_i$ denotes the row vector with components equal to the ith row of A then it is conventional to write

$$A=[\boldsymbol{a}_{ij}]=[\boldsymbol{a}_1 \ \ \boldsymbol{a}_2 \ \ \ldots \ \ \boldsymbol{a}_n]=\begin{bmatrix} \hat{\boldsymbol{a}}_1 \\ \hat{\boldsymbol{a}}_2 \\ \vdots \\ \hat{\boldsymbol{a}}_m \end{bmatrix}$$

Example 2.2.5

(1) The submatrices of $\begin{bmatrix} 2 & 1 \\ 3 & 4 \\ 4 & 1 \end{bmatrix}$ with more than one element are

$[2 \quad 1], [3 \quad 4], [4 \quad 1],$

$\begin{bmatrix} 2 \\ 3 \end{bmatrix}, \begin{bmatrix} 1 \\ 4 \end{bmatrix}, \begin{bmatrix} 3 \\ 4 \end{bmatrix}, \begin{bmatrix} 4 \\ 1 \end{bmatrix}, \begin{bmatrix} 2 \\ 4 \end{bmatrix}, \begin{bmatrix} 1 \\ 1 \end{bmatrix},$

$\begin{bmatrix} 2 & 1 \\ 3 & 4 \end{bmatrix}, \begin{bmatrix} 2 & 1 \\ 4 & 1 \end{bmatrix}, \begin{bmatrix} 3 & 4 \\ 4 & 1 \end{bmatrix}, \begin{bmatrix} 2 \\ 3 \\ 4 \end{bmatrix}, \begin{bmatrix} 1 \\ 4 \\ 1 \end{bmatrix}$ and $\begin{bmatrix} 2 & 1 \\ 3 & 4 \\ 4 & 1 \end{bmatrix}$

(2) If $A = \begin{bmatrix} -4 & 3 \\ 2 & 1 \\ -5 & 0 \end{bmatrix}$ then $A = [\boldsymbol{a}_1 \quad \boldsymbol{a}_2] = \begin{bmatrix} \hat{\boldsymbol{a}}_1 \\ \hat{\boldsymbol{a}}_2 \\ \hat{\boldsymbol{a}}_3 \end{bmatrix}$

where $\boldsymbol{a}_1 = \begin{bmatrix} -4 \\ 2 \\ -5 \end{bmatrix}$, $\boldsymbol{a}_2 = \begin{bmatrix} 3 \\ 1 \\ 0 \end{bmatrix}$, $\hat{\boldsymbol{a}}_1 = [-4 \quad 3]$, $\hat{\boldsymbol{a}}_2 = [2 \quad 1]$ and $\hat{\boldsymbol{a}}_3 = [-5 \quad 0]$.

Matrix Arithmetic

Definition 2.2.6
Let A, B be two $m \times n$ matrices. The *sum of A and B, A + B*, is the matrix obtained by adding corresponding elements of A and B

$$A + B = \begin{bmatrix} a_{11} & a_{12} & \dots & a_{1n} \\ a_{21} & a_{22} & \dots & a_{2n} \\ \vdots & \vdots & & \vdots \\ a_{m1} & a_{m2} & \dots & a_{mn} \end{bmatrix} + \begin{bmatrix} b_{11} & b_{12} & \dots & b_{1n} \\ b_{21} & b_{22} & \dots & b_{2n} \\ \vdots & \vdots & & \vdots \\ b_{m1} & b_{m2} & \dots & b_{mn} \end{bmatrix}$$

$$= \begin{bmatrix} a_{11}+b_{11} & a_{12}+b_{12} & \dots & a_{1n}+b_{1n} \\ a_{21}+b_{21} & a_{22}+b_{22} & \dots & a_{2n}+b_{2n} \\ \vdots & \vdots & & \vdots \\ a_{m1}+b_{m1} & a_{m2}+b_{m2} & \dots & a_{mn}+b_{mn} \end{bmatrix}$$

Example 2.2.6

(1) $\begin{bmatrix} 1 & 2 \\ -3 & 5 \end{bmatrix} + \begin{bmatrix} 3 & 7 \\ 0 & -6 \end{bmatrix} = \begin{bmatrix} 4 & 9 \\ -3 & -1 \end{bmatrix}$

(2) $\begin{bmatrix} 1 & 2 \\ 3 & -2 \\ 5 & 4 \end{bmatrix} + \begin{bmatrix} 3 & 0 & 5 \\ 7 & 8 & -3 \\ 6 & 9 & 5 \end{bmatrix}$

is not defined since the two matrices do not have the same number of columns.

Definition 2.2.7
If A is an $m \times n$ matrix and $\lambda \in \mathbb{R}$ is some scalar, then the *scalar multiplication of A by λ* is the matrix obtained from A by multiplying each of its elements by λ

$$\lambda A = \lambda \begin{bmatrix} a_{11} & a_{12} & \dots & a_{1n} \\ a_{21} & a_{22} & \dots & a_{2n} \\ \vdots & \vdots & & \vdots \\ a_{m1} & a_{m2} & \dots & a_{mn} \end{bmatrix} = \begin{bmatrix} \lambda a_{11} & \lambda a_{12} & \dots & \lambda a_{1n} \\ \lambda a_{21} & \lambda a_{22} & \dots & \lambda a_{2n} \\ \vdots & \vdots & & \vdots \\ \lambda a_{m1} & \lambda a_{m2} & \dots & \lambda a_{mn} \end{bmatrix}$$

If $\lambda = -1$ then $-1A$ is often denoted by $-A$ and then $A+(-B)$ is denoted by $A-B$.

Example 2.2.7

If

$$A = \begin{bmatrix} 2 & 9 \\ -3 & -8 \end{bmatrix}, B = \begin{bmatrix} 1 & -3 \\ 7 & 6 \end{bmatrix}, C = \begin{bmatrix} 2 & 3 \\ -4 & 8 \\ 5 & 0 \end{bmatrix}, D = \begin{bmatrix} 4 & 9 \\ -2 & 6 \\ 0 & -1 \end{bmatrix}$$

then

$$2A = \begin{bmatrix} 4 & 18 \\ -6 & -16 \end{bmatrix}, -B = \begin{bmatrix} -1 & 3 \\ -7 & -6 \end{bmatrix}, 2A-B = \begin{bmatrix} 3 & 21 \\ -13 & -22 \end{bmatrix}$$

$$-3C = \begin{bmatrix} -6 & -9 \\ 12 & -24 \\ -15 & 0 \end{bmatrix}, 2D = \begin{bmatrix} 8 & 18 \\ -4 & 12 \\ 0 & -2 \end{bmatrix}, -3C+2D = \begin{bmatrix} 2 & 9 \\ 8 & -12 \\ -15 & -2 \end{bmatrix}$$

No scalar multiple of either A or B can be added to a scalar multiple of C or D since A and B do not have the same number of rows as C or D.

The operations of matrix addition and scalar multiplication act on matrices in a way very similar to that of ordinary addition and multiplication on real numbers. This fact is summarised in theorem 2.2.1. The proofs of the results are easy but the reader should nevertheless check through each of them carefully. In parts (iii) and (vii), 0 is used to denote a matrix consisting of all zero entries. Such a matrix is called the *null matrix* or *zero matrix*.

Theorem 2.2.1
Let A, B, C be any $m \times n$ matrices and $\lambda_1, \lambda_2 \in \mathbb{R}$ be any scalars. Then

(i) $(A+B)+C=A+(B+C)$;

(ii) $A+B=B+A$;

(iii) $A+0=A=0+A$;

(iv) $\lambda_1(\lambda_2 A)=(\lambda_1\lambda_2)A$;

(v) $(\lambda_1+\lambda_2)A=\lambda_1 A+\lambda_2 A$;

(vi) $\lambda_1(A+B)=\lambda_1 A+\lambda_1 B$;

(vii) $A-A=0$.

Proof In this proof, $A=[a_{ij}]$, $B=[b_{ij}]$ and $C=[c_{ij}]$.

(i)
$$\begin{aligned}(A+B)+C&=([a_{ij}]+[b_{ij}])+[c_{ij}]=[a_{ij}+b_{ij}]+[c_{ij}]\\&=[a_{ij}+b_{ij}+c_{ij}]=[a_{ij}]+[b_{ij}+c_{ij}]=[a_{ij}]+([b_{ij}]+[c_{ij}])\\&=A+(B+C)\end{aligned}$$

(ii)
$$\begin{aligned}A+B&=[a_{ij}]+[b_{ij}]=[a_{ij}+b_{ij}]=[b_{ij}+a_{ij}]=[b_{ij}]+[a_{ij}]\\&=B+A\end{aligned}$$

(iii)
$$A+0=[a_{ij}]+[0]=[a_{ij}+0]=[a_{ij}]=A$$

and $A+0=0+A$ by (ii) above.

(iv)
$$\begin{aligned}\lambda_1(\lambda_2 A)&=\lambda_1(\lambda_2[a_{ij}])=\lambda_1[\lambda_2 a_{ij}]=[\lambda_1\lambda_2 a_{ij}]=\lambda_1\lambda_2[a_{ij}]\\&=(\lambda_1\lambda_2)A\end{aligned}$$

(v)
$$\begin{aligned}(\lambda_1+\lambda_2)A&=(\lambda_1+\lambda_2)[a_{ij}]=[(\lambda_1+\lambda_2)a_{ij}]=[\lambda_1 a_{ij}+\lambda_2 a_{ij}]\\&=[\lambda_1 a_{ij}]+[\lambda_2 a_{ij}]=\lambda_1[a_{ij}]+\lambda_2[a_{ij}]=\lambda_1 A+\lambda_2 A\end{aligned}$$

(vi)
$$\begin{aligned}\lambda_1(A+B)&=\lambda_1([a_{ij}]+[b_{ij}])=\lambda_1[a_{ij}+b_{ij}]=[\lambda_1(a_{ij}+b_{ij})]\\&=[\lambda_1 a_{ij}+\lambda_1 b_{ij}]=[\lambda_1 a_{ij}]+[\lambda_1 b_{ij}]=\lambda_1[a_{ij}]+\lambda_1[b_{ij}]\\&=\lambda_1 A+\lambda_1 B\end{aligned}$$

(vii) $A-A=A+(-1)A=[a_{ij}]+(-1)[a_{ij}]=[a_{ij}]+[-a_{ij}]$
$=[a_{ij}-a_{ij}]=[0]=0$

Having defined matrix addition and multiplication by a scalar, it seems natural to ask: what about multiplication of one matrix by another? In section 2.1, the scalar product of two vectors was defined and, since vectors can be regarded as special cases of matrices, it is natural to try to generalise this type of product.

Consider the row vector $\boldsymbol{a}=[a_1\ a_2\ \dots\ a_n]$ and the column vector

$$\boldsymbol{b}=\begin{bmatrix} b_1 \\ b_2 \\ \vdots \\ b_n \end{bmatrix}$$

The scalar product of these two vectors gives the scalar $c=a_1b_1+a_2b_2+\dots+a_nb_n$.

Now, an $m\times n$ matrix can be regarded as a column of m row vectors. It is natural then to define the product of such a matrix with the vector $\boldsymbol{b}$ to be the column vector consisting of the m scalars $c_1, c_2, \dots, c_m$ obtained by taking the scalar product of each of the m rows of A, in turn, with $\boldsymbol{b}$. Hence

$$\begin{bmatrix} a_{11} & a_{12} & \dots & a_{1n} \\ a_{21} & a_{22} & \dots & a_{2n} \\ \vdots & & & \vdots \\ a_{m1} & a_{m2} & \dots & a_{mn} \end{bmatrix}\begin{bmatrix} b_1 \\ b_2 \\ \vdots \\ b_n \end{bmatrix}=\begin{bmatrix} c_1 \\ c_2 \\ \vdots \\ c_m \end{bmatrix}$$

where $c_i=a_{i1}b_1+a_{i2}b_2+\dots+a_{in}b_n$, $i=1, 2, \dots, m$.

Definition 2.2.8

Multiplying an n-dimensional vector by an $m\times n$ matrix transforms the vector into an m-dimensional vector. Thus, any $m\times n$ matrix defines a mapping $\mathbb{R}^n\rightarrow\mathbb{R}^m$. Such a mapping is called a *linear transformation.*

Linear transformations have the interesting property that they 'preserve' the two basic operations of vector algebra, namely vector addition and scalar multiplication. The proof follows easily from the definitions and is left to the reader.

Theorem 2.2.2

Let A be any $m\times n$ matrix. Then

(i) $\forall\boldsymbol{v}\in\mathbb{R}^n$, $A\boldsymbol{v}\in\mathbb{R}^m$,

(ii) $\forall \boldsymbol{v},\boldsymbol{w} \in \mathbb{R}^n$, $A(\boldsymbol{v}+\boldsymbol{w}) = A\boldsymbol{v} + A\boldsymbol{w}$,

(iii) $\forall \boldsymbol{v} \in \mathbb{R}^n, \forall \lambda \in \mathbb{R}$, $A(\lambda \boldsymbol{v}) = \lambda(A\boldsymbol{v})$.

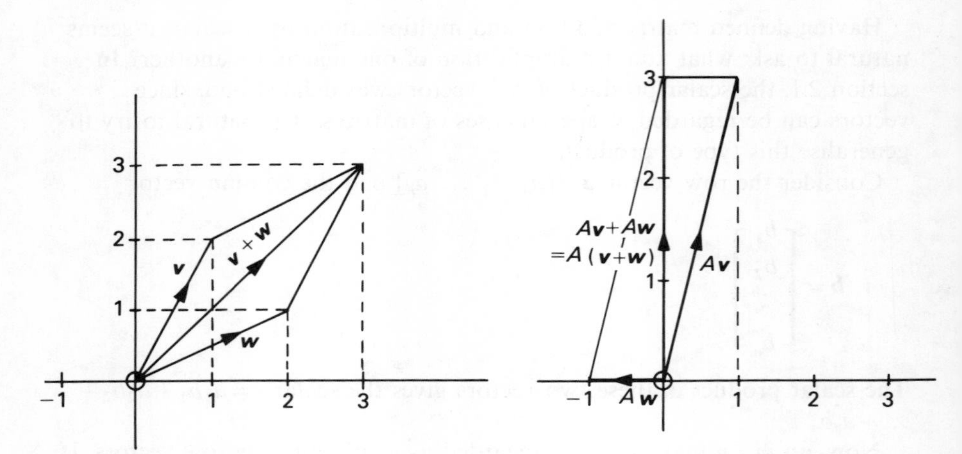

Figure 2.11 Example 2.2.8(1)

Example 2.2.8

(1) In figure 2.11, the linear transformation $\boldsymbol{v} \mapsto A\boldsymbol{v}$, where $A = \begin{bmatrix} -1 & 1 \\ -1 & 2 \end{bmatrix}$, is represented diagrammatically acting on vectors $\boldsymbol{v} = \begin{bmatrix} 1 \\ 2 \end{bmatrix}$, $\boldsymbol{w} = \begin{bmatrix} 2 \\ 1 \end{bmatrix}$ and $\boldsymbol{v}+\boldsymbol{w} = \begin{bmatrix} 3 \\ 3 \end{bmatrix}$

$$A\boldsymbol{v} = \begin{bmatrix} 1 \\ 3 \end{bmatrix}, A\boldsymbol{w} = \begin{bmatrix} -1 \\ 0 \end{bmatrix} \text{ and } A(\boldsymbol{v}+\boldsymbol{w}) = \begin{bmatrix} 0 \\ 3 \end{bmatrix} = A\boldsymbol{v} + A\boldsymbol{w}$$

(2) If $A = \begin{bmatrix} 3 & -4 \\ 2 & 3 \\ 0 & 1 \end{bmatrix}$, $\boldsymbol{v} = \begin{bmatrix} 2 \\ 1 \end{bmatrix}$ and $\boldsymbol{w} = \begin{bmatrix} -2 \\ 5 \end{bmatrix}$, then $A\boldsymbol{v} = \begin{bmatrix} 2 \\ 7 \\ 1 \end{bmatrix}$,

$A\boldsymbol{w} = \begin{bmatrix} -26 \\ 11 \\ 5 \end{bmatrix}$. If $\lambda_1, \lambda_2 \in \mathbb{R}$, then

$$\lambda_1 \boldsymbol{v} + \lambda_2 \boldsymbol{w} = \begin{bmatrix} 2\lambda_1 - 2\lambda_2 \\ \lambda_1 + 5\lambda_2 \end{bmatrix}$$

so

$$A(\lambda_1 \boldsymbol{v}+\lambda_2 \boldsymbol{v})=\begin{bmatrix} 3(2\lambda_1-2\lambda_2)-4(\lambda_1+5\lambda_2) \\ 2(2\lambda_1-2\lambda_2)+3(\lambda_1+5\lambda_2) \\ 0(2\lambda_1-2\lambda_2)+1(\lambda_1+5\lambda_2) \end{bmatrix}=\begin{bmatrix} 2\lambda_1-26\lambda_2 \\ 7\lambda_1+11\lambda_2 \\ \lambda_1+5\lambda_2 \end{bmatrix}$$
$$=\lambda_1 A\boldsymbol{v}+\lambda_2 A\mathbf{w}$$

(3) If $A=[a_{ij}]$ and $\boldsymbol{v}$ is the identity vector, $\boldsymbol{e}_k$, $(1 \leqslant k \leqslant n)$ then $A\boldsymbol{v}$ is a vector equal to the kth column of A, that is if $A=[\boldsymbol{a}_1 \ \boldsymbol{a}_2 \ \dots \ \boldsymbol{a}_n]$ then $A\boldsymbol{e}_k = \boldsymbol{a}_k$.

Having defined the result of multiplying an $m \times n$ matrix and an $n \times 1$ matrix, it now remains to define the result of multiplying an $m \times n$ matrix and an $n \times p$ matrix. The definition arises from regarding the second matrix as a row of p column vectors. Multiplying the first matrix by each of these, in turn, results in a row of p column vectors each of m elements, that is in an $m \times p$ matrix.

Definition 2.2.9
Let A be an $m \times n$ matrix and B be an $n \times p$ matrix then AB, the *matrix product of A and B*, is the $m \times p$ matrix C, where c_{ij} is the scalar product of the ith row of A with the jth column of B. That is, $AB=C=[c_{ij}]$, where $c_{ij}=\sum_{k=1}^{n} a_{ik}b_{kj}=a_{i1}b_{1j}+a_{i2}b_{2j}+\dots+a_{in}b_{nj}$.

Example 2.2.9

(1) If $A=\begin{bmatrix} -2 & 3 \\ 1 & 4 \end{bmatrix}$, $B=\begin{bmatrix} 2 & -1 & 3 \\ 0 & 4 & 1 \end{bmatrix}$ and $C=\begin{bmatrix} 3 & -1 \\ 2 & 0 \\ -1 & 1 \end{bmatrix}$ then

$AB=\begin{bmatrix} -4 & 14 & -3 \\ 2 & 15 & 7 \end{bmatrix}$ but BA is not defined.

$BC=\begin{bmatrix} 1 & 1 \\ 7 & 1 \end{bmatrix}$ but CB is not defined.

$(AB)C=\begin{bmatrix} 19 & 1 \\ 29 & 5 \end{bmatrix}$ and also $A(BC)=\begin{bmatrix} 19 & 1 \\ 29 & 5 \end{bmatrix}$

but neither is equal to $C(AB)$ which is a 3×3 matrix.

This example shows that AB does not in general equal BA. In fact, even if AB is defined it does not follow that BA is defined.

(2) If A is any $m \times n$ matrix and I is the $n \times n$ identity matrix then $AI = A$. This follows immediately from example 2.2.8(3).

(3) If A is any $m \times n$ matrix and 0 is the $n \times n$ zero matrix, then $A0$ is an $m \times n$ matrix with all entries equal to zero.

(4) If $A = \begin{bmatrix} 1 & 2 \\ 4 & 3 \end{bmatrix}$ and

$$B = \begin{bmatrix} -3/5 & 2/5 \\ 4/5 & -1/5 \end{bmatrix}$$

then $AB = \begin{bmatrix} 1 & 0 \\ 0 & 1 \end{bmatrix}$ and $BA = \begin{bmatrix} 1 & 0 \\ 0 & 1 \end{bmatrix}$, that is $AB = BA = I$ and B is said to be the *inverse matrix* of A.

In general, if there exists a matrix B such that $AB = BA = I$ then this matrix B is denoted by A^{-1} and A is said to be an *invertible matrix*.

Although it is not true that $AB = BA$ in general, matrix multiplication does have some useful properties which are summarised below.

Theorem 2.2.3

(i) If A is an $m \times n$ matrix and B an $n \times p$ matrix, then

$\lambda(AB) = (\lambda A)B = A(\lambda B), \ \forall \lambda \in \mathbb{R}$

(ii) If A is an $m \times n$ matrix, B an $n \times p$ matrix and C a $p \times q$ matrix, then $A(BC) = (AB)C$

(iii) If A is an $m \times n$ matrix and B and C are $n \times p$ matrices, then

$A(B + C) = AB + AC$

(iv) If A and B are $m \times n$ matrices and C is an $n \times p$ matrix, then

$(A + B)C = AC + BC$

Matrix addition, subtraction and multiplication are not defined for any two arbitrary matrices. For addition and subtraction to be defined, both matrices must have the same number of rows and columns. For multiplication to be defined, the number of columns of the first matrix must equal the number of rows of the second. However, if attention is confined to $n \times n$ matrices, then all these operations are defined.

Definition 2.2.10
Let $P(x) \equiv a_0 + a_1 x + a_2 x^2 + \ldots + a_m x^m$, $a_i \in \mathbb{R}$, denote a polynomial of degree m. For any $n \times n$ matrix, $P(A)$ is defined to be the matrix $a_0 I + a_1 A + a_2 A^2 + \ldots + a_m A^m$, where I is the $n \times n$ identity matrix, $A^2 = AA$ and $A^i = AA^{i-1}$ for $i = 3, 4, \ldots, m$.

If $P(A) = 0$, A is said to be a *zero of the polynomial P*.

Example 2.2.10

(1) Consider the matrix of order 2, $A = \begin{bmatrix} 1 & 2 \\ -1 & 3 \end{bmatrix}$. $A^2 = \begin{bmatrix} -1 & 8 \\ -4 & 7 \end{bmatrix}$ and $A^3 = \begin{bmatrix} -9 & 22 \\ -11 & 13 \end{bmatrix}$. Hence, if $P(x) = 2 + 2x + x^2 + 3x^3$

$$P(A) = \begin{bmatrix} 2 & 0 \\ 0 & 2 \end{bmatrix} + \begin{bmatrix} 2 & 4 \\ -2 & 6 \end{bmatrix} + \begin{bmatrix} -1 & 8 \\ -4 & 7 \end{bmatrix} + \begin{bmatrix} -27 & 66 \\ -33 & 39 \end{bmatrix} = \begin{bmatrix} -24 & 78 \\ -39 & 54 \end{bmatrix}$$

(2) The matrix of order 2, $J = \begin{bmatrix} 0 & -1 \\ 1 & 0 \end{bmatrix}$ is a zero of the polynomial $x^2 + 1$.

Definition 2.2.11
The transpose, A^T, of an $m \times n$ matrix A is the $n \times m$ matrix obtained by taking the rows of A, in order, as columns, that is, by interchanging the rows and columns of A.

$$\begin{bmatrix} a_{11} & a_{12} & \ldots & a_{1n} \\ a_{21} & a_{22} & \ldots & a_{2n} \\ \vdots & \vdots & & \vdots \\ a_{m1} & a_{m2} & \ldots & a_{mn} \end{bmatrix}^T = \begin{bmatrix} a_{11} & a_{21} & \ldots & a_{m1} \\ a_{12} & a_{22} & \ldots & a_{m2} \\ \vdots & \vdots & & \vdots \\ a_{1n} & a_{2n} & \ldots & a_{mn} \end{bmatrix}$$

A square matrix of order n is said to be *symmetric* if $A = A^T$, that is if $a_{ij} = a_{ji}, \forall i,j$, and *skew-symmetric* if $a_{ij} = -a_{ji}, \forall i,j$.

Example 2.2.11

(1) $$\begin{bmatrix} 2 & 1 \\ -3 & 4 \\ 5 & 1 \end{bmatrix}^T = \begin{bmatrix} 2 & -3 & 5 \\ 1 & 4 & 1 \end{bmatrix}$$

(2) The transpose of a row vector $[v_1 \ v_2 \ \dots \ v_n]$ is the column vector

$$\begin{bmatrix} v_1 \\ v_2 \\ \vdots \\ v_n \end{bmatrix}$$

(3) $\begin{bmatrix} 3 & 1 & 2 \\ 1 & 4 & 7 \\ 2 & 7 & 6 \end{bmatrix}$ is a 3×3 symmetric matrix.

(4) $\begin{bmatrix} 1.1 & -6.2 & 3.0 \\ 6.2 & 7.0 & 4.2 \\ -3 & -4.2 & 5.7 \end{bmatrix}$ is a 3×3 skew-symmetric matrix.

The transpose operation on matrices satisfies the properties of theorem 2.2.4. Again the reader is asked to provide the proof.

Theorem 2.2.4
(i) If A is an $m \times n$ matrix then A^T is an $n \times m$ matrix such that $(A^T)^T = A$ and for any $\lambda \in \mathbb{R}$, $\lambda(A^T) = (\lambda A)^T$.

(ii) If A, B are both $m \times n$ matrices then $(A+B)^T = A^T + B^T$ is an $n \times m$ matrix.

(iii) If A is an $m \times n$ matrix and B is an $n \times p$ matrix then $(AB)^T = B^T A^T$ is a $p \times m$ matrix.

Exercise 2.2

1. If $A = \begin{bmatrix} 4 & -1 \\ 0 & 1 \\ 2 & 3 \end{bmatrix}$, $B = \begin{bmatrix} 2 & 1 \\ -1 & 1 \end{bmatrix}$ and $C = [1 \quad 2]$

evaluate the following (where possible)

(a) $AB + 2B$,

(b) ABC,

(c) $2A^T A - 3B$,

(d) BC^T.

2. If $A=\begin{bmatrix}-2 & 1\\ 1 & 0\end{bmatrix}$ and $B=\begin{bmatrix}-1 & 0\\ 2 & -3\end{bmatrix}$ evaluate

(a) $A^{\mathrm{T}}(2A+AB)$,

(b) $(A-2B)(A+2B)$,

(c) $((A^{\mathrm{T}}A)^{\mathrm{T}}A)^{\mathrm{T}}$.

3. A square $n\times n$ matrix, A, is called *tridiagonal* if $a_{ij}=0$ whenever $|i-j|>1$. Construct a 4×4 tridiagonal matrix. If only the non-zero elements of A were to be stored in a computer, how might this be achieved?

4. Show that $\begin{bmatrix}1 & 2\\ 3 & -4\end{bmatrix}$ is a zero of the polynomial $x^2+3x-10$.

5. Show that the linear transformation $\boldsymbol{v}\mapsto\begin{bmatrix}0 & -1\\ 1 & 0\end{bmatrix}\boldsymbol{v}$ preserves the length of the vector $\boldsymbol{v}$ in $\mathbb{R}^2$. Diagrammatically represent the action of this linear transformation on an arbitrary vector in $\mathbb{R}^2$.

6. How many submatrices are there of a 2×3 matrix? How many of a 3×4 matrix?

7. Prove theorem 2.2.2.

8. Prove theorem 2.2.3.

9. Prove theorem 2.2.4.

2.3 SYSTEMS OF LINEAR EQUATIONS

The solution of a *system of simultaneous linear equations* is a common computational problem. Indeed, it has been estimated that 75 per cent of all scientific problems give rise to such a system of equations.

A system of m linear equations in n unknowns $x_1, x_2, \ldots, x_n$ has the form

$$\begin{aligned}
a_{11}x_1 + a_{12}x_2 + \ldots + a_{1n}x_n &= b_1\\
a_{21}x_1 + a_{22}x_2 + \ldots + a_{2n}x_n &= b_2\\
\vdots \qquad\qquad\qquad\qquad & \ \ \vdots\\
a_{m1}x_1 + a_{m2}x_2 + \ldots + a_{mn}x_n &= b_m
\end{aligned}$$

in which each a_{ij} and each b_i is a given real number. Using vector and matrix notation, the above system may be written in the compact form

$$A\boldsymbol{x} = \boldsymbol{b}$$

where $A = [a_{ij}]$, $\boldsymbol{x} = [x_1\ x_2\ \dots\ x_n]^T$ and $\boldsymbol{b}[b_1\ b_2\ \dots\ b_m]^T$. The $m \times n$ matrix A is called the *coefficient matrix* and $\boldsymbol{b}$ is called the *right-hand-side vector*.

Rather than denote a set of linear equations by

$$A\boldsymbol{x} = \boldsymbol{b}$$

where A is the coefficient matrix and $\boldsymbol{x}$ and $\boldsymbol{b}$ are column vectors, it is possible (by applying the transpose operator to both sides of the equation) to write

$$\boldsymbol{x}^T A^T = \boldsymbol{b}^T$$

where $\boldsymbol{x}^T$ and $\boldsymbol{b}^T$ are row vectors. Although used later in the book, this notation will not be used further in the current chapter.

Definition 2.3.1
If $\boldsymbol{c}$ is an n-component vector such that $A\boldsymbol{c} = \boldsymbol{b}$ then $\boldsymbol{x} = \boldsymbol{c}$ is a *solution* of the system of equations $A\boldsymbol{x} = \boldsymbol{b}$.

Example 2.3.1

(1) The system of three linear equations in three unknowns

$$2x_1 + x_2 + 4x_3 = 12$$

$$8x_1 - 3x_2 + 2x_3 = 20$$

$$4x_1 + 11x_2 - x_3 = 33$$

may be written in the matrix form

$$\begin{bmatrix} 2 & 1 & 4 \\ 8 & -3 & 2 \\ 4 & 11 & -1 \end{bmatrix} \begin{bmatrix} x_1 \\ x_2 \\ x_3 \end{bmatrix} = \begin{bmatrix} 12 \\ 20 \\ 33 \end{bmatrix}$$

The vector

$$\begin{bmatrix} 3 \\ 2 \\ 1 \end{bmatrix}$$

is a solution of this system of equations since

$$2\times3+1\times2+4\times1=12$$
$$8\times3-3\times2+2\times1=20$$
$$4\times3+11\times2-1\times1=33$$

(2) Each of the equations

$$2x_1+4x_2=8$$
$$x_1-3x_2=6$$

may be represented by a straight line in the cartesian plane (figure 2.12). It is from this geometrical interpretation that the term 'linear' equation arises. Clearly, any solution of the given pair of equations must be

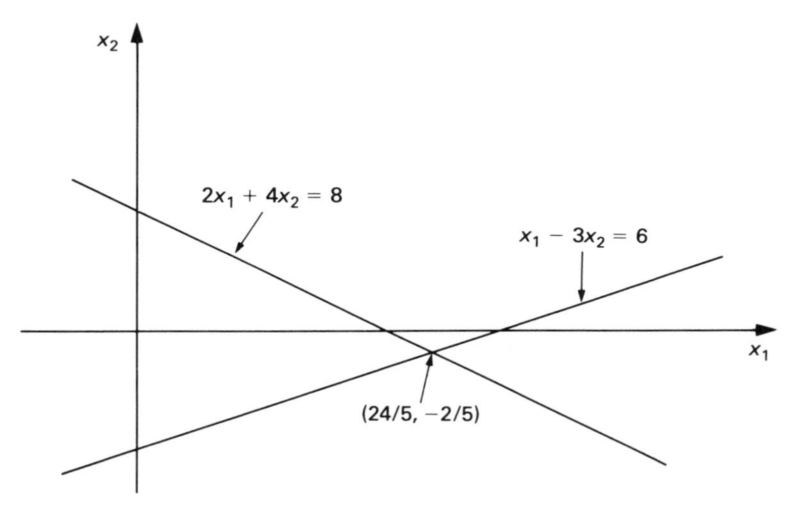

Figure 2.12 Example 2.3.1(2)

represented by a point which lies on both of the corresponding straight lines. Since the lines intersect at the point (24/5, −2/5), the unique solution of the pair of equations is $\begin{bmatrix} 24/5 \\ -2/5 \end{bmatrix}$.

(3) Consider the equations

$$2x_1+4x_2=8$$
$$x_1-3x_2=6$$
$$x_1-x_2=4$$

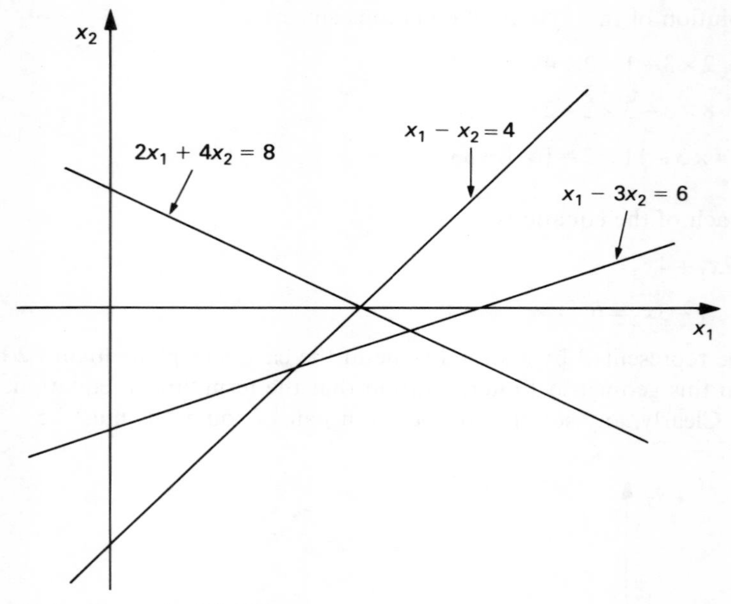

Figure 2.13 Example 2.3.1(3)

There is no point in the cartesian plane at which the straight lines representing these three equations simultaneously intersect. There is thus no solution to the given system. There is, however, a unique solution to each of the three systems consisting of just two of the equations. In figure 2.13, these solutions are represented by the points of intersection of each pair of lines.

(4) The straight lines representing the equations

$$x_1 - x_2 = 3$$
$$2x_1 + 3x_2 = 2$$
$$3x_1 + 2x_2 = 5$$

simultaneously intersect at the point $(11/5, -4/5)$; see figure 2.14. The unique solution of the system is thus $\begin{bmatrix} 11/5 \\ -4/5 \end{bmatrix}$. This is clearly also the unique solution of the system consisting of any pair of the equations.

(5) The straight lines representing the two equations

$$2x_1 + 4x_2 = 5$$
$$x_1 + 2x_2 = 2$$

are parallel (figure 2.15) and hence the system has no solution.

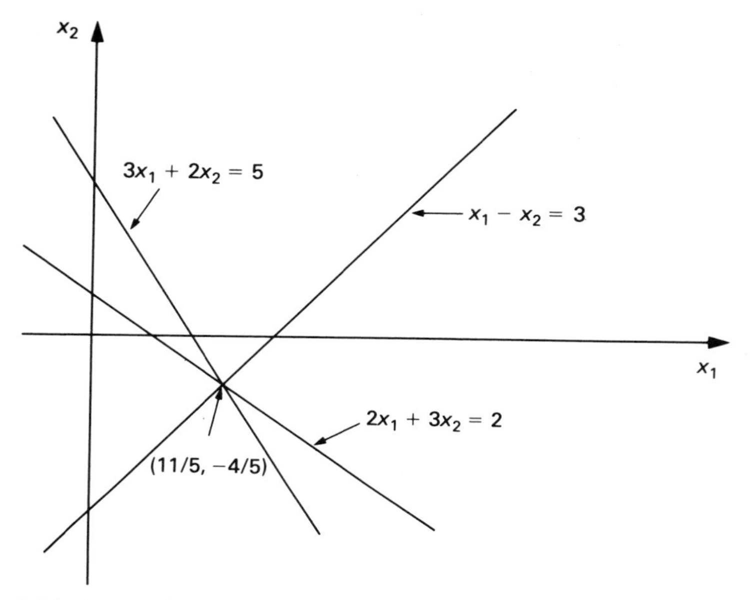

Figure 2.14 Example 2.3.1(4)

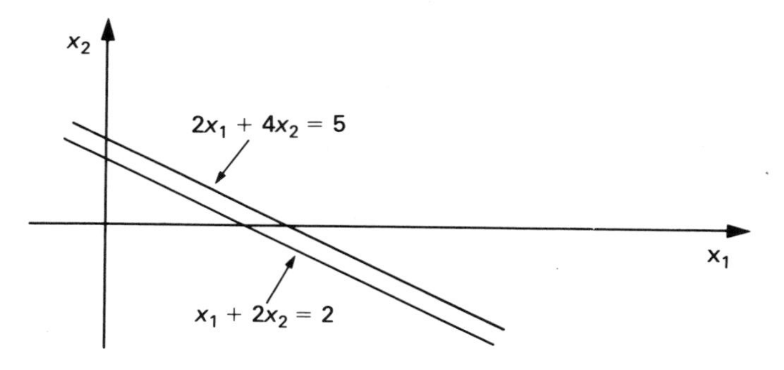

Figure 2.15 Example 2.3.1(5)

(6) The system

$$x_1 + x_2 = 2$$

consists of one equation in two unknowns. If c is any real number then $\begin{bmatrix} c \\ 2-c \end{bmatrix}$ is a solution of the given equation , that is, there are an infinite number of solutions.

Example 2.3.1(3) illustrates that if $m > n$ in a system of m linear equations in n unknowns then, in general, a solution does not exist. The exception to this general rule occurs when $m-n$ of the equations in the system are *redundant equations*. This is illustrated in example 2.3.1(4), in which the third equation may be obtained by adding the first two equations together. The third equation is thus redundant since it may be deleted from the system of equations without affecting the solution of the system.

If $m>n$ and none of the equations is redundant, the system is said to be *over-determined*. For such a system it is often required to determine a vector $\boldsymbol{c}$ such that the difference in 'size' between $A\boldsymbol{c}$ and $\boldsymbol{b}$ is minimised in some sense. If $\|\cdot\|$ denotes a vector norm (see definition 2.1.7), the problem for an over-determined system may be stated as follows

$$\text{determine } \boldsymbol{c} \text{ such that } \|A\boldsymbol{c}-\boldsymbol{b}\| \leqslant \|A\boldsymbol{x}-\boldsymbol{b}\|, \ \forall x \in \mathbb{R}^n$$

The equations in an over-determined system are *inconsistent* in the sense that no vector $\boldsymbol{c}$ can satisfy all of the equations simultaneously. Systems that are not over-determined (that is, for which $m \leqslant n$) may also be inconsistent, as illustrated by the pair of equations in example 2.3.1(5). If $m<n$, the system of equations is said to be *under-determined*. Example 2.1.3(6) illustrates that an under-determined system of equations has an infinite number of solutions, provided the equations are consistent.

Systems of Linearly Independent Equations

Each of the rows in the coefficient matrix A of a system of m linear equations in n unknowns may be regarded as a vector in $\mathbb{R}^n$. If $m>n$, at least $m-n$ of the rows of A may be written as linear combinations of the remaining rows, since no more than n vectors in $\mathbb{R}^n$ can be linearly independent. Hence, either at least $m-n$ of the equations in the system $A\boldsymbol{x}=\boldsymbol{b}$ are redundant or else the system is inconsistent. In example 2.3.1(4)

$$A=\begin{bmatrix} 1 & -1 \\ 2 & 3 \\ 3 & 2 \end{bmatrix}, \ \boldsymbol{b}=\begin{bmatrix} 3 \\ 2 \\ 5 \end{bmatrix}$$

Since the third row of A is equal to the sum of the first two rows and $b_3=b_1+b_2$, the third equation is redundant and may be deleted from the system. In example 2.3.1(3)

$$A=\begin{bmatrix} 2 & 4 \\ 1 & -3 \\ 1 & -1 \end{bmatrix}, \ \boldsymbol{b}=\begin{bmatrix} 8 \\ 6 \\ 4 \end{bmatrix}$$

Now row one of A is equal to $-3 \times$ row two $+ 5 \times$ row three. However,

$-3 \times b_2 + 5 \times b_3 = 2$, which is not equal to b_1. Hence the system is inconsistent.

Definition 2.3.2
If a linear system of equations is consistent and none of the equations is redundant, the system is said to be a *linearly independent system of equations.*

Rank of a Matrix

Definition 2.3.3
The number of linearly independent rows in an $m \times n$ matrix A is called the *row rank* of A and the number of linearly independent columns of A is called the *column rank* of A.

Theorem 2.3.1
The row rank of a matrix is equal to its column rank.

Proof Let $A \equiv [a_{ij}]$ be an $m \times n$ matrix and let r and c denote the row rank of A and the column rank of A, respectively. Without loss of generality, assume that the first r rows of A are linearly independent and the first c columns of A are linearly independent.

Notation Let $\hat{A}$ denote the $r \times n$ matrix obtained by deleting the last $m-r$ rows of A.

Let A^* denote the $(m-r) \times n$ matrix obtained by deleting the first r rows of A.

For $j = 1, 2, \ldots, n$, let $\boldsymbol{a}_j$ denote the m-component vector whose elements form the jth column of A; let $\hat{\boldsymbol{a}}_j$ denote the r-component vector whose elements form the jth column of $\hat{A}$; and let $\boldsymbol{a}_j^*$ denote the $(m-r)$-component vector whose elements form the jth column of A^*.

If $\hat{r}$ and $\hat{c}$ denote the row rank of $\hat{A}$ and the column rank of $\hat{A}$, respectively, then clearly $\hat{r} = r$ and $\hat{c} \leqslant c$.

Assume $\hat{c} < c$, that is, that less than c columns of $\hat{A}$ are linearly independent. It must therefore be possible to express at least one of the first c columns of $\hat{A}$ as a linear combination of the remaining $c-1$ columns. Assume, for the sake of argument, that $\hat{\boldsymbol{a}}_1$ may be expressed as a linear combination of $\hat{\boldsymbol{a}}_2, \hat{\boldsymbol{a}}_3, \ldots, \hat{\boldsymbol{a}}_c$, that is

$$\hat{\boldsymbol{a}}_1 = \sum_{j=2}^{c} \alpha_j \hat{\boldsymbol{a}}_j$$

where $\alpha_2, \alpha_3, \ldots, \alpha_c$ are scalar values.

By assumption, each of the rows of A^* may be written as a linear combination of the r rows of $\hat{A}$, that is, for $i = 1, 2, \ldots, m-r$

$$[a_{r+i,1}a_{r+i,2}\ldots\ a_{r+i,n}]=\sum_{k=1}^{r}\beta_{ik}[a_{k1}\ \ a_{k2}\ \ \ldots\ \ a_{kn}]$$

where β_{i1}, β_{i2}, ..., β_{ir} are scalar values.

If B denotes the $(m-r)\times r$ matrix $[\beta_{ik}]$ then

$$A^*=B\hat{A}$$

Hence

$$\begin{aligned}\boldsymbol{a}_1^*&=B\hat{\boldsymbol{a}}_1\\&=B\left(\sum_{j=2}^{c}\alpha_j\hat{\boldsymbol{a}}_j\right)\\&=\sum_{j=2}^{c}\alpha_jB\hat{\boldsymbol{a}}_j\\&=\sum_{j=2}^{c}\alpha_j\boldsymbol{a}_j^*\end{aligned}$$

Thus

$$\boldsymbol{a}_1=\begin{bmatrix}\hat{\boldsymbol{a}}_1\\ \boldsymbol{a}_1^*\end{bmatrix}=\begin{bmatrix}\sum_{j=2}^{c}\alpha_j\hat{\boldsymbol{a}}_j\\ \sum_{j=2}^{c}\alpha_j\boldsymbol{a}_j^*\end{bmatrix}=\sum_{j=2}^{c}\alpha_j\boldsymbol{a}_j$$

But this contradicts the assumption that the first c columns of A are linearly independent. Hence $\hat{c}\nless c$ and so $\hat{c}=c$. Since at most r vectors in $\mathbb{R}^r$ can be linearly independent, $\hat{c}\leqslant r$ and thus $c\leqslant r$.

Repeating the above argument with A replaced by A^{T} gives $r\leqslant c$. Hence $r=c$, as required.

Definition 2.3.4
The *rank* of a matrix is the number of linearly independent rows or columns of the matrix.

Theorem 2.3.2
Let A be an $n\times n$ matrix. The system of equations $A\boldsymbol{x}=\boldsymbol{b}$ has a unique solution for each $\boldsymbol{b}\in\mathbb{R}^n$ if and only if A has rank n.

Proof Suppose that A has rank n; the columns $\boldsymbol{a}_1$, $\boldsymbol{a}_2$, ..., $\boldsymbol{a}_n$ of A are then linearly independent and so form a basis in $\mathbb{R}^n$. Any vector $\boldsymbol{b}\in\mathbb{R}^n$ may thus be written as a linear combination of these vectors: that is, there exist scalars c_1, c_2, ..., c_n such that

$$\boldsymbol{b}=\sum_{j=1}^{n}c_j\boldsymbol{a}_j$$

But $\sum_{j=1}^{n} c_j \boldsymbol{a}_j = A\boldsymbol{c}$, where

$$\boldsymbol{c} = \begin{bmatrix} c_1 \\ c_2 \\ \vdots \\ c_n \end{bmatrix}$$

Thus $\boldsymbol{x} = \boldsymbol{c}$ is a solution of the system $A\boldsymbol{x} = \boldsymbol{b}$. In order to show that this solution is unique for a given right-hand-side vector $\boldsymbol{b}$, assume that there exist two solutions, $\boldsymbol{c}$ and $\boldsymbol{d}$, such that $\boldsymbol{c} \neq \boldsymbol{d}$. Then

$$A\boldsymbol{c} = \boldsymbol{b}, \; A\boldsymbol{d} = \boldsymbol{b}$$

Hence

$$A(\boldsymbol{c} - \boldsymbol{d}) = \boldsymbol{0}$$

that is

$$\sum_{j=1}^{n} (c_j - d_j)\boldsymbol{a}_j = \boldsymbol{0}$$

Since $\boldsymbol{a}_1, \boldsymbol{a}_2, \ldots, \boldsymbol{a}_n$ are linearly independent vectors, each of the coefficients in the above sum must be zero, that is $c_j - d_j = 0$ for $j = 1, 2, \ldots, n$. But this contradicts the assumption that $\boldsymbol{c} \neq \boldsymbol{d}$.

Hence, there exists a unique solution to the system $A\boldsymbol{x} = \boldsymbol{b}$ if A has rank n.

For the proof in the other direction, assume that the system $A\boldsymbol{x} = \boldsymbol{b}$ has a unique solution for each $\boldsymbol{b} \in \mathbb{R}^n$. Suppose that A does not have rank n. Then there exist scalars $c_1, c_2, \ldots, c_n$, not all of which are zero, such that

$$\sum_{j=1}^{n} c_j \boldsymbol{a}_j = \boldsymbol{0}$$

that is

$$A\boldsymbol{c} = \boldsymbol{0}$$

where $\boldsymbol{c} = [c_1 \; c_2 \; \ldots \; c_n]^{\mathrm{T}}$.

Let $\boldsymbol{b}_1$ be any vector in $\mathbb{R}^n$ and let $\boldsymbol{c}_1$ be the solution of

$$A\boldsymbol{x} = \boldsymbol{b}_1$$

Define $\boldsymbol{c}_2 = \boldsymbol{c} + \boldsymbol{c}_1$. Clearly, $\boldsymbol{c}_2 \neq \boldsymbol{c}_1$ if $\boldsymbol{c} \neq \boldsymbol{0}$. Let

$$\boldsymbol{b}_2 = A\boldsymbol{c}_2$$

Since $\boldsymbol{b}_1$ is an arbitrary vector in $\mathbb{R}^n$, $\boldsymbol{b}_2 \neq \boldsymbol{b}_1$ in general. Hence

$$A\boldsymbol{c}_2 - A\boldsymbol{c}_1 = \boldsymbol{b}_2 - \boldsymbol{b}_1 \neq \boldsymbol{0}$$

But

$$A\boldsymbol{c}_2 - A\boldsymbol{c}_1 = A(\boldsymbol{c}_2 - \boldsymbol{c}_1) = A\boldsymbol{c} = \boldsymbol{0}$$

A contradiction has thus been obtained and so A must have rank n.

Theorem 2.3.3
An $n \times n$ matrix A is invertible if and only if A has rank n.
Proof First assume that A^{-1} exists. The system $A\boldsymbol{x} = \boldsymbol{b}$ then has a unique solution, given by $\boldsymbol{x} = A^{-1}\boldsymbol{b}$, for each $\boldsymbol{b} \in \mathbb{R}^n$. Thus, A has rank n, by the previous theorem.

To establish the proof in the other direction, assume that A has rank n. The system $A\boldsymbol{x} = \boldsymbol{b}$ then has a unique solution for each $\boldsymbol{b} \in \mathbb{R}^n$.
For $j = 1, 2, \ldots, n$, let $\boldsymbol{c}_j$ denote the unique solution of the system $A\boldsymbol{x} = \boldsymbol{e}_j$, where $\boldsymbol{e}_j$ denotes the jth unit vector. If C denotes the $n \times n$ matrix $[\boldsymbol{c}_1 \; \boldsymbol{c}_2 \; \ldots \; \boldsymbol{c}_n]$ then

$$AC = I$$

that is, C is the inverse of A (furthermore, C is unique and $CA = I$; see exercise 2.3).

Equivalent Systems of Equations

One of the most important methods for the computer solution of a system of n linear equations in n unknowns (see section 2.4) begins by reducing the given system to an equivalent system whose solution may readily be obtained.

Definition 2.3.5
Two systems of linear equations are said to be *equivalent systems of equations* if either system can be derived from the other by performing only the following types of operation

(E1) interchanging two equations;

(E2) multiplying both sides of an equation by a non-zero scalar;

(E3) adding a multiple of one equation to another.

If a system of equations has a solution then, clearly, any equivalent system of equations also has that solution.

Performing operations E1, E2 and E3 on the equations in an $m \times n$ system $A\boldsymbol{x} = \boldsymbol{b}$ is equivalent to performing the *elementary row operations* ER1, ER2 and ER3, given below, on an $m \times (n+1)$ matrix, $\tilde{A}$. The first n columns of the latter matrix are the same as the columns of A and the final column consists of the elements of the right-hand-side vector, $\boldsymbol{b}$. It is customary to write $\tilde{A}=[A|\boldsymbol{b}]$ and $\tilde{A}$ is called an *augmented matrix*. For example, if

$$A=\begin{bmatrix}1 & 2 & 3\\ 2 & 4 & 6\\ 5 & 7 & 9\end{bmatrix}, \quad \boldsymbol{b}=\begin{bmatrix}8\\ 3\\ 1\end{bmatrix}$$

then

$$\tilde{A}=\left[\begin{array}{ccc|c}1 & 2 & 3 & 8\\ 2 & 4 & 6 & 3\\ 5 & 7 & 9 & 1\end{array}\right]$$

The elementary row operations are

(ER1) Interchanging two rows, for example, interchanging row i and row k

$$\tilde{a}_{ij} \leftrightarrow \tilde{a}_{kj}, \; j=1, 2, \ldots, n+1$$

(ER2) Multiplying row i, say, by a non-zero scalar, α

$$\tilde{a}_{ij} \leftarrow \alpha \times \tilde{a}_{ij}, \; j=1, 2, \ldots, n+1$$

(ER3) Adding a multiple of one row to another row, for example, adding $\beta \times$ row k to row i

$$\tilde{a}_{ij} \leftarrow \tilde{a}_{ij}+\beta \times \tilde{a}_{kj}, \; j=1, 2, \ldots, n+1$$

Definition 2.3.6
Two matrices are said to be *row equivalent* if either matrix can be obtained from the other matrix through a succession of elementary row operations.

If A and B are row equivalent matrices (written $A \sim B$) then, clearly, they have equal rank.

Example 2.3.2

Let

$$A=\begin{bmatrix}2 & 5 & 0 & -3\\ 3 & 2 & 1 & 2\\ 1 & 2 & 1 & 0\\ 5 & 6 & 3 & 2\end{bmatrix}$$

Then rank $A \leqslant 4$. Adding $-2 \times$ row three to row one, $-3 \times$ row three to row two and $-5 \times$ row three to row four gives

$$A \sim \begin{bmatrix} 0 & 1 & -2 & -3 \\ 0 & -4 & -2 & 2 \\ 1 & 2 & 1 & 0 \\ 0 & -4 & -2 & 2 \end{bmatrix}$$

Now adding $-1 \times$ row two to row four gives

$$A \sim \begin{bmatrix} 0 & 1 & -2 & -3 \\ 0 & -4 & -2 & 2 \\ 1 & 2 & 1 & 0 \\ 0 & 0 & 0 & 0 \end{bmatrix}$$

Row four of this final matrix is clearly redundant and the first three rows are linearly independent. Thus, A has rank three.

Elementary Matrices

Definition 2.3.7
The matrix obtained by performing an elementary row operation on an identity matrix is called an *elementary row matrix*, or simply an *elementary matrix*.

Conventionally, E_{ik} denotes the elementary matrix obtained by interchanging row i and row k of I, while $E_i(\alpha)$ denotes the elementary matrix obtained by multiplying row i of I by α ($\neq 0$) and $E_{ik}(\alpha)$ denotes the elementary matrix obtained by adding $\alpha \times$ row k to row i of I.

Since I_n has rank n, so does every elementary matrix derived from I_n. Hence every elementary matrix has an inverse. The inverse of E_{ik} is clearly E_{ik} itself, the inverse of $E_i(\alpha)$ is $E_i(1/\alpha)$ and the inverse of $E_{ik}(\alpha)$ is $E_{ik}(-\alpha)$. The inverse of an elementary matrix is thus also an elementary matrix.

Example 2.3.3

Let I be the 4×4 identity matrix. Then

$$E_{24} = E_{24}^{-1} = \begin{bmatrix} 1 & 0 & 0 & 0 \\ 0 & 0 & 0 & 1 \\ 0 & 0 & 1 & 0 \\ 0 & 1 & 0 & 0 \end{bmatrix}$$

$$E_3(5)=\begin{bmatrix}1&0&0&0\\0&1&0&0\\0&0&5&0\\0&0&0&1\end{bmatrix},\quad E_3^{-1}(5)=\begin{bmatrix}1&0&0&0\\0&1&0&0\\0&0&\frac{1}{5}&0\\0&0&0&1\end{bmatrix}$$

$$E_{14}(2)=\begin{bmatrix}1&0&0&2\\0&1&0&0\\0&0&1&0\\0&0&0&1\end{bmatrix},\quad E_{14}^{-1}(2)=\begin{bmatrix}1&0&0&-2\\0&1&0&0\\0&0&1&0\\0&0&0&1\end{bmatrix}$$

Theorem 2.3.4
The product of two or more elementary matrices is invertible.

Proof The proof requires the following lemma.

Lemma 2.3.1
If A and B are invertible matrices then

$$(AB)^{-1}=B^{-1}A^{-1}$$

Proof of Lemma

$$\begin{aligned}I&=AA^{-1}\\&=AIA^{-1}\\&=A(BB^{-1})A^{-1}\\&=(AB)(B^{-1}A^{-1})\end{aligned}$$

Hence

$$(AB)^{-1}=B^{-1}A^{-1}$$

The proof of the theorem uses mathematical induction. It follows immediately from the lemma that if F_2 is the product of two elementary matrices then F_2^{-1} exists.

Let F_r be the product of r elementary matrices and assume that it is invertible.

Suppose $F_{r+1}=E_{ik}F_r$. Then, since E_{ik}^{-1} exists and F_r^{-1} is assumed to exist, $F_{r+1}^{-1}=F_r^{-1}E_{ik}^{-1}$, from the lemma. Similarly, if $F_{r+1}=E_{ik}(\alpha)F_r$ or $F_{r+1}=E_i(\alpha)F_r$ then F_{r+1}^{-1} exists.

Hence, by the inductive hypothesis, the product of any number of elementary matrices is invertible.

Performing an elementary row operation on a general $m\times n$ matrix A is equivalent to premultiplying A by an $m\times m$ elementary matrix. Furthermore, if A and B are row equivalent matrices then $A=EB$, where

E is the product of a number of elementary matrices. Clearly, in such a case, $B=E^{-1}A$.

Example 2.3.4

Let

$$A=\begin{bmatrix}1 & 2 & 3 & 4\\2 & 4 & 6 & 8\\1 & 3 & 5 & 7\end{bmatrix}$$

Interchanging row one and row two of A is equivalent to premultiplying A by E_{12}, that is

$$\begin{bmatrix}0 & 1 & 0\\1 & 0 & 0\\0 & 0 & 1\end{bmatrix}\begin{bmatrix}1 & 2 & 3 & 4\\2 & 4 & 6 & 8\\1 & 3 & 5 & 7\end{bmatrix}=\begin{bmatrix}2 & 4 & 6 & 8\\1 & 2 & 3 & 4\\1 & 3 & 5 & 7\end{bmatrix}$$

Multiplying row three of the resulting matrix by -2 is equivalent to premultiplying this matrix by $E_3(-2)$, that is

$$\begin{bmatrix}1 & 0 & 0\\0 & 1 & 0\\0 & 0 & -2\end{bmatrix}\begin{bmatrix}2 & 4 & 6 & 8\\1 & 2 & 3 & 4\\1 & 3 & 5 & 7\end{bmatrix}=\begin{bmatrix}2 & 4 & 6 & 8\\1 & 2 & 3 & 4\\-2 & -6 & -10 & -14\end{bmatrix}$$

Adding $3\times$ row one of this modified matrix to row three is equivalent to premultiplying it by $E_{31}(3)$, that is

$$\begin{bmatrix}1 & 0 & 0\\0 & 1 & 0\\3 & 0 & 1\end{bmatrix}\begin{bmatrix}2 & 4 & 6 & 8\\1 & 2 & 3 & 4\\-2 & -6 & -10 & -14\end{bmatrix}=\begin{bmatrix}2 & 4 & 6 & 8\\1 & 2 & 3 & 4\\4 & 6 & 8 & 10\end{bmatrix}$$

If

$$B=\begin{bmatrix}2 & 4 & 6 & 8\\1 & 2 & 3 & 4\\4 & 6 & 8 & 10\end{bmatrix}$$

then $A\sim B$ and

$$B=EA$$

where

$$E = E_{31}(3)E_3(-2)E_{12}$$
$$= \begin{bmatrix} 0 & 1 & 0 \\ 1 & 0 & 0 \\ 0 & 3 & -2 \end{bmatrix}$$

Also

$$A = E^{-1}B$$

where

$$E^{-1} = E_{12}^{-1}E_3^{-1}(-2)E_{31}^{-1}(3)$$
$$= E_{12}E_3\left(-\frac{1}{2}\right)E_{31}(-3)$$
$$= \begin{bmatrix} 0 & 1 & 0 \\ 1 & 0 & 0 \\ \frac{3}{2} & 0 & -\frac{1}{2} \end{bmatrix}$$

An important special case of row equivalent matrices occurs when two matrices only differ in the ordering of their rows. This has application to the solution of an $n \times n$ system of linear equations on a computer, when it is often necessary to interchange the order in which the equations are taken for stability reasons (see section 2.4). If two matrices A and B differ only in the ordering of their rows then

$$B = PA$$

where P is a permutation matrix.

Definition 2.3.8
An $n \times n$ matrix P is a *permutation matrix* if every row and every column contains exactly one 1 and $(n-1)$ 0s.

Example 2.3.5

If

$$A = \begin{bmatrix} 1 & 2 & 3 \\ 5 & 3 & 1 \\ 2 & 4 & 6 \end{bmatrix}, \quad P = \begin{bmatrix} 0 & 1 & 0 \\ 0 & 0 & 1 \\ 1 & 0 & 0 \end{bmatrix}$$

then

$$B=PA=\begin{bmatrix} 5 & 3 & 1 \\ 2 & 4 & 6 \\ 1 & 2 & 3 \end{bmatrix}$$

Exercise 2.3

1. Show that the following two matrices are row equivalent

$$\begin{bmatrix} 2 & 3 & 4 \\ 1 & 3 & 7 \\ 3 & 7 & 11 \end{bmatrix}, \begin{bmatrix} 1 & 1 & -3 \\ 11 & 24 & 37 \\ 5 & 9 & 5 \end{bmatrix}$$

2. Prove:

(a) that $AB=I$ if and only if $BA=I$;

(b) that the inverse of a matrix is unique.

3. Write down the 5×5 elementary matrices corresponding to each of the following row operations:

(1) multiplying row two by -6;

(2) adding $7 \times$ row three to row five;

(3) interchanging row one and row four.

4. Let A be a 3×3 matrix. Give two elementary row operations that together will reduce the second and third elements of column one to zero. Under what circumstance can this pair of operations not be performed? Write down the corresponding elementary matrices.

5. Corresponding to the elementary row operations ER1, ER2 and ER3 given in the text are the following *elementary column operations*:

(EC1) interchanging two columns;

(EC2) multiplying a column by a non-zero scalar;

(EC3) adding a multiple of one column to another column.

Show that performing any elementary column operation on a general $m \times n$ matrix A is equivalent to post-multiplying A by an elementary matrix.

6. Prove that the following matrix has rank 3

$$\begin{bmatrix} 1 & 2 & 2 & 3 & 2 \\ 2 & 5 & 3 & 10 & 7 \\ 3 & 5 & 7 & 10 & 4 \end{bmatrix}$$

7. Show that any non-null real matrix A can be reduced by a succession of elementary row operations to a *row canonical matrix* (or an *echelon matrix*) B, having the following properties:

(i) each of the first $r(\geqslant 1)$ rows of B has at least one non-zero entry; the remaining rows, if any, consist entirely of zero entries;

(ii) for $i=1, 2, \ldots, r$, the first non-zero entry of row i is equal to one: let j_i denote the column in which this entry occurs;

(iii) for $i=1, 2, \ldots, r$, the only non-zero element in column j_i is a_{ij} (which is equal to 1);

(iv) $j_1 < j_2 < \ldots < j_r$.

2.4 THE SOLUTION OF SYSTEMS OF LINEAR EQUATIONS

From section 2.3 it is clear that, in general, a linear system of m equations in n unknowns will have an infinite number of solutions if $m<n$ and no solution if $m>n$. If A is an $n \times n$ matrix, however, with rank n then the system $A\boldsymbol{x}=\boldsymbol{b}$ has a unique solution for each $\boldsymbol{b} \in \mathbb{R}^n$.

From theorem 2.3.3, if A has rank n then A is invertible. The unique solution of $A\boldsymbol{x}=\boldsymbol{b}$ could therefore be computed directly as $A^{-1}\boldsymbol{b}$. However, unless A^{-1} is already available it is computationally inefficient to solve the system $A\boldsymbol{x}=\boldsymbol{b}$ by first computing A^{-1}.

Before discussing methods for the solution of a general $n \times n$ system of linear equations, the special case of a triangular coefficient matrix is considered.

Solving Triangular Systems of Linear Equations

If A is a lower triangular matrix, that is

$$A = \begin{bmatrix} a_{11} & & & \bigcirc \\ a_{21} & a_{22} & & \\ \vdots & \vdots & \ddots & \\ a_{n1} & a_{n2} & \cdots & a_{nn} \end{bmatrix}$$

then the solution of $A\mathbf{x}=\mathbf{b}$ may be obtained by *forward substitution*, as described in algorithm 2.4.1. This algorithm is valid provided $a_{ii}\neq 0$ for any i, that is, provided A has rank n.

Algorithm 2.4.1 Forward Substitution

$x_1 \leftarrow b_1/a_{11}$;

for i **from** 2 **in steps of** 1 **to** n **do**

$\quad x_i \leftarrow b_i$;

$\quad$ **for** k **from** 1 **in steps of** 1 **to** $i-1$ **do**

$\quad\quad x_i \leftarrow x_i - a_{ik}*x_k$

$\quad$ **endfor**;

$\quad x_i \leftarrow x_i/a_{ii}$

endfor

Example 2.4.1

Solve the following system of equations by forward substitution

$$2x_1=14$$
$$2x_1+3x_2=18$$
$$3x_1-5x_2+2x_3=20$$

By forward substitution

$$x_1=b_1/a_{11}=14/2=7$$
$$x_2=(b_2-a_{21}x_1)/a_{22}=(18-14)/3=4/3$$
$$x_3=(b_3-a_{31}x_1-a_{32}x_2)/a_{33}=(20-21+20/3)/2=17/6$$

If A is upper triangular, that is

$$A=\begin{bmatrix} a_{11} & a_{12} & \cdots & a_{1n} \\ & a_{22} & \cdots & a_{2n} \\ \bigcirc & & \ddots & \vdots \\ & & & a_{nn} \end{bmatrix}$$

then the solution of $A\boldsymbol{x}=\boldsymbol{b}$ may be obtained by *backward substitution*, as described in algorithm 2.4.2. It is again assumed that $a_{ii}\neq 0$ for any i.

Algorithm 2.4.2 Backward Substitution

$x_n \leftarrow b_n/a_{nn}$;

for i **from** $n-1$ **in steps of** -1 **to** 1 **do**

$x_i \leftarrow b_i$;

for k **from** n **in steps of** -1 **to** $i+1$ **do**

$x_i \leftarrow x_i - a_{ik} * x_k$

endfor

$x_i \leftarrow x_i/a_{ii}$

endfor

Example 2.4.2

Solve the following system of equations by backward substitution

$$7x_1+2x_2+3x_3=14$$
$$5x_2-4x_3=-10$$
$$-24x_3=-72$$

By backward substitution

$$x_3=b_3/a_{33}=-72/(-24)=3$$
$$x_2=(b_2-a_{23}x_3)/a_{22}=(-10+12)/5=2/5$$
$$x_1=(b_1-a_{13}x_3-a_{12}x_2)/a_{11}=(14-9-4/5)/7=3/5$$

Both forward substitution and backward substitution require n divisions and $1+2+\ldots+(n-1)=n(n-1)/2$ (see example 1.3.13) additions and multiplications.

Gaussian Elimination

If the coefficient matrix of a system of equations is not triangular, Gaussian elimination may be used to replace the system by an equivalent system in which the coefficient matrix is upper triangular. The latter system may then be solved by backward substitution as described above.

In its simplest form, Gaussian elimination uses only the third operation given in definition 2.3.5, that is, adding a multiple of one equation to another, in order to effect the reduction to a triangular system. The process is illustrated in the following example.

Example 2.4.3

Solve the following system of equations using Gaussian elimination followed by backward substitution

$$5x_1+2x_2+2x_3=3 \quad (1)$$

$$2x_1+3x_2+4x_3=2 \quad (2)$$

$$3x_1+2x_2+x_3=1 \quad (3)$$

First, eliminate x_1 from equations 2 and 3. To do this, subtract $(2/5)\times$ equation 1 from equation 2 and $(3/5)\times$ equation 1 from equation 3. This gives the equivalent system of equations

$$5x_1+2x_2+2x_3=3$$

$$\frac{11}{5}x_2+\frac{16}{5}x_3=\frac{4}{5}$$

$$\frac{4}{5}x_2-\frac{1}{5}x_3=-\frac{4}{5}$$

Equation 2 is now used to eliminate x_2 from equation 3. This is done by subtracting $(4/11)\times$ equation 2 from equation 3, giving the equivalent system

$$5x_1+2x_2+2x_3=3$$

$$\frac{11}{5}x_2+\frac{16}{5}x_3=\frac{4}{5}$$

$$-\frac{15}{11}x_3=-\frac{12}{11}$$

The coefficient matrix in this system is upper triangular. Using backward substitution gives

$$x_3=4/5,\ x_2=-4/5,\ x_1=3/5$$

which is the solution of the original system.

Algorithm 2.4.3 describes the basic Gaussian elimination procedure in terms of the elementary row operations that are performed on the rows of the augmented matrix $\tilde{A}=[A|\boldsymbol{b}]$ (see section 2.3).

Algorithm 2.4.3 Gaussian Elimination

> **for** i **from** 1 **in steps of** 1 **to** $n-1$ **do**
>
> > **for** k **from** $i+1$ **in steps of** 1 **to** n **do**
> >
> > > $m_{ki} \leftarrow \tilde{a}_{ki}/\tilde{a}_{ii}$;
> > >
> > > $\tilde{a}_{ki} \leftarrow 0$;
> > >
> > > **for** j **from** $i+1$ **in steps of** 1 **to** $n+1$ **do**
> > >
> > > > $\tilde{a}_{kj} \leftarrow \tilde{a}_{kj} - m_{ki} * \tilde{a}_{ij}$
> > >
> > > **endfor**
> >
> > **endfor**
>
> **endfor**

The values m_{ki} are called the *multipliers*. The structure of the matrix at a typical stage in the elimination is illustrated in figure 2.16. At the ith stage, the ith row is called the *pivot row* and $\tilde{a}_{ii}$ is called the *pivot element*. The algorithm clearly fails if any pivot element has value zero.

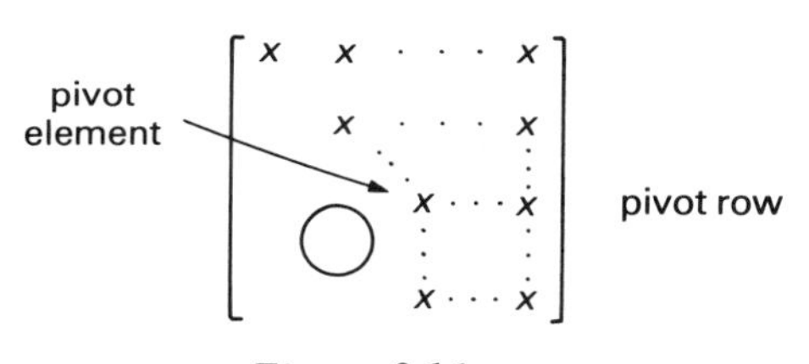

Figure 2.16

Computational Cost of Gaussian Elimination

Gaussian elimination involves $(n-1)+(n-2)+\ldots+2+1=\frac{1}{2}n(n-1)$ divisions. Each time round the j **for** loop, one addition/subtraction and one multiplication are performed. When $i=1$, the j **for** loop is executed for each value of $k=2, 3, \ldots, n$, that is, $n-1$ times. For each of these values of k, j takes values $2, 3, \ldots, n+1$, that is, for each value of k, j takes n values. Hence, when $i=1$, $(n-1)n$ additions/subtractions and $(n-1)n$ multiplications are performed. Similarly, when $i=2$, $(n-2)(n-1)$ additions/subtractions and $(n-2)(n-1)$ multiplications are performed. Continuing in this fashion, the complete elimination requires

$$(n-1)n+(n-2)(n-1)+\ldots+2\times 3+1\times 2$$

additions/subtractions and the same number of multiplications. It is easy to show by mathematical induction (see exercise 2.4) that the latter summation is equal to $(n+1)n(n-1)/3$. The total computational cost of solving an $n \times n$ system of linear equations by Gaussian elimination followed by back substitution is thus

$$\frac{1}{2}n(n-1)+n=\frac{1}{2}n(n+1) \text{ divisions}$$

$$\frac{1}{3}(n+1)n(n-1)+\frac{1}{2}n(n-1)=\frac{1}{6}n(n-1)(2n+5) \text{ multiplications}$$

and

$$\frac{1}{6}n(n-1)(2n+5) \text{ additions/subtractions}$$

Algorithm 2.3.3 is easily modified to facilitate the efficient simultaneous solution of a number of systems of equations having the same coefficient matrix but different right-hand-side vectors. Suppose it is required to solve each of the systems $A\boldsymbol{x}=\boldsymbol{b}_1$, $A\boldsymbol{x}=\boldsymbol{b}_2$. ..., $A\boldsymbol{x}=\boldsymbol{b}_p$. Let B be the $n \times p$ matrix which has the elements of $\boldsymbol{b}_j$ for its jth column. The given problem is thus equivalent to determining an $n \times p$ matrix X satisfying the matrix equation

$$AX=B$$

If $X=C$ is such a matrix, then the jth column of C is the solution of $A\boldsymbol{x}=\boldsymbol{b}_j$.

If $\tilde{A}$ denotes the $n \times (n+p)$ augmented matrix $[A|B]$, then the only modification required in algorithm 2.3.3 is that the j **for** loop should be replaced by

for j **from** $i+1$ **in steps of** 1 **to** $n+p$ **do**

After execution of the resulting algorithm the solution of the jth system is obtained by backward substitution using the upper triangular coefficient matrix in the first n columns of $\tilde{A}$ and the right-hand-side vector in the $(n+j)$th column of $\tilde{A}$.

Example 2.4.4

Solve the two systems of equations $A\boldsymbol{x}=\boldsymbol{b}_1$, $A\boldsymbol{x}=\boldsymbol{b}_2$, where

$$A=\begin{bmatrix} 5 & 2 & 2 \\ 2 & 3 & 4 \\ 3 & 2 & 1 \end{bmatrix}, \boldsymbol{b}_1=\begin{bmatrix} 3 \\ 2 \\ 1 \end{bmatrix}, \boldsymbol{b}_2=\begin{bmatrix} 4 \\ 1 \\ 3 \end{bmatrix}$$

$$\tilde{A}=\left[\begin{array}{ccc|cc} 5 & 2 & 2 & 3 & 4 \\ 2 & 3 & 4 & 2 & 1 \\ 3 & 2 & 1 & 1 & 3 \end{array}\right]$$

$$m_{21}=\tilde{a}_{21}/\tilde{a}_{11}=2/5,\ m_{31}=\tilde{a}_{31}/\tilde{a}_{11}=3/5$$

Subtracting $m_{21}\times$ row one from row two and $m_{31}\times$ row one from row three gives

$$\tilde{A}=\left[\begin{array}{ccc|cc} 5 & 2 & 2 & 3 & 4 \\ 0 & 11/5 & 16/5 & 4/5 & -3/5 \\ 0 & 4/5 & -1/5 & -4/5 & 3/5 \end{array}\right]$$

$$m_{32}=\tilde{a}_{32}/\tilde{a}_{22}=\left(\frac{4}{5}\right)\Big/\left(\frac{11}{5}\right)=4/11$$

Subtracting $m_{32}\times$ row two from row three gives

$$\tilde{A}=\left[\begin{array}{ccc|cc} 5 & 2 & 2 & 3 & 4 \\ 0 & 11/5 & 16/5 & 4/5 & -3/5 \\ 0 & 0 & -15/11 & -12/11 & 9/11 \end{array}\right]$$

Let

$$\begin{bmatrix} x_{11} \\ x_{21} \\ x_{31} \end{bmatrix}$$

denote the solution of $A\boldsymbol{x}=\boldsymbol{b}_1$ and let

$$\begin{bmatrix} x_{12} \\ x_{22} \\ x_{32} \end{bmatrix}$$

denote the solution of $A\boldsymbol{x}=\boldsymbol{b}_2$. Then

$$\begin{bmatrix} 5 & 2 & 2 \\ 0 & 11/5 & 16/5 \\ 0 & 0 & -15/11 \end{bmatrix}\begin{bmatrix} x_{11} \\ x_{21} \\ x_{31} \end{bmatrix}=\begin{bmatrix} 3 \\ 4/5 \\ -12/11 \end{bmatrix}$$

Using backward substitution gives

$$x_{31}=(-12/11)/(-15/11)=4/5$$

$$x_{21}=(4/5-64/25)/(11/5)=-4/5$$

$$x_{11}=(3+8/5-8/5)/5=3/5$$

Similarly

$x_{32} = -3/5$, $x_{22} = 3/5$ and $x_{21} = 4/5$

The solution of the matrix equation

$$AX = B$$

is thus

$$X = \begin{bmatrix} 3/5 & 4/5 \\ -4/5 & 3/5 \\ 4/5 & -3/5 \end{bmatrix}$$

Computing the Inverse of a Matrix

If $B = I_n$ in the matrix equation $AX = B$, then $X = A^{-1}$ is the solution matrix (assuming the $n \times n$ matrix A has rank n). Gaussian elimination may thus be used to determine the inverse of a matrix.

Example 2.4.5

Given

$$A = \begin{bmatrix} 5 & 2 & 2 \\ 2 & 3 & 4 \\ 3 & 2 & 1 \end{bmatrix}$$

determine A^{-1}.

Applying Gaussian elimination

$$\tilde{A} = \left[\begin{array}{ccc|ccc} 5 & 2 & 2 & 1 & 0 & 0 \\ 2 & 3 & 4 & 0 & 1 & 0 \\ 3 & 2 & 1 & 0 & 0 & 1 \end{array}\right]$$

Subtracting $2/5 \times$ row one from row two and $3/5 \times$ row one from row three gives

$$\left[\begin{array}{ccc|ccc} 5 & 2 & 2 & 1 & 0 & 0 \\ 0 & 11/5 & 16/5 & -2/5 & 1 & 0 \\ 0 & 4/5 & -1/5 & -3/5 & 0 & 1 \end{array}\right]$$

Subtracting $4/11 \times$ row two from row three gives

$$\left[\begin{array}{ccc|ccc} 5 & 2 & 2 & 1 & 0 & 0 \\ 0 & 11/5 & 16/5 & -2/5 & 1 & 0 \\ 0 & 0 & -15/11 & -5/11 & -4/11 & 1 \end{array}\right]$$

The inverse could now be determined by using back substitution to determine each of the columns of $X = A^{-1}$: for example, to get the first column of A^{-1}, solve the following upper triangular system by back substitution

$$\begin{bmatrix} 5 & 2 & 2 \\ 0 & 11/5 & 16/5 \\ 0 & 0 & -15/11 \end{bmatrix} \begin{bmatrix} x_{11} \\ x_{21} \\ x_{31} \end{bmatrix} = \begin{bmatrix} 1 \\ -2/5 \\ -5/11 \end{bmatrix}$$

etc. For determining the inverse of a matrix on a computer this is, in fact, the procedure that is generally followed. However, for hand calculations, it is easier to use a modification of Gaussian elimination, called the *Gauss–Jordan method*. In the latter method, A is reduced to a diagonal (usually unit) matrix. This is effected by performing further row operations.

Subtract $(10/11) \times$ row two from row one; then

$$\left[\begin{array}{ccc|ccc} 5 & 0 & -10/11 & 15/11 & -10/11 & 0 \\ 0 & 11/5 & 16/5 & -2/5 & 1 & 0 \\ 0 & 0 & -15/11 & -5/11 & -4/11 & 1 \end{array}\right]$$

Now subtract $(2/3) \times$ row three from row one and $(-176/75) \times$ row three from row two;

$$\left[\begin{array}{ccc|ccc} 5 & 0 & 0 & 5/3 & -2/3 & -2/3 \\ 0 & 11/5 & 0 & -22/15 & 11/75 & 176/75 \\ 0 & 0 & -15/11 & -5/11 & -4/11 & 1 \end{array}\right]$$

Finally, multiplying row one by $(1/5)$, row two by $(5/11)$ and row three by $(-11/15)$ gives

$$\left[\begin{array}{ccc|ccc} 1 & 0 & 0 & 1/3 & -2/15 & -2/15 \\ 0 & 1 & 0 & -2/3 & 1/15 & 16/15 \\ 0 & 0 & 1 & 1/3 & 4/15 & -11/15 \end{array}\right]$$

The original system $AX = I$ has now been reduced to the equivalent system

$$IX = \begin{bmatrix} 1/3 & -2/15 & -2/15 \\ -2/3 & 1/15 & 16/15 \\ 1/3 & 4/15 & -11/15 \end{bmatrix}$$

Hence

$$A^{-1} = \begin{bmatrix} 1/3 & -2/15 & -2/15 \\ -2/3 & 1/15 & 16/15 \\ 1/3 & 4/15 & -11/15 \end{bmatrix}$$

Gaussian Elimination and LU Decomposition

In Gaussian elimination, the coefficient matrix A is reduced by a sequence of elementary row operations to an upper triangular matrix, U say. Each elementary row operation involves subtracting a multiple of one row, the ith say, from another row, the kth say, where $k > i$. Each such operation is equivalent to premultiplying the current matrix by a lower triangular elementary matrix (see section 2.3).

Example 2.4.6

Let

$$A = \begin{bmatrix} 3 & 2 & -1 \\ 1 & 2 & 3 \\ 2 & 4 & 6 \end{bmatrix}$$

Subtracting $(1/3) \times$ row one from row two is equivalent to premultiplying A by an elementary matrix, M_{21}, where

$$M_{21} = \begin{bmatrix} 1 & 0 & 0 \\ -1/3 & 1 & 0 \\ 0 & 0 & 1 \end{bmatrix}$$

This reduces A to the matrix

$$\begin{bmatrix} 3 & 2 & -1 \\ 0 & 4/3 & 10/3 \\ 2 & 4 & 6 \end{bmatrix}$$

Subtracting $(2/3) \times$ row one from row three of this matrix is equivalent to premultiplying it by

$$M_{31} = \begin{bmatrix} 1 & 0 & 0 \\ 0 & 1 & 0 \\ -2/3 & 0 & 1 \end{bmatrix}$$

giving the matrix

$$\begin{bmatrix} 3 & 2 & -1 \\ 0 & 4/3 & 10/3 \\ 0 & 8/3 & 20/3 \end{bmatrix}$$

In general, reducing to zero all of the entries below the diagonal in the ith column of an $n \times n$ matrix is equivalent to premultiplying the matrix by the product of $n-i$ elementary matrices. The row equivalent matrices A and U are therefore related by

$$U = MA$$

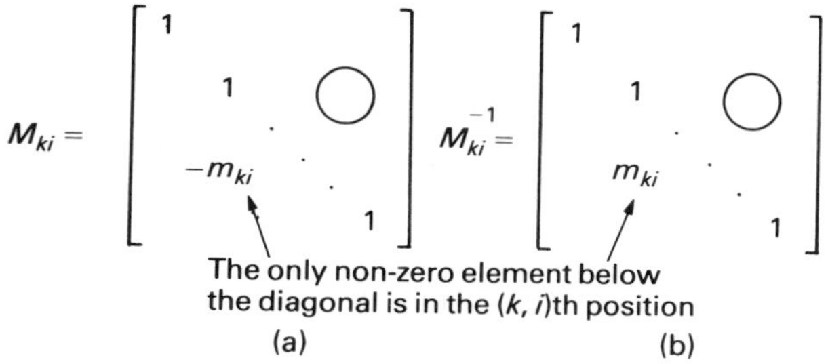

$$M_{ki} = \begin{bmatrix} 1 & & & & \\ & 1 & & \bigcirc & \\ & & \cdot & & \\ & -m_{ki} & & \cdot & \\ & & & & 1 \end{bmatrix} \qquad M_{ki}^{-1} = \begin{bmatrix} 1 & & & & \\ & 1 & & \bigcirc & \\ & & \cdot & & \\ & m_{ki} & & \cdot & \\ & & & & 1 \end{bmatrix}$$

The only non-zero element below the diagonal is in the (k, i)th position

(a) (b)

Figure 2.17

where $M = M_{n,n-1}M_{n,n-2}M_{n-1,n-2}\ldots M_{n1}M_{n-1,1}\ldots M_{31}M_{21}$.

Here M_{ki} denotes the $n \times n$ elementary matrix that reduces to zero the element in the kth row and ith column of the current coefficient matrix. The (k, i)th element of M_{ki} (see figure 2.17a) is simply $-m_{ki}$, where m_{ki} is the appropriate multiplier calculated during Gaussian elimination. It is easy to show that the product of two lower triangular matrices is also lower triangular. It follows by mathematical induction that M is a lower triangular matrix.

From theorem 2.3.4, the inverse of M is given by

$$M^{-1} = M_{21}^{-1}M_{31}^{-1}\ldots M_{n1}^{-1}M_{32}^{-1}\ldots M_{n-1,n-2}^{-1}M_{n,n-2}^{-1}M_{n,n-1}^{-1}$$

From section 2.3, M_{ki}^{-1} is a lower triangular elementary matrix, obtained from M_{ki} by reversing the sign of the value in the (k, i)th position (see figure 2.17b). Thus, M^{-1} is a lower triangular matrix. It is customary to denote this matrix by L. Hence

$$A = LU$$

that is, provided algorithm 2.4.3 can be successfully executed, A may be factorised into the product of a lower and an upper triangular matrix. This is usually called the *LU decomposition theorem.*

Example 2.4.7

Suppose A is a 4×4 matrix. Then

$$L = M_{21}^{-1}M_{31}^{-1}M_{41}^{-1}M_{32}^{-1}M_{42}^{-1}M_{43}^{-1}$$

Now

$$M_{42}^{-1}M_{43}^{-1} = \begin{bmatrix} 1 & 0 & 0 & 0 \\ 0 & 1 & 0 & 0 \\ 0 & 0 & 1 & 0 \\ 0 & m_{42} & 0 & 1 \end{bmatrix} \begin{bmatrix} 1 & 0 & 0 & 0 \\ 0 & 1 & 0 & 0 \\ 0 & 0 & 1 & 0 \\ 0 & 0 & m_{43} & 1 \end{bmatrix} = \begin{bmatrix} 1 & 0 & 0 & 0 \\ 0 & 1 & 0 & 0 \\ 0 & 0 & 1 & 0 \\ 0 & m_{42} & m_{43} & 1 \end{bmatrix}$$

$$M_{32}^{-1}M_{42}^{-1}M_{43}^{-1}=\begin{bmatrix}1&0&0&0\\0&1&0&0\\0&m_{32}&1&0\\0&0&0&1\end{bmatrix}\begin{bmatrix}1&0&0&0\\0&1&0&0\\0&0&1&0\\0&m_{42}&m_{43}&1\end{bmatrix}=\begin{bmatrix}1&0&0&0\\0&1&0&0\\0&m_{32}&1&0\\0&m_{42}&m_{43}&1\end{bmatrix}$$

$$M_{41}^{-1}M_{32}^{-1}M_{42}^{-1}M_{43}^{-1}=\begin{bmatrix}1&0&0&0\\0&1&0&0\\0&0&1&0\\m_{41}&0&0&1\end{bmatrix}\begin{bmatrix}1&0&0&0\\0&1&0&0\\0&m_{32}&1&0\\0&m_{42}&m_{43}&1\end{bmatrix}=\begin{bmatrix}1&0&0&0\\0&1&0&0\\0&m_{32}&1&0\\m_{41}&m_{42}&m_{43}&1\end{bmatrix}$$

$$M_{31}^{-1}M_{41}^{-1}M_{32}^{-1}M_{42}^{-1}M_{43}^{-1}=\begin{bmatrix}1&0&0&0\\0&1&0&0\\m_{31}&0&1&0\\0&0&0&1\end{bmatrix}\begin{bmatrix}1&0&0&0\\0&1&0&0\\0&m_{32}&1&0\\m_{41}&m_{42}&m_{43}&1\end{bmatrix}$$

$$=\begin{bmatrix}1&0&0&0\\0&1&0&0\\m_{31}&m_{32}&1&0\\m_{41}&m_{42}&m_{43}&1\end{bmatrix}$$

Finally

$$L=\begin{bmatrix}1&0&0&0\\m_{21}&1&0&0\\0&0&1&0\\0&0&0&1\end{bmatrix}\begin{bmatrix}1&0&0&0\\0&1&0&0\\m_{31}&m_{32}&1&0\\m_{41}&m_{42}&m_{43}&1\end{bmatrix}=\begin{bmatrix}1&0&0&0\\m_{21}&1&0&0\\m_{31}&m_{32}&1&0\\m_{41}&m_{42}&m_{43}&1\end{bmatrix}$$

The above example illustrates that in the LU decomposition of A, the elements below the unit diagonal in the lower triangular matrix L are simply the multipliers calculated during Gaussian elimination, that is, if $L=[l_{ij}]$, then

$$l_{ij}=\begin{cases}0, & i<j\\ 1, & i=j\\ m_{ij}, & i>j\end{cases}$$

Since the matrix $U=[u_{ij}]$ is just the final version of the coefficient matrix by Gaussian elimination, algorithm 2.4.3 may be modified to determine the LU decomposition of an $n \times n$ matrix, A (assuming that such a decomposition exists).

It is only necessary to store the matrix U together with the subdiagonal elements of L, since each of the diagonal elements of L is known to be one. The LU decomposition of A may thus be conveniently stored in a single $n \times n$ matrix, T, as illustrated in figure 2.18.

$$\begin{bmatrix} u_{11} & u_{12} & u_{13} & \cdots & u_{1n-1} & u_{1n} \\ l_{21} & u_{22} & u_{23} & \cdots & u_{2n-1} & u_{2n} \\ l_{31} & l_{32} & u_{33} & \cdots & u_{3n-1} & u_{3n} \\ \vdots & \vdots & \vdots & & \vdots & \vdots \\ l_{n1} & l_{n2} & l_{n3} & \cdots & l_{nn-1} & u_{nn} \end{bmatrix}$$

Figure 2.18

The procedure for the triangular decomposition of an $n \times n$ matrix A is summarised in algorithm 2.4.4. The algorithm begins by initialising T to A. Clearly, if A is not subsequently required, there is no need to use an additional matrix T since A may itself be overwritten by its LU factorisation.

Algorithm 2.4.4 LU Decomposition

$T \leftarrow A$;

for i **from** 1 **in steps of** 1 **to** $n-1$ **do**

  for k **from** $i+1$ **in steps of** 1 **to** n **do**

    $t_{ki} \leftarrow t_{ki}/t_{ii}$,

    for j **from** $i+1$ **in steps of** 1 **to** n **do**

      $t_{kj} \leftarrow t_{kj} - t_{ki} * t_{ij}$

    endfor

  endfor

endfor

Given the LU decomposition of a matrix A, the solution of the system of equations $A\mathbf{x} = \mathbf{b}$ may be obtained as follows. First solve the lower triangular system

$L\mathbf{x} = \mathbf{b}$

by forward substitution. Suppose $\mathbf{x} = \mathbf{c}$ is the solution of this system. The solution of the system $A\boldsymbol{x}=\boldsymbol{b}$ is then obtained by solving the upper triangular system

$$U\boldsymbol{x}=\boldsymbol{c}$$

by backward substitution.

The vector $\boldsymbol{c}$ is the same as the final version of the right-hand-side vector determined by Gaussian elimination. Clearly, the LU decomposition of A followed by the forward substitution $L\boldsymbol{x}=\boldsymbol{b}$ is equivalent to the Gaussian elimination procedure for replacing the system $A\boldsymbol{x}=\boldsymbol{b}$ by the equivalent system $U\boldsymbol{x}=\boldsymbol{c}$. The computational cost of the LU decomposition algorithm is

$$\frac{1}{2}n(n-1) \quad \text{divisions}$$

$$\frac{1}{6}n(n-1)(2n-1) \quad \text{multiplications}$$

and

$$\frac{1}{6}n(n-1)(2n-1) \quad \text{additions/subtractions}$$

The total cost of solving a system of equations $A\boldsymbol{x}=\boldsymbol{b}$ by the LU decomposition of A followed by a forward substitution and a backward substitution is exactly the same as for Gaussian elimination followed by a backward substitution.

The advantage of the LU decomposition approach over Gaussian elimination occurs when it is required to solve a number of systems

$$A\boldsymbol{x}=\boldsymbol{b}_1,\ A\boldsymbol{x}=\boldsymbol{b}_2,\ \ldots$$

where only $\boldsymbol{b}_1$ is specified in advance and each successive right-hand-side vector $\boldsymbol{b}_i$ depends upon the solution $\boldsymbol{x}_{i-1}$ of the previous system. In such a case, the matrix A can be factorised once and for all into its L and U factors. The solution of each system is then obtained by executing a forward substitution followed by a backward substitution, the two substitutions costing a total of n divisions, $n(n-1)$ multiplications and $n(n-1)$ additions/subtractions.

Example 2.4.8

Let

$$A=\begin{bmatrix} 3 & -5 & 4 \\ -8 & 4 & 1 \\ 5 & -6 & 2 \end{bmatrix},\ \boldsymbol{b}=\begin{bmatrix} 2 \\ -3 \\ 1 \end{bmatrix}$$

Find the LU decomposition of A and hence solve the system $A\boldsymbol{x}=\boldsymbol{b}$.

Now

$$m_{21}=-8/3,\ m_{31}=5/3$$

Subtracting $m_{21}\times$row one from row two of A and $m_{31}\times$row one from row three gives

$$\begin{bmatrix} 3 & -5 & 4 \\ 0 & -28/3 & 35/3 \\ 0 & 7/3 & -14/3 \end{bmatrix}$$

$$m_{32}=-1/4$$

Subtracting $m_{32}\times$row two from row three now gives

$$\begin{bmatrix} 3 & -5 & 4 \\ 0 & -28/3 & 35/3 \\ 0 & 0 & -7/4 \end{bmatrix}$$

Hence

$$\begin{bmatrix} 3 & -5 & 4 \\ -8 & 4 & 1 \\ 5 & -6 & 2 \end{bmatrix}=\begin{bmatrix} 1 & 0 & 0 \\ -8/3 & 1 & 0 \\ 5/3 & -1/4 & 1 \end{bmatrix}\begin{bmatrix} 3 & -5 & 4 \\ 0 & -28/3 & 35/3 \\ 0 & 0 & -7/4 \end{bmatrix}$$

For this example, the matrix T in algorithm 2.4.4 is finally equal to

$$\begin{bmatrix} 3 & -5 & 4 \\ -8/3 & -28/3 & 35/3 \\ 5/3 & -1/4 & -7/4 \end{bmatrix}$$

To solve $A\boldsymbol{x}=\boldsymbol{b}$, first solve $L\boldsymbol{x}=\boldsymbol{b}$

$$\begin{bmatrix} 1 & 0 & 0 \\ -8/3 & 1 & 0 \\ 5/3 & -1/4 & 1 \end{bmatrix}\begin{bmatrix} x_1 \\ x_2 \\ x_3 \end{bmatrix}=\begin{bmatrix} 2 \\ -3 \\ 1 \end{bmatrix}$$

Forward substitution gives $x_1=2$, $x_2=-3+(8/3)\times 2=7/3$, $x_3=1-(5/3)\times 2+(1/4)\times(7/3)=-(7/4)$.

Next solve $U\boldsymbol{x}=\boldsymbol{c}$, where $\boldsymbol{c}$ is the solution just determined of the lower triangular system

$$\begin{bmatrix} 3 & -5 & 4 \\ 0 & -28/3 & 35/3 \\ 0 & 0 & -7/4 \end{bmatrix}\begin{bmatrix} x_1 \\ x_2 \\ x_3 \end{bmatrix}=\begin{bmatrix} 2 \\ 7/3 \\ -7/4 \end{bmatrix}$$

Backward substitution gives

$$x_3=1,\ x_2=\left(\frac{7}{3}-\frac{35}{3}\right)\Big/\left(-\frac{28}{3}\right)=1,\ x_1=(2-4+5)/3=1$$

The solution of the original system is thus $[1 \quad 1 \quad 1]^T$.

Gaussian Elimination and Row Interchanges

Algorithms 2.4.3 and 2.4.4 can always be executed provided no pivot element has value zero. If at the ith stage of Gaussian elimination, $\tilde{a}_{ii}=0$ then the statement $m_{ki}\leftarrow\tilde{a}_{ki}/\tilde{a}_{ii}$ cannot be successfully executed. However, if $\tilde{a}_{li}\neq 0$ for some $l>i$, then the ith and lth rows may be interchanged and the elimination process continued as before. Such an interchange of rows of A does not affect the solution of the underlying system, $A\boldsymbol{x}=\boldsymbol{b}$.

Although in theory it is only necessary to interchange rows in order to avoid pivoting on a zero element, in practice it is also necessary to interchange rows in order to avoid pivoting on an element that has a relatively small absolute value. Failure to do so can lead to severe stability problems when working with arithmetic of limited precision (as, for example, on a computer).

Definition 2.4.1
An algorithm is said to be *numerically unstable* if relatively small errors in the input data give rise to relatively large errors in the output data.

Example 2.4.9

Using four-digit decimal floating-point arithmetic, apply algorithms 2.4.3 and 2.4.2 to solve the following system of equations

$$\begin{bmatrix} 1.62 & 1.10 & 0.65 \\ 6.18 & 4.20 & -3.04 \\ 4.65 & -3.05 & 2.10 \end{bmatrix}\begin{bmatrix} x_1 \\ x_2 \\ x_3 \end{bmatrix}=\begin{bmatrix} 3.37 \\ 7.34 \\ 3.70 \end{bmatrix}$$

Initially

$$\tilde{A}=\left[\begin{array}{ccc|c} 0.1620\times 10^1 & 0.1100\times 10^1 & 0.6500\times 10^0 & 0.3370\times 10^1 \\ 0.6180\times 10^1 & 0.4200\times 10^1 & -0.3040\times 10^1 & 0.7340\times 10^1 \\ 0.4650\times 10^1 & -0.3050\times 10^1 & 0.2100\times 10^1 & 0.3700\times 10^1 \end{array}\right]$$

Pivoting on $\tilde{a}_{11}=0.1620\times 10^1$

$$m_{21}=(0.6180\times 10^1)/(0.1620\times 10^1)=0.38148148\ldots\times 10^1\approx 0.3815\times 10^1$$

Similarly, $m_{31}\approx 0.2870\times 10^1$. Hence

$$\begin{aligned}\tilde{a}_{22}&=0.4200\times 10^1-(0.3815\times 10^1)\times(0.1100\times 10^1)\\&=0.4200\times 10^1-0.41965\times 10^1\\&\approx 0.4200\times 10^1-0.4197\times 10^1\\&=0.3000\times 10^{-2}\end{aligned}$$

Similarly computing $\tilde{a}_{23}$, $\tilde{a}_{24}$, $\tilde{a}_{32}$, $\tilde{a}_{33}$, $\tilde{a}_{34}$ and setting $\tilde{a}_{21}=\tilde{a}_{31}=0$ gives

$$\left[\begin{array}{ccc|c}0.1620\times 10^1 & 0.1100\times 10^1 & 0.6500\times 10^0 & 0.3370\times 10^1\\ 0 & 0.3000\times 10^{-2} & -0.5520\times 10^1 & -0.5520\times 10^1\\ 0 & -0.6207\times 10^1 & 0.2340\times 10^0 & -0.5972\times 10^1\end{array}\right]$$

Pivoting on $\tilde{a}_{22}=0.3000\times 10^{-2}$ gives

$m_{32}=-0.2069\times 10^4$ resulting in

$$\left[\begin{array}{ccc|c}0.1620\times 10^1 & 0.1100\times 10^1 & 0.6500\times 10^0 & 0.3370\times 10^1\\ 0 & 0.3000\times 10^{-2} & -0.5520\times 10^1 & -0.5520\times 10^1\\ 0 & 0 & -0.1142\times 10^5 & -0.1143\times 10^5\end{array}\right]$$

Backward substitution now gives

$$x_3=0.1001\times 10^1,\ x_2=0.2000\times 10^1,\ x_1=0.3204\times 10^0$$

that is, the computed solution, rounded to the number of significant digits given in the data, is $x_1=0.32$, $x_2=2.00$, $x_3=1.00$. The reader may easily verify for himself that the exact solution of the given system is

$$x_1=x_2=x_3=1$$

Thus, although the computed value of x_3 is accurate, the results for x_1 and x_2 are badly in error.

Suppose now that the second and third equations in the given system are interchanged. Then, after pivoting on $\tilde{a}_{11}=0.1620\times 10^1$, the matrix is

$$\left[\begin{array}{ccc|c}0.1620\times 10^1 & 0.110\times 10^1 & 0.6500\times 10^0 & 0.3370\times 10^1\\ 0 & -0.6207\times 10^1 & 0.2340\times 10^0 & -0.5972\times 10^1\\ 0 & 0.3000\times 10^{-2} & -0.5520\times 10^1 & -0.5520\times 10^1\end{array}\right]$$

Pivoting on $\tilde{a}_{22}=-0.6207\times 10^1$ gives

$$m_{32}=-0.4833\times 10^{-3}$$

$$\left[\begin{array}{ccc|c}0.1620\times 10^1 & 0.1100\times 10^1 & 0.6500\times 10^0 & 0.3370\times 10^1\\ 0 & -0.6207\times 10^1 & 0.2340\times 10^0 & -0.5972\times 10^1\\ 0 & 0 & -0.5520\times 10^1 & -0.5523\times 10^1\end{array}\right]$$

Backward substitution now gives $x_3 = 0.1001 \times 10^1$, $x_2 = 0.9998 \times 10^0$, $x_1 = 0.9994 \times 10^0$. The computed solution, rounded to the number of significant digits given in the data, is thus

$$x_1 = 0.999, \; x_2 = 1.00, \; x_3 = 1.00$$

In the first attempt to solve the above system, the multiplier m_{32} has a very large absolute value, while in the second attempt it has a very small absolute value. By applying floating-point rounding-error analysis (Forsythe and Moler, 1967), it may be shown that it is generally advisable to choose as the ith pivot the element $\tilde{a}_{li}$ satisfying

$$|\tilde{a}_{li}| = \max_{i \leqslant k \leqslant n} |\tilde{a}_{ki}|$$

If $l \neq i$, then it is necessary to interchange rows i and l. This strategy for selecting pivot elements is known as *partial pivoting.*

In the computer solution of a linear system of equations, it is not necessary to perform actual row interchanges. It is sufficient instead to interchange the elements of a *permutation vector* $\boldsymbol{p} = [p_1 \; p_2 \; \dots \; p_n]^{\mathrm{T}}$. Initially, p_k is assigned the value k, for each $k = 1, 2, \dots, n$. If it is subsequently required to 'interchange' rows i and j, say, then this is effected by interchanging the values of p_i and p_j. At any stage in the elimination, the rows of $\tilde{A}$ may be obtained in the interchanged order by referencing rows $p_1, p_2, \dots, p_n$, respectively, instead of rows $1, 2, \dots, n$. Let P denote the permutation matrix (see definition 2.3.8) with 1s in positions (i, r_i), $i = 1, 2, \dots, n$, where r_i denotes the final value of p_i. Then using Gaussian elimination with partial pivoting to solve the system $A\boldsymbol{x} = \boldsymbol{b}$ is equivalent to using ordinary Gaussian elimination to solve the system $PA\boldsymbol{x} = P\boldsymbol{b}$.

Given a matrix A of rank n, algorithm 2.4.5 obtains the LU decomposition of the matrix $C = PA$. After execution of the algorithm, $t_{p_i j}$ contains the (i, j)th element of L if $i > j$ and the (i, j)th element of U if $i \leqslant j$.

Algorithm 2.4.5

$T \leftarrow A$;

for k **from** 1 **in steps of** 1 **to** n **do**

$p_k \leftarrow k$

endfor;

for i **from** 1 **in steps of** 1 **to** $n-1$ **do**

$\max \leftarrow |t_{p_i i}|$; $l \leftarrow i$;

```
        for k from i+1 in steps of 1 to n do
            if |t_{p_k i}| > max then
                max <- |t_{p_k i}|; l <- k
            endif;
        endfor;
        p_i <-> p_l;
        for k from i+1 in steps of 1 to n do
                t_{p_k i} <- t_{p_k i}/t_{p_i i};
            for j from i+1 in steps of 1 to n do
                t_{p_k j} <- t_{p_k j} - t_{p_k i} * t_{p_i j}
            endfor
        endfor
    endfor
```

Exercise 2.4

1. Solve the following system of equations by Gaussian elimination followed by back substitution

$$3x_1 - 5x_2 + 4x_3 = 2$$
$$-8x_1 + 4x_2 + x_3 = -3$$
$$5x_1 - 6x_2 + 2x_3 = 1$$

Write down the *LU* decomposition of the coefficient matrix.

2. Given

$$A = \begin{bmatrix} 6 & -4 & -1 \\ -4 & 11 & 7 \\ -1 & 7 & 5 \end{bmatrix}$$

compute A^{-1}.

3. Show that the number of multiplications performed in Gaussian elimination is equal to $(n+1)n(n-1)/3$, where n is the number of equations in the system.

4. Solve the following system of equations:

(i) using Gaussian elimination without partial pivoting:

(ii) using Gaussian elimination with partial pivoting.

In both cases, use four-digit decimal floating-point arithmetic

$$\begin{aligned} -1.41x_1+2x_2&=2\\ x_1-1.41x_2+x_3&=2\\ 3x_2-3x_3&=4 \end{aligned}$$

5. An $n\times n$ matrix is said to be *tridiagonal* if it has the form

$$\begin{bmatrix} a_1 & c_1 & & & \bigcirc \\ b_2 & a_2 & c_2 & & \\ & \cdot & \cdot & \cdot & \\ & & \cdot & \cdot & \cdot \\ & \bigcirc & & \cdot & \cdot & c_{n-1} \\ & & & & b_n & a_n \end{bmatrix}$$

Show that if the LU decomposition of such a matrix exists then it can be written in the form

$$\begin{bmatrix} 1 & & & \bigcirc \\ \beta_2 & 1 & & \\ & \beta_3 & \cdot & \\ & & \cdot & \cdot \\ \bigcirc & & & \cdot & \cdot \\ & & & & \beta_n & 1 \end{bmatrix}\begin{bmatrix} \alpha_1 & c_1 & & & \bigcirc \\ & \alpha_2 & c_2 & & \\ & & \cdot & \cdot & \\ & & & \cdot & \cdot \\ & \bigcirc & & & \cdot & c_{n-1} \\ & & & & & \alpha_n \end{bmatrix}$$

and give recursive formulae for determining the αs and βs. Hence compute the LU decomposition of the matrix

$$\begin{bmatrix} 1 & 1 & & & \bigcirc \\ 1 & 2 & 1 & & \\ & 1 & 3 & 1 & \\ \bigcirc & & 1 & 4 & 1 \\ & & & 1 & 5 \end{bmatrix}$$

6. The solutions of some systems of equations are extremely sensitive to relatively small changes in the elements of the coefficient matrix and the right-hand-side vector. Such systems are said to be *ill-conditioned.*

Illustrate this phenomenon by computing the solutions of $A\boldsymbol{x}=\boldsymbol{b}_1$ and $A\boldsymbol{x}=\boldsymbol{b}_2$, where

$$A=\begin{bmatrix}\frac{1}{2} & \frac{1}{3} & \frac{1}{4}\\ \frac{1}{3} & \frac{1}{4} & \frac{1}{5}\\ \frac{1}{4} & \frac{1}{5} & \frac{1}{6}\end{bmatrix},\ \boldsymbol{b}_1=\begin{bmatrix}0.95\\0.67\\0.52\end{bmatrix},\ \boldsymbol{b}_2=\begin{bmatrix}0.96\\0.66\\0.53\end{bmatrix}$$

2.5 DETERMINANTS

Introduction

The simplest linear equations involve just one variable, x (say), and are of the form

$$ax=b$$

where a, b are constants. Such equations have a solution $x=b/a$, provided $a\neq 0$.

Linear equations in two variables, x and y (say), are of the form

$$a_{11}x+a_{12}y=b_1$$

$$a_{21}x+a_{22}y=b_2$$

where a_{11}, a_{12}, a_{21}, a_{22}, b_1 and b_2 are constants. Such equations have a solution $x=(b_1a_{22}-b_2a_{12})/(a_{11}a_{22}-a_{21}a_{12})$, $y=(b_2a_{11}-b_1a_{21})/(a_{11}a_{22}-a_{21}a_{12})$, provided $a_{11}a_{22}-a_{21}a_{12}\neq 0$.

In both these cases, whether solutions exist depends on the values of the constants appearing on the left-hand sides of the equation. This is the case in general. The n linear equations in n variables $x_1, x_2, \ldots, x_n$, of the form

$$\begin{matrix}a_{11}x_1+ & a_{12}x_2+ & \ldots+a_{1n}x_n= & b_1\\ a_{21}x_1+ & a_{22}x_2+ & \ldots+a_{2n}x_n= & b_2\\ \vdots & \vdots & \vdots & \vdots\\ a_{n1}x_1+ & a_{n2}x_2+ & \ldots+a_{nn}x_n= & b_n\end{matrix}$$

only have a solution providing some function of the a_{ij}s is non-zero. This function is called the *determinant*.

Writing the linear equations in the form

$$A\boldsymbol{x}=\boldsymbol{b}$$

the determinant is a property of the $n\times n$ coefficient matrix, A. This section is devoted to defining the determinant of a square matrix and to showing that linear systems of the form $A\boldsymbol{x}=\boldsymbol{b}$ have a solution if and only if the determinant of A is non-zero. From theorem 2.3.2, the determinant of A must be zero if and only if the rank of A is less than n.

The determinant of a matrix A is denoted by $|A|$ or by det A. The former notation will be used throughout this section.

From the above discussion, it is clear that the determinant of a 1×1 matrix $[a_{11}]$ should be defined as a_{11} and that the determinant of a 2×2 matrix

$$\begin{bmatrix} a_{11} & a_{12} \\ a_{21} & a_{22} \end{bmatrix}$$

should be $a_{11}a_{22} - a_{12}a_{21}$.

The definition of the determinant of an $n \times n$ matrix is recursive in the sense that it is assumed that the determinant of an $(n-1) \times (n-1)$ matrix is already defined. Before giving this recursive definition, some preliminary concepts are required.

Definition 2.5.1

The *minor*, $|M_{ij}|$, of any element a_{ij} in an $n \times n$ matrix, A, is the determinant of M_{ij}, the $(n-1) \times (n-1)$ submatrix of A obtained from A by omitting the ith row and jth column.

Example 2.5.1

(1) If $A = \begin{bmatrix} 2 & 1 \\ 4 & 3 \end{bmatrix}$, $M_{11} = [3]$, $M_{12} = [4]$, $M_{21} = [1]$ and $M_{22} = [2]$. Hence the minors of A are $|M_{11}| = 3$, $|M_{12}| = 4$, $|M_{21}| = 1$ and $|M_{22}| = 2$.

(2) If $A = \begin{bmatrix} 3 & -4 & 2 \\ 0 & 5 & 7 \\ 1 & 0 & 6 \end{bmatrix}$

$|M_{23}|$ is the determinant of the matrix $\begin{bmatrix} 3 & -4 \\ 1 & 0 \end{bmatrix}$ which is equal to $3 \times 0 - (-4) \times 1 = 4$.

Definition 2.5.2

The *cofactor*, A_{ij}, of any element a_{ij} in an $n \times n$ matrix A is $(-1)^{i+j}|M_{ij}|$. By definition, A_{ij} only differs from $|M_{ij}|$ in its sign. These signs differ as illustrated in figure 2.19.

The *adjoint* of A, Adj(A), is the transpose of the matrix $[A_{ij}]$.

i values \ *j* values	1	2	3	4	5	...
1	+	−	+	−	+	...
2	−	+	−	+	−	...
3	+	−	+	−	+	...
4	−	+	−	+	−	...
5	+	−	+	−	+	...
.	.	.	.	.	.	
.	.	.	.	.	.	
.	.	.	.	.	.	

Figure 2.19

Example 2.5.2

(1) If $A=\begin{bmatrix}2 & 1\\ 4 & 3\end{bmatrix}$, $A_{11}=3$, $A_{12}=-4$, $A_{21}=-1$ and $A_{22}=2$. Hence

$$\text{Adj}(A)=\begin{bmatrix}3 & -4\\ -1 & 2\end{bmatrix}^{\text{T}}=\begin{bmatrix}3 & -1\\ -4 & 2\end{bmatrix}$$

(2) If $A=\begin{bmatrix}3 & -4 & 2\\ 0 & 5 & 7\\ 1 & 0 & 6\end{bmatrix}$, $\text{Adj}(A)=\begin{bmatrix}30 & 7 & -5\\ 24 & 16 & -4\\ -38 & -21 & 15\end{bmatrix}^{\text{T}}$

$$=\begin{bmatrix}30 & 24 & -38\\ 7 & 16 & -21\\ -5 & -4 & 15\end{bmatrix}$$

Since the determinant has only been defined for 1×1 and 2×2 matrices, definitions 2.5.1 and 2.5.2 would seem to be applicable to at most 3×3 matrices. This situation can now be remedied by giving the definition of the determinant for an arbitrary $n\times n$ matrix.

Definition 2.5.3
Let $A=[a_{ij}]$ denote an $n\times n$ matrix. The *determinant* of A, denoted by $|A|$, is defined recursively as

(i) if $n=1$, then $|A|=a_{11}$;

(ii) if $n>1$, then $|A|=a_{11}A_{11}+a_{12}A_{12}+\ldots a_{1n}A_{1n}$.

Example 2.5.3

(1) If $A=\begin{bmatrix} a_{11} & a_{12} \\ a_{21} & a_{22} \end{bmatrix}$ is an arbitrary 2×2 matrix, then

$$|A|=\begin{vmatrix} a_{11} & a_{12} \\ a_{21} & a_{22} \end{vmatrix}=a_{11}A_{11}+a_{12}A_{12}=a_{11}a_{22}-a_{12}a_{21}$$

This agrees with our initial definition of the determinant of a 2×2 matrix.

(2) If

$$A=\begin{bmatrix} a_{11} & a_{12} & a_{13} \\ a_{21} & a_{22} & a_{23} \\ a_{31} & a_{32} & a_{33} \end{bmatrix}$$

is an arbitrary 3×3 matrix, then

$$|A|=\begin{vmatrix} a_{11} & a_{12} & a_{13} \\ a_{21} & a_{22} & a_{23} \\ a_{31} & a_{32} & a_{33} \end{vmatrix} =a_{11}\begin{vmatrix} a_{22} & a_{23} \\ a_{32} & a_{33} \end{vmatrix}-a_{12}\begin{vmatrix} a_{21} & a_{23} \\ a_{31} & a_{33} \end{vmatrix}+a_{13}\begin{vmatrix} a_{21} & a_{22} \\ a_{31} & a_{32} \end{vmatrix}$$

$$=a_{11}(a_{22}a_{33}-a_{23}a_{32})-a_{12}(a_{21}a_{33}-a_{23}a_{31})$$
$$+a_{13}(a_{21}a_{32}-a_{22}a_{31})$$
$$=a_{11}a_{22}a_{33}-a_{11}a_{23}a_{32}-a_{12}a_{21}a_{33}$$
$$+a_{12}a_{23}a_{31}+a_{13}a_{21}a_{32}-a_{13}a_{22}a_{31}$$

(3) If

$$A=\begin{bmatrix} 3 & -4 & 2 \\ 0 & 5 & 7 \\ 1 & 0 & 6 \end{bmatrix}$$

then

$$|A|=\begin{vmatrix} 3 & -4 & 2 \\ 0 & 5 & 7 \\ 1 & 0 & 6 \end{vmatrix}$$
$$=3(30)+4(-7)+2(-5)$$
$$=90-28-10=52$$

(4) If

$$A=\begin{bmatrix} 1 & 3 & 0 & 1 \\ -1 & 0 & 2 & 3 \\ 2 & -1 & 0 & 3 \\ 1 & 0 & 1 & 0 \end{bmatrix}$$

then

$$|A|=\begin{vmatrix} 1 & 3 & 0 & 1 \\ -1 & 0 & 2 & 3 \\ 2 & -1 & 0 & 3 \\ 1 & 0 & 1 & 0 \end{vmatrix}$$

$$=1\begin{vmatrix} 0 & 2 & 3 \\ -1 & 0 & 3 \\ 0 & 1 & 0 \end{vmatrix} -3\begin{vmatrix} -1 & 2 & 3 \\ 2 & 0 & 3 \\ 1 & 1 & 0 \end{vmatrix} +0\begin{vmatrix} -1 & 0 & 3 \\ 2 & -1 & 3 \\ 1 & 0 & 0 \end{vmatrix}$$

$$-1\begin{vmatrix} -1 & 0 & 2 \\ 2 & -1 & 0 \\ 1 & 0 & 1 \end{vmatrix}$$

$$=1\left(0\begin{vmatrix} 0 & 3 \\ 1 & 0 \end{vmatrix}-2\begin{vmatrix} -1 & 3 \\ 0 & 0 \end{vmatrix}+3\begin{vmatrix} -1 & 0 \\ 0 & 1 \end{vmatrix}\right)$$

$$-3\left(-1\begin{vmatrix} 0 & 3 \\ 1 & 0 \end{vmatrix}-2\begin{vmatrix} 2 & 3 \\ 1 & 0 \end{vmatrix}+3\begin{vmatrix} 2 & 0 \\ 1 & 1 \end{vmatrix}\right)$$

$$-1\left(-1\begin{vmatrix} -1 & 0 \\ 0 & 1 \end{vmatrix}-0\begin{vmatrix} 2 & 0 \\ 1 & 1 \end{vmatrix}+2\begin{vmatrix} 2 & -1 \\ 1 & 0 \end{vmatrix}\right)$$

$$=(0+0-3)-3(3+6+6)-(1+2)$$

$$=-3-45-3=-51$$

Evaluating the determinant of a large matrix using the definition of a determinant given above can be a time-consuming activity. To evaluate a 4×4 determinant may require over 50 simple arithmetic operations and a 10×10 determinant may require several million such operations! The determinant of a matrix is required sufficiently frequently in practice for it to be important to seek techniques which will simplify the task of evaluating a determinant. A useful start is the observation that the determinant of a lower triangular matrix is simply the product of the elements on its leading diagonal. The result also holds for an upper triangular matrix but the proof of this fact is left until theorem 2.5.4.

Theorem 2.5.1
If $A=[a_{ij}]$ is an $n\times n$ lower triangular matrix then $|A|=a_{11}a_{22}\ldots a_{nn}$.

Proof The proof is by induction on n. If $n=1$, $A=[a_{11}]$ and $|A|=a_{11}$.

Assume the result holds for all lower triangular matrices of order less than $k\times k$. Let A be any $k\times k$ lower triangular matrix; then

$$|A|=\begin{vmatrix} a_{11} & & & \bigcirc \\ a_{21} & a_{22} & & \\ \vdots & \vdots & \ddots & \\ a_{k1} & a_{k2} & \cdots & a_{kk} \end{vmatrix}$$

so $|A|=a_{11}A_{11}$, but by the induction hypothesis

$$A_{11}=\begin{vmatrix} a_{22} & & \bigcirc \\ \vdots & \ddots & \\ a_{k2} & \cdots & a_{kk} \end{vmatrix}=a_{22}a_{33}\ldots a_{kk}$$

Thus $|A|=a_{11}a_{22}\ldots a_{kk}$ and the result holds for A. Hence the theorem is proved by induction.

From the discussion in section 2.4, it follows that any $n\times n$ matrix can be converted to an upper (or lower) triangular matrix by a number of simple row operations, that is

(1) interchanging two rows;

(2) adding a multiple of one row to another.

Alternatively, the matrix can be converted to an upper (or lower) triangular matrix by similar operations which act on the columns. The next step in the study of determinants is thus to look at the effect that these and similar elementary row and column operations have on the value of the determinant.

Properties of Determinants

Let $A=[a_{ij}]$ be an $n\times n$ matrix. An alternative definition of the determinant, $|A|$, can be given by observing that $|A|$ is the sum of $n!$ different terms, each of which is of the form $\pm a_{1\sigma_1}a_{2\sigma_2}\ldots a_{n\sigma_n}$, where $\sigma_1\sigma_2\ldots\sigma_n$ is an arrangement (or, in mathematical terms, a *permutation*; see definition 4.1.7) of the n symbols 1, 2, ..., n.

Example 2.5.4

(1) If $n=2$, there are two permutations of 1, 2 namely 12 and 21

$$\begin{vmatrix} a_{11} & a_{12} \\ a_{21} & a_{22} \end{vmatrix} = a_{11}a_{22} - a_{12}a_{21}$$

(2) If $n=3$, there are $3!=6$ permutations of 1, 2, 3 namely 123, 132, 213, 231, 312 and 321. The determinant of a 3×3 matrix $A=[a_{ij}]$ should thus involve terms in $a_{11}a_{22}a_{33}$, $a_{11}a_{23}a_{32}$, $a_{12}a_{21}a_{33}$, $a_{12}a_{23}a_{31}$, $a_{13}a_{21}a_{32}$ and $a_{13}a_{22}a_{31}$. This is indeed the case — see example 2.5.3(2).

The sign of the term $a_{1\sigma_1}\sigma_{2\sigma_2}\ldots a_{n\sigma_n}$ in $|A|$ is determined by the parity of the permutation $\sigma_1\sigma_2\ldots\sigma_n$.

Definition 2.5.4
Let $\sigma=\sigma_1\sigma_2\ldots\sigma_n$ be a permutation of 1, 2, ..., n. Then, σ is called *even* if there is an even number of pairs (i, j) such that $i>j$ and i precedes j in σ. Otherwise, the permutation is called *odd*.

The *parity* of σ, par(σ), is defined as follows

$$\text{par}(\sigma)=\begin{cases} 1, & \text{if } \sigma \text{ is even} \\ -1, & \text{if } \sigma \text{ is odd} \end{cases}$$

Example 2.5.5

(1) The identity permutation $\varepsilon=123\ldots n$ is even since there are no pairs (i, j) such that $i>j$ and i precedes j in ε.

(2) By definition 2.5.4, 123, 231 and 312 are the even permutations of 1, 2, 3, while 132, 213 and 321 are the odd permutations.

(3) Let τ be a permutation of 1, 2, ..., n which interchanges two numbers r and s but leaves the others fixed. Such a permutation is called a *transposition*. If $r<s$, then $\tau=12\ldots(r-1)s(r+1)\ldots(s-1)r(s+1)\ldots n$. There are $2(s-r-1)+1$ pairs (i, j) such that $i>j$ and i precedes j in τ. These comprise pairs (x, r) and (s, x) where $x=r+1, \ldots, s-1$ together with the pair (s, r). Thus, every transposition is odd.

Definition 2.5.5
If σ and τ are two permutations of 1, 2, ..., n, then their *composition* $\sigma_0\tau$ is the permutation $\rho_1\rho_2\ldots\rho_n$ where $\rho_i=\sigma_{\tau_i}$.

If σ and τ are such that $\sigma_0\tau=\tau_0\sigma=\varepsilon$, then τ is the *inverse permutation* of σ and is denoted by σ^{-1}.

Example 2.5.6

(1) If $\sigma = 12435$ and $\tau = 52314$, then $\sigma_0\tau = 52413$ and $\tau_0\sigma = 52134$.

(2) If $\sigma = 12435$, then $\sigma^{-1} = 12435$.

(3) If $\sigma = 52314$, then $\sigma^{-1} = 42351$.

The proof of the following theorem, which follows reasonably easily from the above definitions, is left to the reader.

Theorem 2.5.2
(i) If σ, τ are permutations, then $\text{par}(\sigma_0\tau) = \text{par}(\tau_0\sigma) = \text{par}(\sigma) \times \text{par}(\tau)$.

(ii) If σ is a permutation, then $\text{par}(\sigma) = \text{par}(\sigma^{-1})$.

Using the concepts of permutation and parity, an alternative definition of the determinant can now be given.

Definition 2.5.6

The *determinant*, $|A|$, of an $n \times n$ matrix $A = [a_{ij}]$ is $\sum_{\sigma} \text{par}(\sigma)\ a_{1\sigma_1}a_{2\sigma_2}\ldots a_{n\sigma_n}$, where the sum is taken over all possible permutations $\sigma = \sigma_1\sigma_2\ldots\sigma_n$ of 1, 2, ..., n.

Example 2.5.7

Using the above definition, if $A = [a_{ij}]$ is a 3×3 matrix

$$|A| = a_{11}a_{22}a_{33} + a_{12}a_{23}a_{31} + a_{13}a_{21}a_{32} - a_{11}a_{23}a_{32} - a_{12}a_{21}a_{33} - a_{13}a_{22}a_{31}$$

which agrees with the result obtained from the previous definition. For a 3×3 matrix, which terms in the expression for the determinant are positive and which are negative can be remembered from figure 2.20.

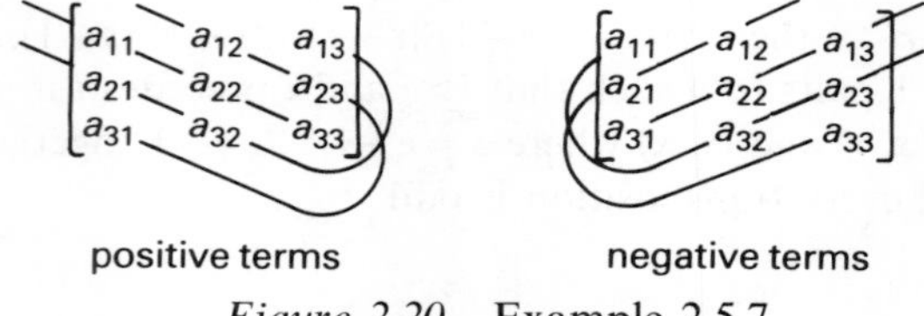

Figure 2.20 Example 2.5.7

In the expansion of a determinant, any term $a_{1\sigma_1}a_{2\sigma_2}\ldots a_{n\sigma_n}$ involves precisely one term from any given row (or column). An immediate corollary to this is the following.

Theorem 2.5.3
Let A be an $n \times n$ matrix.

(i) If A has a row (or column) of zeros then $|A|=0$.

(ii) If B is obtained from A by multiplying any row (or column) by $\lambda \in \mathbb{R}$, then $|B|=\lambda|A|$.

Example 2.5.8

(1)
$$\begin{vmatrix} 2 & 0 & 3 \\ 1 & 0 & -1 \\ 5 & 0 & 2 \end{vmatrix} = 2\times 0\times 2+0\times(-1)\times 5+3\times 1\times 0-2\times 0\times(-1) -0\times 1\times 2-3\times 0\times 5=0$$

(2)
$$\begin{vmatrix} 40 & 8 & 12 \\ 90 & 3 & 6 \\ 0 & 5 & 10 \end{vmatrix} = 10\begin{vmatrix} 4 & 8 & 12 \\ 9 & 3 & 6 \\ 0 & 5 & 10 \end{vmatrix} = 10\times 4\begin{vmatrix} 1 & 2 & 3 \\ 9 & 3 & 6 \\ 0 & 5 & 10 \end{vmatrix}$$

$$= 40\times 3\begin{vmatrix} 1 & 2 & 3 \\ 3 & 1 & 2 \\ 0 & 5 & 10 \end{vmatrix} = 120\times 5\begin{vmatrix} 1 & 2 & 3 \\ 3 & 1 & 2 \\ 0 & 1 & 2 \end{vmatrix}$$

$$=600(2+0+9-2-12-0)=-1800$$

Theorem 2.5.4
Let A be an $n \times n$ matrix.

(i) If $B=A^{\mathrm{T}}$, then $|B|=|A|$.

(ii) If A is an upper triangular matrix, then $|A|=a_{11}a_{22}\ldots a_{nn}$.

Proof (i) If $A=[a_{ij}]$ then $A^{\mathrm{T}}=B=[b_{ij}]$ where $b_{ij}=a_{ji}$. Hence

$$|B|=\sum_{\sigma}\operatorname{par}(\sigma)b_{1\sigma_1}b_{2\sigma_2}\ldots b_{n\sigma_n}$$

$$=\sum_{\sigma}\operatorname{par}(\sigma)a_{\sigma_1 1}a_{\sigma_2 2}\ldots a_{\sigma_n n}$$

Now $a_{\sigma_1 1}a_{\sigma_2 2}\ldots a_{\sigma_n n}=a_{1\tau_1}a_{2\tau_2}\ldots a_{n\tau_n}$ where τ is the inverse of σ. As σ ranges over all permutations, so does $\tau=\sigma^{-1}$. Thus

$$|B|=\sum_{\tau}\operatorname{par}(\tau^{-1})a_{1\tau_1}a_{2\tau_2}\ldots a_{n\tau_n}$$

$$=\sum_{\tau}\text{par}(\tau)a_{1\tau_1}a_{2\tau_2}\ldots a_{n\tau_n}=|A|$$

(ii) The proof follows immediately from theorem 2.5.1 and part (i).

Example 2.5.9

(1) $\begin{vmatrix}3 & 1 & 2\\0 & 4 & 6\\5 & 1 & 2\end{vmatrix}=3\times4\times2+1\times6\times5+2\times0\times1-3\times1\times6-1\times0\times2-2\times4\times5=-4$

$\begin{vmatrix}3 & 0 & 5\\1 & 4 & 1\\2 & 6 & 2\end{vmatrix}=3\times4\times2+0\times1\times2+5\times1\times6-3\times6\times1-0\times1\times2-5\times4\times2=-4$

(2) $\begin{vmatrix}3 & 2 & 9\\0 & 5 & 6\\0 & 0 & 2\end{vmatrix}=3\times5\times2=30$

Theorem 2.5.5
Let A be an $n\times n$ matrix.

(i) If B is obtained from A by interchanging two rows (or columns) of A then $|B|=-|A|$.

(ii) If two rows (or columns) of A are equal then $|A|=0$.

(iii) If B is obtained from A by adding a multiple of one row (or column) to another row (or column) then $|B|=|A|$.

(iv) If B is obtained from A by adding to one row (or column) a linear combination of the remaining rows (or columns) then $|B|=|A|$.

Proof (i) Assume column r is interchanged with column s and $r<s$. Let τ denote the permutation of 1, 2, ..., n which interchanges r and s but leaves the others fixed. From example 2.5.5(3), τ is an odd permutation so $\text{par}(\tau)=-1$.

If $A=[a_{ij}]$ then $B=[b_{ij}]=[a_{i\tau_j}]$. Thus

$$|B|=\sum_{\sigma}\text{par}(\sigma)b_{1\sigma_1}b_{2\sigma_2}\ldots b_{n\sigma_n}$$

$$=\sum_{\sigma}\text{par}(\sigma)a_{1(\sigma_0\tau)_1}a_{2(\sigma_0\tau)_2}\ldots a_{n(\sigma_0\tau)_n}$$

As σ ranges over all permutations, so too does $\sigma_0\tau$. Thus

$$|A| = \sum_{\sigma} \text{par}(\sigma_0\tau) a_{1(\sigma_0\tau)_1} a_{2(\sigma_0\tau)_2} \ldots a_{n(\sigma_0\tau)_n}$$

$$= \text{par}(\tau) \times \sum_{\sigma} \text{par}(\sigma) a_{1(\sigma_0\tau)_1} a_{2(\sigma_0\tau)_2} \ldots a_{n(\sigma_0\tau)_n}$$

$$= -|B|$$

The proof for rows follows immediately from this result and theorem 2.5.4(i).

(ii) If the two identical rows of A are interchanged, the matrix A remains unchanged. By part (i), it follows that $|A| = -|A|$. Hence $|A| = 0$.

(iii) and (iv) are left as exercises for the reader. They both follow from applications of result (ii).

The above results give a number of techniques which can be used to ease the evaluation of determinants.

Example 2.5.10

(1) $$\begin{vmatrix} 3 & 0 & 1 \\ 1 & 2 & 3 \\ 5 & 0 & 0 \end{vmatrix} = -\begin{vmatrix} 5 & 0 & 0 \\ 1 & 2 & 3 \\ 3 & 0 & 1 \end{vmatrix} = -5\begin{vmatrix} 2 & 3 \\ 0 & 1 \end{vmatrix} = -10.$$

(2) $$\begin{vmatrix} 1 & 3 & 5 \\ 2 & 0 & 1 \\ 1 & 3 & 5 \end{vmatrix} = 0$$ since the first and third rows are equal.

(3) $$\begin{vmatrix} 2 & -1 & 4 \\ 3 & 5 & -9 \\ 7 & 3 & -1 \end{vmatrix} = \begin{vmatrix} 2 & -1 & 4 \\ 3 & 5 & -9 \\ 0 & 0 & 0 \end{vmatrix}$$ (3rd row = 3rd row − 2nd row − 2 × 1st row)

(4) Any matrix can be converted to upper [lower] triangular form by interchanging columns or rows and adding multiples of columns or rows to other columns or rows. None of these operations alters the absolute value of the determinant of the matrix and since the determinant of a triangular matrix can be so easily found, converting a matrix to triangular form is a useful technique in the evaluation of the determinant.

$$\begin{vmatrix} 4 & 3 & 1 \\ 11 & 2 & 3 \\ 6 & 1 & 2 \end{vmatrix} = -\begin{vmatrix} 1 & 3 & 4 \\ 3 & 2 & 11 \\ 2 & 1 & 6 \end{vmatrix}$$ (interchange 1st and 3rd column)

$$= -\begin{vmatrix} 1 & 0 & 0 \\ 3 & -7 & -1 \\ 2 & -5 & -2 \end{vmatrix}$$ (2nd column = 2nd column − 3 × 1st column, 3rd column = 3rd column − 4 × 1st column)

$$= \begin{vmatrix} 1 & 0 & 0 \\ 3 & -1 & -7 \\ 2 & -2 & -5 \end{vmatrix}$$ (interchange 2nd and 3rd column)

$$= \begin{vmatrix} 1 & 0 & 0 \\ 3 & -1 & 0 \\ 2 & -2 & 9 \end{vmatrix}$$ (3rd column = 3rd column − 7 × 2nd column)

$$= -9$$

As described in section 2.4, both Gaussian elimination and its variant, triangular decomposition, involve the reduction of a matrix A to an upper triangular matrix, U. The determinant of A is given by

$$|A| = u_{11}u_{22}\ldots u_{nn}$$

Most computer library routines for solving systems of linear equations also compute the determinant of the coefficient matrix in this way.

Theorem 2.5.6
Let A be an $n \times n$ matrix.

(i) For any $n \times n$ elementary matrix E, $|EA| = |E|\,|A|$.

(ii) If $A \sim B$ (that is, A and B are row equivalent) then $|A| = 0$ iff $|B| = 0$.

Proof (i) Consider the three possible types of elementary matrix.
If E is E_{ik} then by theorem 2.5.5(i), $|E_{ik}| = -|I| = -1$ and $|EA| = -|A|$. Hence, $|E|\,|A| = |EA|$.
If E is $E_i(\alpha)$ then by theorem 2.5.3(ii), $|E_i(\alpha)| = \alpha|I| = \alpha$ and $|EA| = \alpha|A|$. Hence, $|E|\,|A| = |EA|$.
If E is $E_{ik}(\alpha)$ then by theorem 2.5.5(iii), $|E_{ik}(\alpha)| = |I| = 1$ and $|EA| = |A|$. Hence, $|E|\,|A| = |EA|$.

(ii) If $A \sim B$ then $B = E_1E_2\ldots E_mA$, where $E_1, E_2, \ldots, E_m$ are elementary matrices. Using part (i) and induction, it is easy to show that
$|B| = |E_1||E_2|\ldots|E_m||A|$. Since $|E_i| \neq 0$ for any elementary matrix, it follows that $|A| = 0$ iff $|B| = 0$.

Definition 2.5.7
A square matrix is *singular* iff its determinant is zero. If a matrix is not singular then it is said to be *non-singular*.

Theorem 2.5.7
Let A be an $n \times n$ matrix. The following four conditions are equivalent

(i) A has rank n;

(ii) A is invertible;

(iii) for any $\boldsymbol{b} \in \mathbb{R}^n$, the set of equations $A\boldsymbol{x} = \boldsymbol{b}$ has a unique solution;

(iv) A is non-singular.

Proof Conditions (i), (ii) and (iii) are equivalent by theorems 2.3.2 and 2.3.3.

However, condition (ii) implies condition (iv) since, if A is invertible, then, from the discussion in section 2.4, it is row equivalent to the identity matrix. Hence $|A| \neq 0$ by theorem 2.5.6(ii).

Also, condition (iv) implies condition (ii) since if A is not invertible, then it must be row equivalent to a matrix with a row of zeros. Hence $|A| = 0$ by theorem 2.5.6(ii).

Hence statement (iv) is equivalent to statement (ii) which proves all the statements are equivalent.

Example 2.5.11

Theorem 2.5.7 has many applications, for example it can be used to determine whether a set of vectors is linearly dependent or not.

(1) The vectors [3 0 1], [1 2 3] and [5 0 0] are linearly independent since the matrix

$$\begin{bmatrix} 3 & 0 & 1 \\ 1 & 2 & 3 \\ 5 & 0 & 0 \end{bmatrix}$$

is non-singular (see example 2.5.10) and therefore has rank 3.

(2) The vectors [2 −1 4], [3 5 −9] and [7 3 −1] are linearly dependent since

$$\begin{vmatrix} 2 & -1 & 4 \\ 3 & 5 & -9 \\ 7 & 3 & -1 \end{vmatrix} = 0$$

(3) The equations

$$3x+2y=5$$
$$2x-3y+z=4$$
$$ky+2z=1$$

have a unique solution provided

$$\begin{vmatrix} 3 & 2 & 0 \\ 2 & -3 & 1 \\ 0 & k & 2 \end{vmatrix} \neq 0$$

that is, provided $3(-6-k)-2(4)\neq 0$, that is, $k\neq -26/3$.

Theorem 2.5.8
For any square matrices A and B, $|AB|=|A||B|$.

Proof If A is non-singular then it is invertible and hence row equivalent to the identity matrix. Hence $A=E_1E_2\ldots E_mI$ for some elementary matrices $E_1, E_2, \ldots, E_m$. Thus $|A|=|E_1|\,|E_2|\ldots|E_m|$ and $|AB|=|E_1E_2\ldots E_mB|$ $=|E_1|\,|E_2|\ldots|E_m|\,|B|=|A|\,|B|$.

If AB is invertible then so is A since $A(B(AB)^{-1})=(AB)(AB)^{-1}=I$. Thus if A is singular, then AB is singular, that is, $|A|=0$ implies $|AB|=0$. Hence if A is singular, $|AB|=|A|\,|B|$.

The theorem holds whether A is non-singular or singular.

Example 2.5.12

(1) $A=\begin{bmatrix} 2 & 0 & 3 \\ 1 & -4 & 0 \\ 0 & 8 & 3 \end{bmatrix}$

has rank two and $|A|=0$. Hence, for any matrix $B=[b_{ij}]$, $|AB|=0$.

(2) If

$$A=\begin{bmatrix} 3 & -4 & 2 \\ 0 & 5 & 7 \\ 1 & 0 & 6 \end{bmatrix}$$

then $|A|=52$ [by example 2.5.3(3)]. If

$$B=\begin{bmatrix} 3 & 0 & 5 \\ 1 & 4 & 1 \\ 2 & 6 & 2 \end{bmatrix}$$

then $|B| = -4$. In this case

$$|AB| = \begin{vmatrix} 9 & -4 & 15 \\ 19 & 62 & 19 \\ 15 & 36 & 17 \end{vmatrix}$$

which, as the reader should verify, is equal to -4×52.

The Inverse of a Matrix

In definition 2.5.3, the determinant of $[a_{ij}]$ was defined as $a_{11}A_{11} + a_{12}A_{12} + \ldots + a_{1n}A_{1n}$ $(n > 1)$, that is, it was defined in terms of the elements and cofactors of the first row. It is possible to evaluate the determinant using the elements and cofactors of any row or column.

Theorem 2.5.9
If $A = [a_{ij}]$ is an $n \times n$ matrix $(n > 1)$, then for any $i, j = 1, 2, \ldots, n$,
$|A| = a_{i1}A_{i1} + a_{i2}A_{i2} + \ldots + a_{in}A_{in} = a_{1j}A_{1j} + a_{2j}A_{2j} + \ldots + a_{nj}A_{nj}$.

Proof First, prove that $|A| = a_{i1}A_{i1} + a_{i2}A_{i2} + \ldots + a_{in}A_{in}$ for $i = 1, 2, \ldots, n$. The result holds for $i = 1$ by definition.

If $i > 1$, exchange row i with each preceding row until it is the first row. For $j = 1, 2, \ldots, n$, the determinant $|M_{ij}|$ is not affected since the relative position of the other rows is not altered by these exchanges. Now, each row interchange changes the sign of the determinant by theorem 2.5.5(i). Thus the determinant of the rearranged matrix is $(-1)^{i-1}|A|$. Hence

$$(-1)^{i-1}|A| = a_{i1}|M_{i1}| - a_{i2}|M_{i2}| + \ldots + a_{in}(-1)^{n-1}|M_{in}|$$

$$|A| = a_{i1}(-1)^{i+1}|M_{i1}| + a_{i2}(-1)^{i+2}|M_{i2}| + \ldots$$

$$+ a_{in}(-1)^{i+n}|M_{in}|$$

$$= a_{i1}A_{i1} + a_{i2}A_{i2} + \ldots + a_{in}A_{in}$$

The proof that $|A| = a_{1j}A_{1j} + a_{2j}A_{2j} + \ldots + a_{nj}A_{nj}$ follows from the result of theorem 2.5.4, that $|A^{\mathrm{T}}| = |A|$.

Example 2.5.13

(1) Let

$$A = \begin{bmatrix} 3 & 0 & 1 \\ 1 & 2 & 3 \\ 5 & 0 & 0 \end{bmatrix}$$

Expanding along the third row gives

$$|A| = a_{31}A_{31} + a_{32}A_{32} + a_{33}A_{33}$$

$$= 5\begin{vmatrix} 0 & 1 \\ 2 & 3 \end{vmatrix} = -10$$

Expanding along the second column gives

$$|A| = a_{12}A_{12} + a_{22}A_{22} + a_{32}A_{32}$$

$$= 2\begin{vmatrix} 3 & 1 \\ 5 & 0 \end{vmatrix} = -10$$

(2) Let

$$A = \begin{bmatrix} 4 & 3 & 1 \\ 11 & 2 & 3 \\ 6 & 1 & 2 \end{bmatrix}$$

Expanding along the second row gives

$$|A| = a_{21}A_{21} + a_{22}A_{22} + a_{23}A_{23}$$

$$= 11 \times -\begin{vmatrix} 3 & 1 \\ 1 & 2 \end{vmatrix} + 2 \times \begin{vmatrix} 4 & 1 \\ 6 & 2 \end{vmatrix} + 3 \times -\begin{vmatrix} 4 & 3 \\ 6 & 1 \end{vmatrix}$$

$$= -11(6-1) + 2(8-6) - 3(4-18)$$

$$= -55 + 4 + 42$$

$$= -9$$

Theorem 2.5.10
If $A = [a_{ij}]$ is an $n \times n$ matrix, then

$$a_{i1}A_{j1} + a_{i2}A_{j2} + \ldots + a_{in}A_{jn} = \begin{cases} |A| & \text{if } i = j \\ 0 & \text{if } i \neq j \end{cases}$$

Proof If $i = j$, the result is simply a restatement of theorem 2.5.9. If $i \neq j$, replace the jth row of A by the elements of the ith row. A now has two rows equal so by theorem 2.5.5(ii) the determinant is zero. But by expanding along the jth row of the altered matrix, the determinant is $a_{i1}A_{j1} + a_{i2}A_{j2} + \ldots + a_{in}A_{jn}$.

Theorem 2.5.11
Provided $|A| \neq 0$

$$A^{-1} = \frac{1}{|A|}\text{Adj}(A)$$

Proof The i,jth element of $A \times \text{Adj}(A)$ is $a_{i1}A_{j1} + a_{i2}A_{j2} + \ldots + a_{in}A_{jn}$. Thus all the elements of $A \times \text{Adj}(A)$ are zero except those on the leading diagonal which are all equal to $|A|$. Hence

$$A \times \text{Adj}(A) = |A|I, \text{ so } A^{-1} = \frac{1}{|A|}\text{Adj}(A)$$

Example 2.5.14

In example 2.5.2(2), the adjoint of

$$A = \begin{bmatrix} 3 & -4 & 2 \\ 0 & 5 & 7 \\ 1 & 0 & 6 \end{bmatrix}$$

was shown to be

$$\begin{bmatrix} 30 & 24 & -38 \\ 7 & 16 & -21 \\ -5 & -4 & 15 \end{bmatrix}$$

In example 2.5.3(3), it is shown that $|A| = 52$. Hence

$$A^{-1} = \frac{1}{52}\begin{bmatrix} 30 & 24 & -38 \\ 7 & 16 & -21 \\ -5 & -4 & 15 \end{bmatrix}$$

Check

$$AA^{-1} = \frac{1}{52}\begin{bmatrix} 3 & -4 & 2 \\ 0 & 5 & 7 \\ 1 & 0 & 6 \end{bmatrix}\begin{bmatrix} 30 & 24 & -38 \\ 7 & 16 & -12 \\ -5 & -4 & 15 \end{bmatrix} = \frac{1}{52}\begin{bmatrix} 52 & 0 & 0 \\ 0 & 52 & 0 \\ 0 & 0 & 52 \end{bmatrix} = I$$

The reader should note that in practice the inverse of a matrix is generally computed by the technique described in section 2.4.

Exercise 2.5

1. Evaluate the following.

(a) $\begin{vmatrix} 2 & -1 \\ 3 & 4 \end{vmatrix}$

(b) $\begin{vmatrix} 3 & -2 & 4 \\ 5 & -12 & 2 \\ 2 & 3 & 5 \end{vmatrix}$ (c) $\begin{vmatrix} 12 & 9 & 6 \\ 6 & 5 & 8 \\ 2 & 0 & 4 \end{vmatrix}$

(d) $\begin{vmatrix} a & b & c \\ c & a & b \\ b & c & a \end{vmatrix}$ (e) $\begin{vmatrix} a & b & c \\ b & c & a \\ c & a & b \end{vmatrix}$

2. By applying elementary row operations, convert the following matrix to upper triangular form. Hence, evaluate its determinant.

$$\begin{bmatrix} 1 & 2 & -1 & 4 \\ 2 & 1 & -1 & 1 \\ 4 & -1 & 3 & 1 \\ -1 & 1 & 2 & 1 \end{bmatrix}$$

3. List all the permutations of 1, 2, 3, 4. State the parity of each of them.

4. Prove theorem 2.5.2.

5. Prove theorem 2.5.5(iii) and (iv).

6. Prove *Cauchy's formula*: $|\text{Adj}(A)| = |A|^{n-1}$ for any $n \times n$ matrix, A.

7. Prove the equivalence of definitions 2.5.3 and 2.5.6.

8. Two square matrices A, B are said to be *similar* if $A = P^{-1}BP$ for some non-singular matrix P. Show that similar matrices have identical determinants.

9. Let σ be a permutation of 1, 2, …, n. If $A = [a_{ij}]$ is an $n \times n$ matrix, let $\sigma(A)$ denote the matrix $[a_{\sigma_i \sigma_j}]$. Show that $|\sigma(A)| = |A|$.

10. The *eigenvalues* of an $n \times n$ matrix, A, are the solutions of the equation $|A - \lambda I| = 0$. Find the four eigenvalues of

$$\begin{bmatrix} 0 & 1 & 2 & 1 \\ 1 & 0 & 1 & 2 \\ 2 & 1 & 0 & 1 \\ 1 & 2 & 1 & 0 \end{bmatrix}$$

11. Calculate the adjoint of

$$A=\begin{bmatrix} 1 & 2 & 3 \\ -1 & 0 & 2 \\ 0 & 5 & 1 \end{bmatrix}$$

Hence, find A^{-1}.

12. Show that, provided $|A| \neq 0$, the unique solution of $A\boldsymbol{x}=\boldsymbol{b}$ is given by

$$[|A_1| \; |A_2| \; \ldots \; |A_n|]^{\mathrm{T}}/|A|$$

where A_i is obtained from A by replacing the ith column of A by $\boldsymbol{b}$. (This result is known as *Cramer's rule.* Because of the cost in evaluating determinants, this is not an efficient way of solving equations.)

Use Cramer's rule to find the solution of

$$3x-2y+z=11$$

$$x+y-z=-2$$

$$2x-5y=9$$

13. Prove that a matrix A has rank k iff the largest square non-singular submatrix of A has dimension $k \times k$.

Hence, find the rank of

$$\begin{bmatrix} 3 & 2 & -2 & 1 \\ 6 & 4 & 3 & 2 \\ 6 & 4 & -4 & 2 \end{bmatrix}$$

3 Calculus

3.1 SEQUENCES

Introduction

Most people have an intuitive feel for many of the concepts that calculus seeks to make precise — convergence, continuity and rate of change are just some of the common ideas with which calculus concerns itself.

Calculus provides a mathematical basis for many important areas of computing, for example, numerical analysis and computer-aided design.

This section considers sequences and introduces the most fundamental concept in calculus — the notion of a limit.

Definition 3.1.1
A *finite sequence* is a function whose domain consists of the first p ($\geqslant 1$) natural numbers. The sequence may be written as a succession of numbers

$$a_1, a_2, a_3, \ldots, a_p$$

in which a_n denotes the nth *term* in the sequence.

An *infinite sequence* is a function whose domain is the set of all natural numbers; thus, an infinite sequence has an infinite number of terms. In the sequel, an infinite sequence with nth term a_n will be denoted by (a_n).

It is often convenient to describe a sequence simply by listing the first few terms. However, if there is to be no possibility of ambiguity, all of the terms must be determined according to a definite rule. Such a rule will often have the form $a_n = f(n)$.

Example 3.1.1

(1) 1, 3, 5, 9, 11, 13 is a finite sequence which has six terms.

(2) If $a_n = (3n-1)/(4n+5)$ then the first five terms of the infinite sequence (a_n) are

$$\frac{2}{9}, \frac{5}{13}, \frac{8}{17}, \frac{11}{21}, \frac{14}{25}$$

Recurrence Relations

In practice, a sequence is sometimes defined *recursively*. This means that the nth term can be computed from previous terms in the sequence, assuming these have already been determined.

Definition 3.1.2
A sequence may be defined recursively if there exists an expression of the form $a_n = F(a_{n-1}, a_{n-2}, \ldots, a_{n-r})$ connecting a_n with previous terms. An expression that connects a_n with the previous r terms in the sequence is called a *recurrence relation of order r*, or a *difference equation of order r*. A recurrence relation of the form $a_n = c_1 a_{n-1} + c_2 a_{n-2} + \ldots + c_r a_{n-r}$, where the coefficients c_i are given real numbers, is called a *linear recurrence relation*.

Given a recursive definition for a sequence (a_n), it may or may not be possible to obtain an explicit formula of the form $a_n = f(n)$, that is, to 'solve' the recurrence relation for a_n.

Example 3.1.2

(1) The sequence defined by the first-order non-linear recurrence relation

$$a_1 = \sqrt{2};\ a_n = \sqrt{(2a_{n-1})},\ n \geqslant 2$$

has the first three terms $\sqrt{2}$, $\sqrt{(2\sqrt{2})}$, $\sqrt{[2\sqrt{(2\sqrt{2})}]}$.

It may be shown by induction (see exercise 3.1) that an explicit formula for defining this sequence is $a_n = 2^{1-2^{-n}}$.

(2) The *Fibonacci sequence* 1, 1, 2, 3, 5, 8, 13, ... may be defined by the second-order linear recurrence relation

$$a_1 = 1;\ a_2 = 1;\ a_n = a_{n-1} + a_{n-2},\ n \geqslant 3$$

In order to solve this recurrence relation, first notice that $a_n = c\alpha^n$ will be a solution if

$$c\alpha^{n-2}(\alpha^2 - \alpha - 1) = 0$$

The solutions of interest are the roots of the *indicial equation*

$$\alpha^2 - \alpha - 1 = 0$$

that is

$$\alpha = \frac{1}{2}(1 \pm \sqrt{5})$$

The general solution of the recurrence relation $a_n = a_{n-1} + a_{n-2}$ is therefore

$$a_n = c_1\left(\frac{1+\sqrt{5}}{2}\right)^n + c_2\left(\frac{1-\sqrt{5}}{2}\right)^n$$

Finally, the coefficients c_1, c_2 may be determined by using the initial conditions $a_1 = 1$, $a_2 = 1$, to give

$$a_n = \frac{1}{\sqrt{5}}\left[\left(\frac{1+\sqrt{5}}{2}\right)^n - \left(\frac{1-\sqrt{5}}{2}\right)^n\right]$$

This example illustrates a general technique for the solution of linear recurrence relations. A further technique is discussed in section 3.3.

(3) The first-order non-linear recurrence relation

$$a_1 = 1; \quad a_n = \sqrt{(a_{n-1} + 1)}, \; n \geqslant 2$$

defines a sequence (a_n) having the first four terms

$$1, \sqrt{2}, \sqrt{(\sqrt{2}+1)}, \sqrt{[\sqrt{(\sqrt{2}+1)}+1]}$$

The above technique is not applicable in this case and the authors are not aware of an explicit formula for defining this sequence.

Limit of a Sequence

Some sequences *converge* to a *limit*. Consider the sequence 0.3, 0.33, 0.333, 0.3333, ... The terms in this sequence are successively better approximations to 1/3 and, by taking sufficiently many terms, the difference between 1/3 and the last term taken can be made as small as one likes. The sequence converges to the limit of 1/3. On the other hand, the terms in the sequence 1, 2, 3, 4, ... steadily increase and the terms in the sequence $-1, 1, -1, 1, \ldots$ *oscillate* between ± 1.

Definition 3.1.3
(a) A sequence (a_n) *converges* to the limit l if, for any given $\varepsilon > 0$ (however small), there is an integer M (which depends on ε) such that $|a_n - l| < \varepsilon$, for all $n \geqslant M$.

(b) A sequence (a_n) *diverges*, or tends to infinity, if for any given C (however large) there is an integer M such that

$$a_n > C$$

for all $n \geqslant M$. Similarly, (a_n) tends to minus infinity if for any C (however large and negative) $a_n < C$ for all n greater than some M.

(c) If the sequence (a_n) does not tend to a limit l nor to $\pm\infty$ then it is an *oscillating* sequence.

If a sequence (a_n) converges to the limit l, then this may be denoted by writing

$$\lim_{n\to\infty} a_n = l, \quad \text{or} \quad a_n \to l \text{ as } n \to \infty$$

and if it tends to infinity, by writing

$$\lim_{n\to\infty} a_n = \infty, \quad \text{or} \quad a_n \to \infty \text{ as } n \to \infty$$

Graphical Representation of a Limit

A sequence is a function and can therefore be represented by a graph, which will consist of a set of isolated points, one corresponding to each integral value $n = 1, 2, 3, \ldots$ (figure 3.1). Obviously, only a finite number of terms of an infinite sequence can be represented as points on a page. Nevertheless, such a pictorial representation is useful for illustrating the idea of limit. Figure 3.1a illustrates a sequence that converges to the limit l. For any given $\varepsilon > 0$, only a finite number of terms lie without the interval bounded by the lines $a = l - \varepsilon$ and $a = l + \varepsilon$.

Figure 3.1b illustrates a sequence tending to infinity. For any given C, only a finite number of terms have value less than C.

Example 3.1.3

(1) The sequence (a_n) defined by $a_n = 1/n$ has limit zero. For, given $\varepsilon > 0$ there is an integer M such that $1/M < \varepsilon$; for example, if $\varepsilon = 0.00001$ then a suitable M is $1/0.00001 + 1 = 100\,001$. Thus, for all $n \geqslant M$

$$a_n = \frac{1}{n} \leqslant \frac{1}{M} < \varepsilon$$

that is, $|a_n - 0| < \varepsilon$.

(2) The sequence $(1.1)^n \to \infty$ as $n \to \infty$. For, given $C > 0$, $(1.1)^n > C$ for all $n > \log C/\log 1.1$.

(3) The sequence (a_n) defined by $a_n = (-1)^n$ oscillates between ± 1; note, however, that the sequence with nth term $(-1)^n/n$ is *not* an oscillating sequence, but converges to the limit zero.

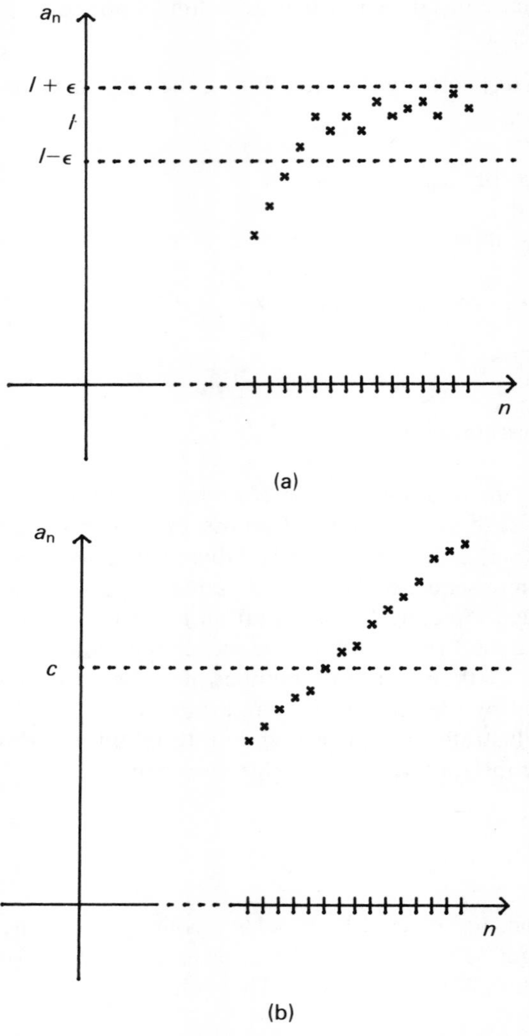

Figure 3.1 (a) A sequence converging to the limit l; (b) a sequence tending to infinity

Limits and Arithmetic Operations

Theorem 3.1.1
If the limit of a sequence exists, it is unique.

Proof The proof is by contradiction. Suppose $\lim_{n\to\infty} a_n = l_1$ and $\lim_{n\to\infty} a_n = l_2$,

$l_1 \neq l_2$. Given $\varepsilon > 0$, there exist integers M_1, M_2 such that

$$|a_n - l_1| < \frac{1}{2}\varepsilon \text{ for all } n \geqslant M_1$$

and

$$|a_n - l_2| < \frac{1}{2}\varepsilon \text{ for all } n \geqslant M_2$$

Let $M = \max\{M_1, M_2\}$ then

$$|a_n - l_1| < \frac{1}{2}\varepsilon \text{ and } |a_n - l_2| < \frac{1}{2}\varepsilon \text{ for all } n \geqslant M$$

Now

$$\begin{aligned} |l_1 - l_2| &= |-(a_n - l_1) + (a_n - l_2)| \\ &\leqslant |a_n - l_1| + |a_n - l_2| \text{ (by the triangle inequality)} \\ &< \frac{1}{2}\varepsilon + \frac{1}{2}\varepsilon = \varepsilon \text{ for all } n \geqslant M \end{aligned}$$

This is true for any $\varepsilon > 0$. If $l_1 \neq l_2$ then $|l_1 - l_2| > 0$ and ε may be given this value, implying

$$|l_1 - l_2| < |l_1 - l_2|$$

which is impossible. Hence l_1 must equal l_2.

By applying the rules given in the following theorem, it is often possible to determine the behaviour of a given sequence by relating it to 'simpler' sequences of known behaviour.

Theorem 3.1.2
Let the sequences (a_n) and (b_n) have limits a and b, respectively

(i) $\lim_{n \to \infty} (a_n \pm b_n) = a \pm b$

(ii) $\lim_{n \to \infty} a_n b_n = ab$

(iii) $\lim_{n \to \infty} \frac{a_n}{b_n} = \frac{a}{b}$ provided $b \neq 0$

(iv) $\lim_{n \to \infty} a_n^m = a^m$ provided $a \geqslant 0$

Proof (i) Given $\varepsilon > 0$, let M be sufficiently large that

$$|a_n - a| < \frac{1}{2}\varepsilon \text{ and } |b_n - b| < \frac{1}{2}\varepsilon \text{ for all } n \geqslant M$$

Then

$$|(a_n \pm b_n)-(a \pm b)| = |(a_n - a) \pm (b_n - b)|$$
$$\leqslant |a_n - a| + |b_n - b|$$
$$< \varepsilon$$

Hence $\lim_{n\to\infty}(a_n \pm b_n) = a \pm b$.

(ii) See exercise 3.1.

(iii) It is sufficient to prove that

$$\lim_{n\to\infty}\frac{1}{b_n} = \frac{1}{b}, \; b \neq 0$$

The more general result then follows by applying the rule for products (ii).

If $b_n \to b$ as $n \to \infty$, it may be shown (see exercise 3.1) that for all n sufficiently large $|b_n| > \frac{1}{2}|b|$, that is, $1/|b_n| < 2/|b|$. Now

$$\left|\frac{1}{b_n} - \frac{1}{b}\right| = \left|\frac{b - b_n}{b_n b}\right| = \frac{|b_n - b|}{|b_n|\,|b|}$$

$$< \frac{2|b_n - b|}{|b|^2} \text{ for } n \text{ sufficiently large}$$

Thus, given $\varepsilon > 0$ let M be sufficiently large such that both

$$|b_n - b| < \frac{|b|^2\varepsilon}{2} \text{ and } |b_n| > \frac{1}{2}|b| \text{ for all } n \geqslant M$$

Then

$$\left|\frac{1}{b_n} - \frac{1}{b}\right| < \frac{2|b_n - b|}{|b|^2} < \varepsilon$$

Hence

$$\lim_{n\to\infty}\frac{1}{b_n} = \frac{1}{b} \text{ provided } b \neq 0$$

(iv) See exercise 3.1.

From definition 1.4.1, a *polynomial of degree m in n* is an expression of the form

$$c_0 + c_1 n + c_2 n^2 + \ldots + c_m n^m$$

where $c_0, c_1, \ldots, c_m$ are constants independent of n, with $c_m \neq 0$.

Definition 3.1.4
A *rational function* is the quotient of one polynomial by another.

Let $r(n)$ denote the rational function

$$\frac{c_0+c_1n+c_2n^2+\ldots+c_mn^m}{d_0+d_1n+d_2n^2+\ldots+d_kn^k}$$

in which $c_m \neq 0$, $d_k \neq 0$.

This expression may be rearranged to give

$$r(n)=\frac{n^m(c_0/n^m+c_1/n^{m-1}+\ldots+c_{m-1}/n+c_m)}{n^k(d_0/n^k+d_1/n^{k-1}+\ldots+d_{k-1}/n+d_k)}$$

The behaviour of $r(n)$ as $n\to\infty$ may be determined by first observing that $1/n^p\to 0$ as $n\to\infty$, for any $p>0$; by applying the rules given in theorem 3.1.2, it then follows that

$$\lim_{n\to\infty} r(n)=\frac{c_m}{d_k}\lim_{n\to\infty} n^{m-k}$$

Thus

if $m>k$, $r(n)\to\pm\infty$, depending on the sign of c_m/d_k;

if $m=k$, $r(n)\to c_m/d_k$;

if $m<k$, $r(n)\to 0$ as $n\to\infty$.

Example 3.1.4

$$\lim_{n\to\infty}\frac{3n^3+9n^2+4}{4n^3-11n^2+7n+3}=\frac{3}{4}$$

Bounded Sequences and Monotone Sequences

Definition 3.1.5
A sequence (a_n) is *bounded above* if there exists a number K such that $a_n\leqslant K$ for all n; it is *bounded below* if there exists a number L such that $a_n\geqslant L$ for all n.

A sequence (a_n) is *monotonically increasing* if $a_{n+1}\geqslant a_n$ for all n, but *monotonically decreasing* if $a_{n+1}\leqslant a_n$ for all n. A sequence which is either monotonically increasing or decreasing is called a *monotone sequence.*

Theorem 3.1.3
If a sequence is monotonically increasing and bounded above, or if it is monotonically decreasing and bounded below, then it is convergent.

Proof Suppose that the sequence (a_n) is monotonically increasing and bounded above. Since (a_n) has an upper bound, it has a *least upper bound*, that is, there exists a number l such that $a_n \leqslant l$ for all n but for any $\varepsilon > 0$ there is some a_M satisfying $a_M > l - \varepsilon$. Since the sequence is monotonically increasing, $a_n \geqslant a_M$ for all $n \geqslant M$. This means that given $\varepsilon > 0$ there is an integer M such that

$$0 \leqslant l - a_n < \varepsilon \text{ for all } n \geqslant M$$

Thus, the limit of a monotonically increasing sequence which is bounded above is its least upper bound. Similarly, it may be shown that the limit of a monotonically decreasing sequence which is bounded below is its greatest lower bound.

Theorem 3.1.4
If $a_r > 0$ and there exists a number $K > 1$ such that $a_{n+1} \geqslant K a_n$ for all $n \geqslant r$, then $a_n \to \infty$.

Proof See exercise 3.1.

Example 3.1.5 The Sequence (x^n)

If $x = 1$, $x^n = 1$ for all n and the limit is also 1.

If $0 < x < 1$, (x^n) is a decreasing sequence, since $x^{n+1} = x x^n < x^n$. It is also bounded below by 0 and must therefore have a limit. Now, if a sequence (a_n) is convergent, a_n and a_{n+1} both approach the same value as $n \to \infty$, that is, $\lim_{n\to\infty} a_{n+1} = \lim_{n\to\infty} a_n$. For the sequence (x^n) therefore, $\lim_{n\to\infty} x^{n+1} = \lim_{n\to\infty} x^n = X$, say. Since x is a constant, $\lim_{n\to\infty} x^{n+1} = x \lim_{n\to\infty} x^n$, that is, $X = xX$; but x is not equal to 0 or 1 and so X must equal 0. Thus, $x^n \to 0$ as $n \to \infty$, if $0 < x < 1$.

Similarly, it may be shown that $x^n \to 0$ as $n \to \infty$, if $-1 < x < 0$. By theorem 3.1.4, $x^n \to \infty$ as $n \to \infty$, if $x > 1$.

Finally, the sequence (x^n) oscillates if $x \leqslant -1$.

Cauchy's Criterion of Convergence

Definition 3.1.6
A sequence (a_n) is a *Cauchy sequence* if for any $\varepsilon > 0$ there is an integer M such that

$$|a_n - a_m| < \varepsilon \text{ for all } n,\ m \geqslant M$$

Theorem 3.1.5
A sequence is convergent if and only if it is a Cauchy sequence.

Proof It is left as an exercise for the reader to show that any convergent sequence is a Cauchy sequence.

If (a_n) is a Cauchy sequence then, given $\varepsilon > 0$ there is an integer M such that for all $n \geqslant M$

$$a_M - \varepsilon < a_n < a_M + \varepsilon$$

that is, the sequence is bounded both above and below and the greatest lower bound and least upper bound differ by an amount equal to 2ε at most. The sequence is therefore convergent.

Iterative Solution of Non-linear Equations

Sequences generated by first-order recurrence relations of the form $a_{n+1} = g(a_n)$ frequently occur in the numerical solution of non-linear equations.

Definition 3.1.7
A first-order recurrence relation of the form $a_{n+1} = g(a_n)$ is called a *one-point iteration formula*, or a *one-point iterative scheme*. The generation of each successive term in the sequence is called a *step in the iteration* and a_n is called the nth *iterate*. If the resulting sequence is convergent, the iterative scheme is also said to be convergent.

The progress of the iterates can be illustrated graphically. Figure 3.2 depicts the behaviour of a convergent one-point iteration formula.

First, the graphs of $y = g(x)$ and $y = x$ are sketched. Then a vertical line is drawn through the point $x = a_1$ to meet the curve $y = g(x)$ in the point A_1. A horizontal line through A_1 is then drawn to meet the line $y = x$ in the point B_1. The vertical through B_1 meets the x axis at $x = a_2$, the second iterate. The procedure is then repeated starting with a_2 in place of a_1, leading to the third iterate a_3.

Example 3.1.6

The square root of a given number $c > 0$ satisfies the equation $x^2 = c$. From this it follows that $2x^2 = x^2 + c$. Dividing throughout by $2x$ gives $x = (x + c/x)/2$, an equation of the form $x = g(x)$, leading to the one-point iteration formula

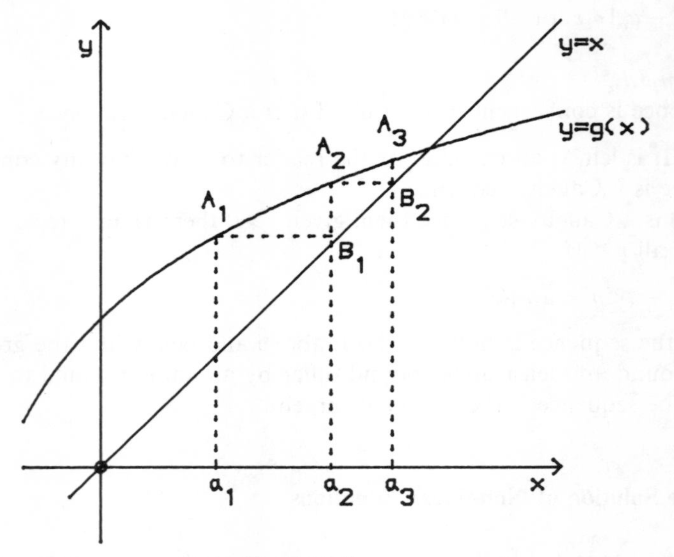

Figure 3.2

$$a_{n+1}=\frac{1}{2}\left(a_n+\frac{c}{a_n}\right)$$

for determining an approximation of the square root of a given number c.

This formula is convergent to the positive square root of $c>0$ for any starting value $a_1>0$. For any $n \in N$, a_n and c/a_n lie on opposite sides of $\sqrt{c}$. For example, if $a_n<\sqrt{c}$ then $c/a_n>\sqrt{c}$, that is, $a_n<\sqrt{c}<c/a_n$. Furthermore, the interval bracketing $\sqrt{c}$ at the nth step in the iteration properly contains the interval bracketing $\sqrt{c}$ at the $(n+1)$th step in the iteration. For example, if $a_n<\sqrt{c}<c/a_n$, then

$$a_{n+1}=\frac{1}{2}\left(a_n+\frac{c}{a_n}\right)<\frac{1}{2}\left(\frac{c}{a_n}+\frac{c}{a_n}\right)=\frac{c}{a_n}$$

and

$$a_{n+1}=\frac{1}{2}\left(a_n+\frac{c}{a_n}\right)>\frac{1}{2}(a_n+a_n)=a_n$$

that is, $a_n<a_{n+1}<c/a_n$. Furthermore, $a_{n+1}<c/a_n$ implies $a_n<c/a_{n+1}$ and $a_n<a_{n+1}$ implies $1/a_{n+1}<1/a_n$ and hence $c/a_{n+1}<c/a_n$, that is, $a_n<c/a_{n+1}<c/a_n$. Thus, both a_{n+1} and c/a_{n+1} lie within the interval $(a_n, c/a_n)$. All four possible arrangements of a_n, c/a_n, a_{n+1} and c/a_{n+1} are shown in figure 3.3. Each step in the iteration leads to a smaller interval bracketing $\sqrt{c}$.

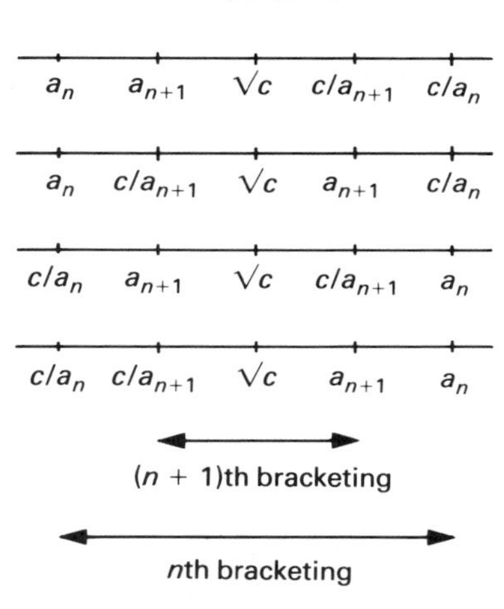

Figure 3.3 Example 3.1.6

If the square-root iteration formula is used to determine $\sqrt{2}$, starting with $a_1 = 1$, the following results are obtained.

n	a_n
1	1
2	1.5
3	1.4166667
4	1.4142157
5	1.4142136

In practice, it is better to use iteration formulae giving rise to rapidly convergent sequences. The above formula satisfies this requirement and for the given numerical example, the value of a_5 is correct to all decimal places shown.

Example 3.1.7

Some early computers did not have a machine divide operation. For such a machine, one way of calculating $1/c$ given $c > 0$ is by using the one-point iteration formula

$$a_{n+1} = a_n(2 - ca_n)$$

If a_1 is a sufficiently good initial approximation to $1/c$, this scheme is convergent to $1/c$ (see exercise 3.1).

If the scheme is used to calculate 1/19.7, say, starting with an initial guess $a_1 = 0.05$, the following results are obtained.

n	a_n
1	0.05
2	0.05075
3	0.0507614
4	0.0507614

The final iterate is correct to all decimal places shown. If a_1 is given the value 0.1, a further seven iterations are required for the same degree of accuracy, while if $a_1 = 0.11$, the scheme is divergent!

The reader is asked in exercise 3.1 to show that the above scheme is convergent provided $0 < a_1 < 2/c$. Suppose that c is equal to $b \times 2^n$, where $1/2 \leqslant b < 1$. Then $1/b > 1$, that is, $2 < 2/b$. The iterative scheme

$$a_{n+1} = a_n(2 - ba_n)$$

with $a_1 = 2$, is thus convergent to $1/b$ for any $b \in [\frac{1}{2}, 1)$. To determine $1/c$ for a given value of c, it is therefore sufficient to determine $1/b$, where b is equal to the mantissa in the binary floating-point representation of c, and shift the result appropriately.

Exercise 3.1

1. Determine, to five-decimal-place accuracy, the first ten terms of each of the sequences defined as follows

(a) $a_n = 1/n^4$;

(b) $a_1 = 1$, $a_{n+1} = \sqrt{(3a_n)}$, $n \geqslant 1$;

(c) $a_n = \sqrt{(n+1)} - \sqrt{n}$.

2. A sequence (a_n) is defined by the recurrence relation

$$a_1 = \sqrt{2},\ a_{n+1} = \sqrt{(2a_n)},\ n \geqslant 1$$

Prove by induction that $a_n = 2^{1-2^{-n}}$.

3. If $\lim_{n\to\infty} a_n = a$ and $\lim_{n\to\infty} b_n = b$, show that $\lim_{n\to\infty} a_n b_n = ab$. [Hint: $a_n b_n - ab = (a_n - a)\{(b_n - b) + b\} + a(b_n - b)$.]

4. If $\lim_{n\to\infty} a_n = a \geqslant 0$, prove by induction on m that $\lim_{n\to\infty} a_n^m = a^m$. [Hint: $\lim_{n\to\infty} a_n^{k+1} = \lim_{n\to\infty} a_n \lim_{n\to\infty} a_n^k$.]

5. Show that if $\lim_{n\to\infty} b_n = b$, then $|b_n| > |b/2|$ for all n sufficiently large. [Hint: $|b| \leqslant |b_n| + |b - b_n|$.]

6. Determine whether each of the sequences defined below is convergent and if so determine the limit.

(a) $a_n = (4^n + 3^n)^{1/n}$ [hint: $(4^n + 3^n)^{1/n} = 4\{1 + (0.75)^n\}^{1/n}$];

(b) $a_n = (n!)^{1/n}$;

(c) $a_1 = 1$, $a_{n+1} = \sqrt{(3a_n)}$, $n \geqslant 1$.

7. Prove theorem 3.1.4

8. Using the iteration formula given in example 3.1.6, determine $\sqrt{29}$, stopping when successive iterates agree to six decimal places.

9. Show that, if the iterative scheme given in example 3.1.7 is convergent, its limit is $1/c$. [Hint: if the scheme is convergent, $\lim_{n\to\infty} a_{n+1} = \lim_{n\to\infty} a_n = a$, say.]

Show that in order for the scheme to be convergent, a_1 must be chosen to satisfy $0 < a_1 < 2/c$. [Hint: show that $r_{n+1} = r_n^2$, where $r_n = 1 - ca_n$ is the *residual*.]

10. Using the iterative formula given in example 3.1.7, compute the value of 1/12.23 correct to six decimal places, if possible, using

(a) $a_1 = 0.1$;

(b) $a_1 = 0.5$.

Explain why convergence occurs in one case but not in the other.

3.2 SERIES

Definition 3.2.1
A *series* is the sum of the terms of a sequence.

Summing the terms of a finite sequence is a straightforward process and gives rise to a *finite* series. The series corresponding to the finite sequence

$a_1, a_2, a_3, \ldots, a_p$ is simply the sum $a_1+a_2+a_3+\ldots+a_p$. As mentioned in chapter 1, this sum is often denoted by

$$\sum_{n=1}^{p} a_n$$

More generally, the sum $a_m+a_{m+1}+\ldots+a_p$ may be denoted by

$$\sum_{n=m}^{p} a_n$$

Example 3.2.1

$$\sum_{n=1}^{6} n(n+1)=1\times 2+2\times 3+3\times 4+4\times 5+5\times 6+6\times 7=112$$

It is not so obvious what is meant by an *infinite series*, that is, by the 'sum' of infinitely many numbers $a_1+a_2+a_3+\ldots$. In fact, it is not clear at first sight that such a sum can be meaningful. However, if an infinite sequence is truncated after p terms ($p\geqslant 1$), that is, if just the first p terms are considered, then this finite *subsequence* can be summed as before.

Definition 3.2.2
The pth *partial sum* s_p of an infinite sequence (a_n) is the sum of the first p terms of the sequence, that is, $s_p=a_1+a_2+a_3+\ldots+a_p$.

If it is to be meaningful to talk of the 'sum to infinity'

$$\sum_{n=1}^{\infty} a_n$$

of an infinite sequence then, intuitively, one would expect that the sequence (s_n) of partial sums should converge to a limit if the sum to infinity of the given infinite sequence is to be defined.

Definition 3.2.3
The *sum to infinity* of the sequence (a_n) is the limit, if it exists, of the sequence (s_n) of partial sums. It this limit does not exist, the sum to infinity of the sequence is undefined. The infinite series

$$\sum_{n=1}^{\infty} a_n$$

is said to be *convergent* if the sequence (s_n) is convergent, *divergent* if (s_n) is divergent and *oscillatory* if (s_n) is an oscillating sequence.

For some series, it is possible to obtain a formula of the form $s_p=f(p)$, expressing the pth partial sum explicity in terms of p. There are several approaches that can sometimes be used to this end but, before considering these, a number of important special types of series are presented.

Some Important Special Types of Series

Definition 3.2.4
An *arithmetic series* has the form

$$a+[a+d]+[a+2d]+\ldots+[a+(p-1)d]$$

that is, the nth term is given by the formula $a+(n-1)d$, where a is the first term of the series and d is the *common difference.*

Thus, in an arithmetic series, each successive term may be obtained by adding a fixed value d to the previous term.

Example 3.2.2

The sum of the first p natural numbers $1+2+3+\ldots+p$ is an arithmetic series, with first term 1 and common difference 1.

If s_p denotes the sum of the first p terms of an arithmetic series, then

$$s_p=a+[a+d]+[a+2d]+\ldots+[a+(p-1)d]$$

Writing the terms in reverse order gives

$$s_p=[a+(p-1)d]+[a+(p-2)d]+[a+(p-3)d]+\ldots+a$$

Adding these two expressions for s_p, term by term, gives

$$\begin{aligned}2s_p&=[2a+(p-1)d]+[2a+(p-1)d]+[2a+(p-1)d]+\ldots+[2a+(p-1)d]\\&=p[2a+(p-1)d]\end{aligned}$$

Hence, a general expression for the sum of the first p terms of an arithmetic series is

$$s_p=\frac{1}{2}p[2a+(p-1)d]$$

Example 3.2.3

Substituting $a=1$, $d=1$ into this expression leads to the formula $p(p+1)/2$ for the sum of the first p natural numbers. Thus, irrespective of the size of

p, the sum of the first p natural numbers can be evaluated by performing just three arithmetic operations.

Definition 3.2.5
A *geometric series* has the form

$$a+ax+ax^2+\ldots+ax^{p-1}$$

that is, the nth term is ax^{n-1}, where a is the first term and x is the *common ratio.*

Thus, in a geometric series, each successive term is obtained by multiplying the previous term by a constant value x.

Example 3.2.4

The series $3+6+12+24+48+96$ is a geometric series with first term 3 and common ratio 2.

If s_p denotes the sum of the first p terms of a geometric series, then

$$s_p=a+ax+ax^2+\ldots+ax^{p-1}$$

Hence

$$xs_p=ax+ax^2+\ldots+ax^{p-1}+ax^p$$

Subtracting the second equation from the first gives

$$(1-x)s_p=a-ax^p$$

Hence, assuming $x\neq 1$, a general expression for the sum of the first p terms of a geometric series is

$$s_p=\frac{a(1-x^p)}{1-x}$$

An infinite geometric series will be convergent if the sequence of partial sums tends to a limit. Now

$$\lim_{n\to\infty} s_n=\lim_{n\to\infty}\frac{a(1-x^n)}{1-x}=\frac{a}{1-x}\left(1-\lim_{n\to\infty}x^n\right)$$

If $x\leqslant -1$ or $x>1$, the sequence (x^n) is not convergent (see example 3.1.5). If $|x|<1$, $x^n\to 0$ and so $s_n\to a/(1-x)$; if $x=1$, $a/(1-x)$ is undefined. Hence, provided $|x|<1$, the infinite geometric series

$$\sum_{n=1}^{\infty} ax^{n-1}$$

is convergent and its sum is $a/(1-x)$.

Example 3.2.5

The series

$$x+\frac{x}{1+x}+\frac{x}{(1+x)^2}+\ldots$$

is a geometric series with first term x and common ratio $1/(1+x)$. The sum to p terms is

$$x\left[1-\left(\frac{1}{1+x}\right)^p\right]\Big/\left(1-\frac{1}{1+x}\right)=(1+x)\left[1-\left(\frac{1}{1+x}\right)^p\right]$$

If $|1/(1+x)|<1$, that is, if $x\notin[-2, 0]$, then the series can be summed to infinity and its sum is $1+x$. (Note that if $x=0$, the sum to infinity is clearly also 0.)

Definition 3.2.6
An *arithmetic–geometric* series has the form

$$ab+(a+d)bx+(a+2d)bx^2+\ldots+[a+(p-1)d]bx^{p-1}$$

that is, the nth term is the product of the nth term of an arithmetic series and the nth term of a geometric series.

A formula for the sum s_p of p terms of an arithmetic–geometric series can always be obtained by deriving an expression for $(1-x)s_p$. The procedure is best illustrated by an example.

Example 3.2.6

The series $1+2x+3x^2+\ldots$ is an arithmetic–geometric series. The nth term is the product of the nth term of the arithmetic series $1+2+3+\ldots$ and the nth term of the geometric series $1+x+x^2+\ldots$ If s_p denotes the sum of the first p terms, then

$$s_p=1+2x+3x^2+\ldots+(p-1)x^{p-2}+px^{p-1}$$

and

$$xs_p=x+2x^2+\ldots+(p-2)x^{p-2}+(p-1)x^{p-1}+px^p$$

Hence

$$(1-x)s_p=1+x+x^2+\ldots+x^{p-2}+x^{p-1}-px^p$$

The series $1+x+x^2+\ldots+x^{p-1}$ is a geometric series and has the sum

$(1-x^p)/(1-x)$, provided $x \neq 1$ (if $x=1$, the given series reverts to the arithmetic series $1+2+3+\ldots$). Thus, if $x \neq 1$

$$s_p = \frac{1-x^p}{(1-x)^2} - \frac{px^p}{1-x}$$

The sum to infinity of the series will exist if the sequence (s_n) of partial sums is convergent. If (s_n) is convergent, then

$$\lim_{n\to\infty} s_n = \frac{1}{(1-x)^2}\left(1 - \lim_{n\to\infty} x^n - (1-x) \lim_{n\to\infty} nx^n\right)$$

Now, if the nth term of any sequence tends to a limit, the $(n+1)$th term tends to the same limit. Hence

$$\lim_{n\to\infty} nx^n = \lim_{n\to\infty} (n+1)x^{n+1} = x\left(\lim_{n\to\infty} nx^n + \lim_{n\to\infty} x^n\right)$$

that is

$$(1-x)\lim_{n\to\infty} nx^n = x\lim_{n\to\infty} x^n$$

Thus

$$\lim_{n\to\infty} s_n = \frac{1}{(1-x)^2}\left(1 - \lim_{n\to\infty} x^n - x\lim_{n\to\infty} x^n\right) = \frac{1}{(1-x)^2}$$

if $|x|<1$, since $x^n \to 0$ if $|x|<1$. Hence, provided $|x|<1$, the sum to infinity of the given series exists and is equal to $1/(1-x)^2$.

Summation by Induction

It is often possible to use mathematical induction to obtain a formula for the sum of the first p terms of a given series. In example 1.3.13, the formula for the sum of the first p natural numbers is proved in this way. A further example is now given.

Example 3.2.7

Consider the series

$$\frac{1}{1\times 4} + \frac{1}{4\times 7} + \frac{1}{7\times 10} + \ldots + \frac{1}{(3p-2)(3p+1)}$$

The sum of the first two terms is 2/7, of the first three terms, 3/10 and of the first four terms, 4/13. This suggests that the sum of the first p terms is

$p/(3p+1)$. Assume that this is true. Then

$$\text{sum of the first } (p+1) \text{ terms} = \frac{p}{3p+1} + (p+1)\text{th term}$$

$$= \frac{p}{3p+1} + \frac{1}{[3(p+1)-2][3(p+1)+1]}$$

$$= \frac{3p^2+4p+1}{(3p+1)(3p+4)}$$

$$= \frac{(3p+1)(p+1)}{(3p+1)(3p+4)}$$

$$= \frac{p+1}{3p+4}$$

Thus, if the sum of the first p terms is $p/(3p+1)$, the sum of the first $(p+1)$ terms is $(p+1)/[3(p+1)+1]$. Furthermore, the sum of the first term is 1/4, which is $p/(3p+1)$ with $p=1$. Hence, by induction, the sum of the first p terms of the given series is equal to $p/(3p+1)$, for all $p \in N$.

Summation by the Method of Differences

Definition 3.2.7
The difference $f_{n+1} - f_n$ between two consecutive terms of a sequence (f_n) is called the *first forward difference* of f_n and is denoted by Δf_n.

If it is possible to express the general term a_n of a given series as the first forward difference of the nth term of a sequence (f_n), then it is very easy to obtain an explicit formula for the sum to p terms of the series. For, if $a_n = f_{n+1} - f_n$ then

$$\sum_{n=1}^{p} a_n = \sum_{n=1}^{p} (f_{n+1} - f_n) = (f_2 - f_1) + (f_3 - f_2) + (f_4 - f_3) + \ldots + (f_{p+1} - f_p)$$

$$= -f_1 + f_{p+1}$$

that is, the sum of the first p terms of the series is simply the difference between the $(p+1)$th and first terms of the sequence (f_n).

Example 3.2.8

Given the series

$$\sum_{n=1}^{p} \frac{1}{n(n+1)}$$

let $f_n = -1/n$. Then

$$\frac{1}{n(n+1)} = f_{n+1} - f_n$$

and so the sum of the series is

$$f_{p+1} - f_1 = -\frac{1}{p+1} + 1$$

Since

$$\lim_{n \to \infty} \frac{1}{n+1} = 0$$

the sum to infinity of the given series is 1.

Definition 3.2.8
The *factorial polynomial* of degree n is given by the formula

$${}^nP_m = n(n-1)(n-2)\ldots(n-m+1) = \frac{n!}{(n-m)!}$$

(${}^nP_m$ denotes the number of permutations of n objects taken m at a time, see definition 4.1.7.)

Example 3.2.9

Let

$$f_n = \frac{{}^nP_{m+1}}{m+1}$$

then

$$f_{n+1} - f_n = \frac{1}{m+1}\{(n+1)n(n-1)\ldots[n+1-(m+1)+1] - n(n-1)(n-2)\ldots[n-(m+1)+1]\}$$

$$= \frac{1}{m+1}[(n+1)-(n-m)]n(n-1)\ldots(n-m+1)$$

$$= {}^nP_m$$

Hence, if $p \geqslant q > m$

$$\sum_{n=q}^{p} {}^nP_m = f_{p+1} - f_q = \frac{{}^{p+1}P_{m+1} - {}^qP_{m+1}}{m+1}$$

For example

$$\sum_{n=4}^{7} {}^{n}P_3 = \sum_{n=4}^{7} n(n-1)(n-2) = \frac{{}^{8}P_4 - {}^{4}P_4}{4}$$

$$= \frac{8 \times 7 \times 6 \times 5 - 4 \times 3 \times 2 \times 1}{4} = 414$$

Summation by the Application of Identities

The method of differences discussed above is a particular example of the application of identities, whereby the general term of a given series is replaced by an equivalent expression which is easier to sum.

Example 3.2.10

An explicit formula for the sum of the series

$$\sum_{n=1}^{p} n^2$$

may be obtained by using the identity

$$n^2 = \frac{1}{3}[(n+1)^3 - n^3 - 3n - 1]$$

Using the method of differences with $f_n = n^3$ gives

$$\sum_{n=1}^{p} [(n+1)^3 - n^3] = (p+1)^3 - 1$$

Furthermore, from example 3.2.3

$$\sum_{n=1}^{p} n = \frac{1}{2}p(p+1)$$

Hence

$$\sum_{n=1}^{p} n^2 = \frac{1}{3}\left\{ \sum_{n=1}^{p} [(n+1)^3 - n^3] - 3\sum_{n=1}^{p} n - \sum_{n=1}^{p} 1 \right\}$$

$$= \frac{1}{3}\left[(p+1)^3 - 1 - \frac{3}{2}p(p+1) - p \right]$$

$$= \frac{1}{6}p(2p+1)(p+1)$$

after simplification.

Summation by the Method of Indeterminate Coefficients

This method is applicable when the nth term in the given series is a polynomial in n. It has been shown above that the sum of the first p natural numbers

$$\sum_{n=1}^{p} n = p(p+1)/2$$

and that the sum of the squares of the first p natural numbers

$$\sum_{n=1}^{p} n^2 = p(p+1)(2p+1)/6$$

In the first case, the general term is a polynomial in n of degree 1 and the sum is a polynomial in p of degree 2; in the second case, the general term is a polynomial in n of degree 2 and the sum is a polynomial in p of degree 3. In general, it may be shown that if the nth term of a series is a polynomial in n of degree r, then the sum to p terms is a polynomial in p of degree $r+1$, with zero constant term. Thus, if

$$a_n = c_0 + c_1 n + c_2 n^2 + \ldots + c_r n^r$$

where $c_0, c_1, \ldots, c_r$ are given constants independent of n, then

$$\sum_{n=1}^{p} a_n = d_1 p + d_2 p^2 + \ldots + d_{r+1} p^{r+1}$$

The coefficients $d_1, \ldots, d_{r+1}$ are constants independent of p. Now

$$a_p = \sum_{n=1}^{p} a_n - \sum_{n=1}^{p-1} a_n$$

$$= d_1 p + d_2 p^2 + \ldots + d_{r+1} p^{r+1} - d_1(p-1) - d_2(p-1)^2$$

$$- \ldots - d_{r+1}(p-1)^{r+1}$$

but

$$a_p = c_0 + c_1 p + c_2 p^2 + \ldots + c_r p^r$$

and so the d_js may be determined by equating coefficients of powers of p.

Example 3.2.11

Since n^3 is a polynomial of degree 3

$$\sum_{n=1}^{p} n^3 = d_1 p + d_2 p^2 + d_3 p^3 + d_4 p^4$$

Now

$$a_p = \sum_{n=1}^{p} a_n - \sum_{n=1}^{p-1} a_n$$

$$= d_1 p + d_2 p^2 + d_3 p^3 + d_4 p^4 - d_1(p-1) - d_2(p-1)^2 - d_3(p-1)^3 - d_4(p-1)^4$$

that is

$$p^3 = 4d_4 p^3 + 3(d_3 - 2d_4)p^2 + (2d_2 - 3d_3 + 4d_4)p + (d_1 - d_2 + d_3 - d_4)$$

Equating coefficient gives rise to the triangular system of equations

$$\begin{aligned} d_1 - d_2 + d_3 - d_4 &= 0 \\ 2d_2 - 3d_3 + 4d_4 &= 0 \\ d_3 - 2d_4 &= 0 \\ 4d_4 &= 1 \end{aligned}$$

Back-substitution then gives $d_4 = 1/4$, $d_3 = 1/2$, $d_2 = 1/4$, $d_1 = 0$. Hence

$$\sum_{n=1}^{p} n^3 = \frac{1}{4}p^2 + \frac{1}{2}p^3 + \frac{1}{4}p^4 = \frac{1}{4}p^2(p+1)^2$$

Tests for the Convergence of Infinite Series

By definition 3.2.3, the infinite series $\sum_{n=1}^{\infty} a_n$ is convergent if the sequence (s_n) of partial sums is convergent. If $s_n \to s$ as $n \to \infty$ then $a_n = s_n - s_{n-1} \to s - s = 0$ as $n \to \infty$. Thus, a *necessary* condition for the convergence of an infinite series is that the nth term should tend to zero. This provides a useful first test for convergence. However it is *not* sufficient for convergence and

$$\sum_{n=1}^{\infty} a_n$$

may be divergent even though a_n tends to zero.

Example 3.2.12

(1) Since the nth term of the series

$$\sum_{n=1}^{\infty} \frac{n}{n+1}$$

tends to 1 as $n \to \infty$, the series is divergent.

(2) The series

$$\sum_{n=1}^{\infty} \frac{1}{n}$$

is called the *harmonic series.* It is easy to show that the sum of the first 2^m terms of this series is not less than $1+m/2$. The harmonic series is therefore divergent, even though $1/n \rightarrow 0$ as $n \rightarrow \infty$.

From theorem 3.1.3, it follows that if

$$\sum_{n=1}^{\infty} a_n$$

is a series of *non-negative* terms, that is $a_n \geqslant 0$ for every n, a necessary *and* sufficient condition for convergence is that there exists $K > 0$ such that

$$s_p = \sum_{n=1}^{p} a_n \leqslant K \text{ for all } p \in N.$$

This result is central to the proof of the following important theorem, through which the convergence or divergence of a given series may be established by comparing it with a series of known behaviour. The details of the proof are left as an exercise for the reader.

Theorem 3.2.1 Comparison Test
Let $a_n \geqslant 0$, $b_n \geqslant 0$ for all $n \in N$; then

(i) if

$$\sum_{n=1}^{\infty} b_n$$

is convergent and there exists $K > 0$ such that $a_n < Kb_n$ for every n, then

$$\sum_{n=1}^{\infty} a_n$$

is also convergent;

(ii) if

$$\sum_{n=1}^{\infty} b_n$$

is divergent and there exists $K > 0$ such that $b_n \leqslant Ka_n$ for every n, then

$$\sum_{n=1}^{\infty} a_n$$

is also divergent.

Example 3.2.13

The series

$$\sum_{n=1}^{\infty} n^{-k},\ k \in N,$$

is a very important series. It is known as the *Riemann zeta function* and is denoted by $\zeta(k)$. It has been shown above that $\zeta(1)$ (the *harmonic series*) is divergent. Consider now $\zeta(k)$, $k \geqslant 2$. Since $1/n^2 \leqslant 3/n(n+1)$ for all $n \in N$ and

$$\sum_{n=1}^{\infty} \frac{1}{n(n+1)}$$

is convergent (see example 3.2.8),

$$\sum_{n=1}^{\infty} \frac{1}{n^2}$$

is also convergent, by the comparison test. Furthermore, since $1/n^k \leqslant 1/n^2$ for all $n \in N$ and for any $k > 2$

$$\sum_{n=1}^{\infty} n^{-k}$$

is also convergent when $k > 2$.

Another useful test for the convergence of a series of positive terms, which is often very easy to apply, is the following.

Theorem 3.2.2 Ratio Test
If $a_n > 0$ for all n and $a_{n+1}/a_n \to l$ as $n \to \infty$ then

(i) if $l < 1$, $\sum_{n=1}^{\infty} a_n$ is convergent;

(ii) if $l > 1$, $\sum_{n=1}^{\infty} a_n$ is divergent.

Proof (i) For any number k satisfying $0 \leqslant l < k < 1$, there is an integer M such that $a_{n+1}/a_n < k$ for all $n \geqslant M$. From this it follows that $a_{M+1} < ka_M$, $a_{M+2} < ka_{M+1} < k^2 a_M$ and in general $a_{M+m} < k^m a_M$, for any $m \in N$. Thus

$$\sum_{n=1}^{\infty} a_n < a_1 + a_2 + \ldots + a_{M-1} + a_M(1 + k + k^2 + \ldots)$$

$$= \sum_{n=1}^{M-1} a_n + \sum_{n=1}^{\infty} a_M k^{n-1}$$

$$= s_{M-1} + \frac{a_M}{1-k}$$

since $k<1$. Hence

$$\sum_{n=1}^{\infty} a_n$$

is convergent.

(ii) This is straightforward and is left as an exercise for the reader.

Example 3.2.14

The series

$$\sum_{n=1}^{\infty} 2^n/n!$$

is convergent by the ratio test, for $a_{n+1}/a_n = 2^{n+1}n!/2^n(n+1)! = 2/(n+1) \to 0$ as $n \to \infty$.

The tests for convergence given above are for series of positive terms. For a series

$$\sum_{n=1}^{\infty} a_n$$

having an infinite number of both positive and negative terms, the given tests can be applied to the related series

$$\sum_{n=1}^{\infty} |a_n|$$

If this series converges, then so does the given series.

Definition 3.2.9
The series

$$\sum_{n=1}^{\infty} a_n$$

is said to be *absolutely convergent* if

$$\sum_{n=1}^{\infty} |a_n|$$

is convergent. If

$$\sum_{n=1}^{\infty} a_n$$

is convergent but

$$\sum_{n=1}^{\infty} |a_n|$$

is divergent

$$\sum_{n=1}^{\infty} a_n$$

is said to be *conditionally convergent.*

The reader should satisfy himself that if

$$\sum_{n=1}^{\infty} |a_n|$$

is convergent, then so is

$$\sum_{n=1}^{\infty} a_n$$

(see exercise 3.1).

Example 3.2.15

(1) Since $\sum_{n=1}^{\infty} 1/n^2$ is convergent, the series $\sum_{n=1}^{\infty} (-1)^{n+1}(1/n^2)$ is absolutely convergent.

(2) The series $\sum_{n=1}^{\infty} (-1)^{n+1}(1/n)$ is convergent, but only conditionally since $\sum_{n=1}^{\infty} 1/n$ is divergent.

An important type of series of positive and negative terms is the alternating series.

Definition 3.2.10
An *alternating series* is one in which positive and negative terms occur alternately, that is, $s = a_1 - a_2 + a_3 - a_4 + \ldots$, where $a_1, a_2, a_3, \ldots$ are non-negative numbers.

Theorem 3.2.3
A necessary *and* sufficient condition for the convergence of an alternating series is that (a_n) is monotonically decreasing.

Proof This is established by observing that the 'odd' partial sums s_1, s_3, s_5, ... form a decreasing sequence which is bounded below, and therefore convergent, and that the 'even' partial sums s_2, s_4, s_6, ... form an increasing sequence which is bounded above, and therefore also convergent. The limits of these two sequences must be the same since $s_{2n-1}-s_{2n}=a_{2n}\to 0$ as $n\to\infty$. Hence, the alternating series is convergent. It is left to the reader to fill in the details of this proof.

Summing a Finite Series on a Computer

Suppose that a computer is to be used to sum a finite series and that no simple formula is available for the sum. Since on a computer real numbers are represented in floating-point form (see section 1.3), this simple mathematical problem may not be straightforward in practice. Summing a set of floating-point numbers may give different results depending on the order in which the numbers are taken. In general, it is preferable to sum a set of floating-point numbers *in order of increasing magnitude*. This is illustrated in the following example, using three-digit decimal floating-point arithmetic, with rounding.

Example 3.2.16

$$\begin{aligned}
&(0.831\times 10^2+0.324\times 10^1)+0.247\times 10^0\\
&=(0.831\times 10^2+0.0324\times 10^2)+0.247\times 10^0\\
&\approx 0.863\times 10^2+0.00247\times 10^2\\
&\approx 0.865\times 10^2
\end{aligned}$$

The final digit in the mantissa of the result is not correct, that is, the result is only correct to two significant digits. However, if the given numbers are summed in reverse order, then

$$\begin{aligned}
&(0.247\times 10^0+0.324\times 10^1)+0.831\times 10^2\\
&=(0.0247\times 10^1+0.324\times 10^1)+0.831\times 10^2\\
&\approx 0.0349\times 10^2+0.831\times 10^2\\
&\approx 0.866\times 10^2
\end{aligned}$$

which is correct to three significant digits.

In addition to the above problem, there is a further possible complication if the series has both positive and negative terms. The subtraction of one floating-point number from another of approximately equal magnitude may result in some of the leading digits of the two mantissae cancelling each other out. If this happens, the resulting floating-point number will have a reduced number of significant digits.

Example 3.2.17

$$0.263 \times 10^2 - 0.198 \times 10^2$$
$$= 0.065 \times 10^2$$
$$= 0.650 \times 10^1$$

Now suppose that the given numbers had been rounded from 0.2628×10^2 and 0.1975×10^2, respectively

$$0.2628 \times 10^2 - 0.1975 \times 10^2$$
$$= 0.0653 \times 10^2$$
$$= 0.653 \times 10^1$$

Thus, 0.650×10^1 is correct to two significant digits only, the final zero in the mantissa having no significance. However, any further operations on this result will act as though the zero were significant.

Algorithm 3.2.1 sums p positive numbers $a_1, a_2, \ldots, a_p$ in order of increasing magnitude. In effect, the numbers are sorted into ascending order, using a sorting algorithm called an *exchange sort.*

For a series having both positive and negative terms, the sum of the positive terms should first be obtained, followed by the sum of the negative terms. Both of these summations should be in order of increasing magnitude and should be carried out using double-length arithmetic if the two sums are expected to be of approximately equal magnitude. The sum of the given series is then the difference between the sum of the positive terms and the sum of the negative terms.

Algorithm 3.2.1

sum ← 0;

for n **from** 1 **in steps of** 1 **to** $p-1$ **do**

nmin ← n;

min ← a_n;

for i **from** $n+1$ **in steps of** 1 **to** p **do**

```
            if a_i < min then
                nmin ← i;
                min ← a_i
            endif
        endfor;
        sum ← sum + min;
        a_nmin ← a_n
    endfor
```

Summing Infinite Series on a Computer

Summing a convergent infinite series is an important practical problem and one that can present severe difficulties. The crux of the problem lies with the speed of convergence of the series. A series may converge so quickly that relatively few terms need to be summed in order to obtain the required accuracy. On the other hand, a series may converge so slowly that thousands of terms may be needed to obtain the same degree of accuracy. In the latter case, the cumulative effect of rounding errors may be such that the required accuracy cannot be achieved by direct summation.

Example 3.2.18

If

$$s_p = \sum_{n=1}^{p} \frac{(-1)^{n+1}}{2n+19}$$

is evaluated directly for increasing values of p, s_{950} and s_{1000} are calculated to be 0.0246778 and 0.0246907, respectively.

It may be shown that

$$\sum_{n=1}^{\infty} \frac{(-1)^{n+1}}{(2n+19)}$$

is 0.02493825... Thus, the values of s_{950} and s_{1000} are correct to only two significant digits, even though they agree to three. More terms could be taken, but the propagation of rounding errors makes it impossible to obtain high accuracy.

The sum of the series in example 3.2.18 may be evaluated quickly and accurately by using a technique which is applicable to alternating series. In order to describe this technique conveniently, it is necessary to extend definition 3.2.7 to higher-order forward differences.

Definition 3.2.11
Given a sequence (f_n), the rth *forward difference* of f_n is denoted by $\Delta^r f_n$ and is defined recursively by

$$\Delta^r f_n = \Delta(\Delta^{r-1} f_n)$$

that is, the rth forward difference of f_n is the first forward difference of the $(r-1)$th forward difference of f_n.

Example 3.2.19

Suppose $f_1 = 1, f_2 = 4$ and $f_3 = 9$. Then $\Delta f_1 = 3$, $\Delta f_2 = 5$ and $\Delta^2 f_1 = \Delta(\Delta f_1) = \Delta(f_2 - f_1) = \Delta f_2 - \Delta f_1 = 5 - 3 = 2$.

For an alternating series, it may be shown that

$$a_1 - a_2 + a_3 - a_4 + \ldots = \frac{1}{2}\left[a_1 - \frac{1}{2}\Delta a_1 + \frac{1}{4}\Delta^2 a_1 - \frac{1}{8}\Delta^3 a_1 + \ldots\right]$$

Example 3.2.20

Using the above formula

$$\begin{aligned}\frac{1}{21} - \frac{1}{23} + \frac{1}{25} - \ldots &\approx \frac{1}{2}\left(\frac{1}{21} + \frac{1}{2} \times 0.0041408 + \frac{1}{4} \times 0.0006625\right.\\ &\quad + \frac{1}{8} \times 0.0001472 + \frac{1}{16} \times 0.0000406\\ &\quad \left. + \frac{1}{32} \times 0.0000131 + \frac{1}{64} \times 0.00000048\right)\\ &\approx 0.0249383\end{aligned}$$

which is correct to all digits shown. This is to be compared with the poor results given in example 3.2.18, and illustrates the principle that one good thought can be worth more than a great deal of computer time.

Another approach that is often useful for speeding up the summation process is essentially a practical application of the comparison test. Suppose that it is required to evaluate

$$s = \sum_{n=1}^{\infty} a_n$$

to a specified degree of accuracy. If

$$s' = \sum_{n=1}^{\infty} b_n$$

is known, then

$$s = s' - (s' - s)$$

and it is only necessary to evaluate

$$s' - s = \sum_{n=1}^{\infty} (b_n - a_n)$$

The *comparison series*

$$\sum_{n=1}^{\infty} b_n$$

is chosen such that the series

$$\sum_{n=1}^{\infty} (b_n - a_n)$$

converges faster than the original series

$$\sum_{n=1}^{\infty} a_n$$

Example 3.2.21

Evaluate

$$s = \sum_{n=1}^{\infty} \frac{n}{(n^3+5)},$$

with an error less than 10^{-4}. This series converges as $1/n^2$, in the sense that $n/(n^3+5) \approx 1/n^2$ for n sufficiently large. The partial sums s_{19} and s_{20} are 0.639448 and 0.641946, to six decimal places, respectively. Thus, the sum is still unsettled in the second decimal place. Now, values of the Riemann zeta function

$$\zeta(k) = \sum_{n=1}^{\infty} n^{-k}$$

are known to a very high accuracy. If $\zeta(2)$ is taken as a comparison series then

$$s = \zeta(2) - [\zeta(2) - s]$$

$$=\zeta(2)-\sum_{n=1}^{\infty}\left(\frac{1}{n^2}-\frac{n}{n^3+5}\right)$$

$$=1.644934067-5\sum_{n=1}^{\infty}\frac{1}{n^2(n^3+5)}$$

The series on the right-hand side converges as $1/n^5$, and its 14th and 15th partial sums are 0.190839 and 0.190841, to six decimal places, respectively. Thus, the 14th and 15th successive approximations to s are 0.690734 and 0.690731, respectively, and it may be shown that the error in the latter of these two numbers is approximately 2.15×10^{-5}, which is less than 10^{-4}, as required.

Multiplication of Infinite Series

The product of two finite series consists of the sum of the products, in any order, of every term in one series with each term in the other series and this sum is equal to the product of the sums of the two given series.

Now consider the product of two absolutely convergent infinite series

$$\sum_{n=1}^{\infty} a_n \text{ and } \sum_{n=1}^{\infty} b_n,$$

with sums S and T, respectively. As in the finite case, every term in one series is multiplied by each term in the other series and the sum of the resulting product pairs is clearly independent of the order in which they are taken. The possible product pairs can be written in a table, as illustrated in figure 3.4.

$$\begin{matrix} a_1b_1 & a_1b_2 & a_1b_3 & a_1b_4 & \cdots \\ a_2b_1 & a_2b_2 & a_2b_3 & a_2b_4 & \cdots \\ a_3b_1 & a_3b_2 & a_3b_3 & a_3b_4 & \cdots \\ a_4b_1 & a_4b_2 & a_4b_3 & a_4b_4 & \cdots \\ \vdots & \vdots & \vdots & \vdots & \end{matrix}$$

Figure 3.4

The $n\times n$ array $[a_ib_j]$ contains all of the products for which the suffix of each factor is less than or equal to n, that is, the sum of the terms in the array is equal to $(a_1+a_2+\ldots+a_n)(b_1+b_2+\ldots+b_n)$, which has the limit ST, as $n\to\infty$.

The following theorem has thus been established.

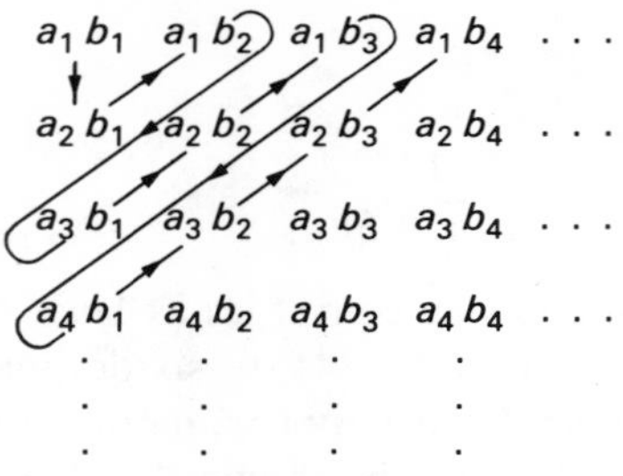

$$\begin{array}{ccccc} a_1 b_1 & a_1 b_2 & a_1 b_3 & a_1 b_4 & \dots \\ a_2 b_1 & a_2 b_2 & a_2 b_3 & a_2 b_4 & \dots \\ a_3 b_1 & a_3 b_2 & a_3 b_3 & a_3 b_4 & \dots \\ a_4 b_1 & a_4 b_2 & a_4 b_3 & a_4 b_4 & \dots \\ \vdots & \vdots & \vdots & \vdots & \end{array}$$

Figure 3.5

Theorem 3.2.4
Given two absolutely convergent series, with sums S and T, respectively, the series consisting of the sum of the products, in any order, of every term in one of the given series with each term in the other series is absolutely convergent and its sum is equal to ST.

If the product pairs $a_i b_j$ are summed in the order indicated by the arrows in figure 3.5 (that is, diagonally) then the following series is obtained

$$\sum_{n=1}^{\infty} [a_n b_1 + a_{n-1} b_2 + \ldots + a_1 b_n]$$

The following corollary to theorem 3.2.4 follows immediately.

Corollary

$$\sum_{n=1}^{\infty} a_n \sum_{n=1}^{\infty} b_n = \sum_{n=1}^{\infty} [a_n b_1 + a_{n-1} b_2 + \ldots + a_1 b_n]$$

Exercise 3.2

1. Evaluate $\sum_{n=10}^{50} (2n+3)$.

2. Evaluate $\sum_{n=1}^{\infty} \left(\frac{2}{3}\right)^n$.

3. Sum to infinity the series

$$1 + \frac{4}{5} + \frac{7}{5^2} + \frac{10}{5^3} + \ldots$$

4. Evaluate

$$\sum_{n=1}^{10} \frac{1}{(n+3)(n+4)}$$

by the method of differences.

5. Prove each of the following results, using mathematical induction:

(i) $\displaystyle\sum_{n=1}^{p} n^2 = \frac{1}{6}p(2p+1)(p+1)$

(ii) $\displaystyle\sum_{n=1}^{p} n^3 = \frac{1}{4}p^2(p+1)^2$

(iii) $\displaystyle\sum_{n=1}^{p} \frac{n}{(n+2)(n+3)(n+4)} = \frac{1}{6} - \frac{1}{(p+3)} + \frac{2}{(p+3)(p+4)}$

6. Use the method of indeterminate coefficients to obtain an expression for the sum to p terms of the series

$$1 \times 3 \times 4 + 4 \times 5 \times 5 + 7 \times 7 \times 6 + 10 \times 9 \times 7 + \ldots$$

7. Use the ratio test to determine whether or not each of the following series is convergent or divergent:

(i) $\displaystyle\sum_{n=1}^{\infty} \frac{n^4}{2^n}$

(ii) $\displaystyle\sum_{n=1}^{\infty} \frac{3^n}{n^5}$

(iii) $\displaystyle\sum_{n=1}^{\infty} \frac{(n!)^2}{(2n)!}$

8. Show that if

$$\sum_{n=1}^{\infty} |a_n|$$

is convergent, then so is

$$\sum_{n=1}^{\infty} a_n.$$

[Hint: define

$$u_n = \begin{cases} a_n, & \text{if } a_n \geqslant 0 \\ 0, & \text{otherwise} \end{cases}$$

$$v_n = \begin{cases} -a_n, & \text{if } a_n < 0 \\ 0, & \text{otherwise} \end{cases}$$

and consider the convergence of

$$\sum_{n=1}^{\infty} (u_n - v_n).]$$

9. Using a calculator, determine the first ten partial sums of the series

$$\sum_{n=1}^{\infty} \frac{1}{(n^2+1)}.$$

Show that

$$\sum_{n=1}^{\infty} \frac{1}{n^2+1} = \zeta(2) - \zeta(4) + \zeta(6) - \sum_{n=1}^{\infty} \frac{1}{n^6(n^2+1)}$$

where $\zeta(k)$ denotes the Riemann zeta function

$$\sum_{n=1}^{\infty} \frac{1}{n^k}.$$

Given that $\zeta(2) = \pi^2/6$, $\zeta(4) = \pi^4/90$, $\zeta(6) = \pi^6/945$, evaluate

$$\sum_{n=1}^{\infty} \frac{1}{(n^2+1)}$$

to the accuracy of your calculator.

3.3 CONTINUOUS REAL FUNCTIONS

As stated in section 1.5, a real function $f\colon X \to \mathbb{R}$, $X \subset \mathbb{R}$, assigns to each $x \in X$ a unique real number y, where $y = f(x)$. Sketches of the graphs of a number of commonly occurring real functions were given in example 1.5.6. On reconsidering these examples, the reader should distinguish two types of curve: those that can be drawn without taking the pen off the paper, for example, those in figures 1.25, 1.26 and 1.29, and those that are broken at one or more points, such as those in figures 1.28 and 1.34. If the graph of a function is not broken at any point, the function is said to be *continuous*. If, on the other hand, the graph is broken at some point, the function is said to be *discontinuous* at that point. These first, intuitive definitions of continuity and discontinuity must now be made rigorous. As a first step, the notion of limit introduced in section 3.1 must be extended. For a sequence, the behaviour as the integer variable n tends to plus infinity is the only concern. An immediate and straightforward extension for a real function is the behaviour as x tends to *minus* infinity.

Definition 3.3.1
(a) $f(x)\to l$ as $x\to\infty$ if, given $\varepsilon>0$, there exists $W>0$ such that $|f(x)-l|<\varepsilon$ for all $x>W$;

(b) $f(x)\to l$ as $x\to-\infty$ if, given $\varepsilon>0$, there exists $W>0$ such that $|f(x)-l| <\varepsilon$ for all $x<-W$;

(c) $f(x)\to\infty$ as $x\to\infty$ if, given $C>0$, no matter how large, there exists W such that $f(x)>C$ for all $x>W$;

(d) $f(x)\to-\infty$ as $x\to-\infty$ if, given $C>0$, there exists $W>0$ such that $f(x)<-C$ for all $x<-W$.

In addition to the behaviour of $y=f(x)$ as $x\to\pm\infty$, an awareness of the behaviour of $f(x)$ as x approaches any feasible value c is important. In order for it to be feasible, c does not have to be an element of the domain X of the function, provided x can take values arbitrarily close to c whilst remaining in X. Often, x will be able to approach c either from the left or from the right, that is, through values less than c or through values greater than c, and the limiting value of $f(x)$ need not necessarily be the same in both cases. These two limit situations are distinguished by writing $x\to c^-$ and $x\to c^+$, respectively.

The curves referred to in the following examples are again those sketched in example 1.5.6.

Example 3.3.1

(1) Figure 1.25 is a sketch of the graph $\{(x, y)|y=ax+b\}$. Consider $x=1$. $f(1)$ exists and is equal to $a+b$. Furthermore, as x approaches one from either the left or the right, y tends to $a+b$, that is $f(x)\to f(1)$ as $x\to 1^-$ and $f(x)\to f(1)$ as $x\to 1^+$.

(2) Figure 1.28 is a sketch of the graph $\{(x, y)|y=1/(x-k)\}$. As x approaches the value k from the left, the value of y tends to minus infinity, that is, $y\to-\infty$ as $x\to k^-$, but as x approaches k from the right, y tends to plus infinity, that is, $y\to\infty$ as $x\to k^+$. Quite clearly, this function is not defined for $x=k$. If the definition is completed by assigning an arbitrary finite value to $f(k)$, then $f(x)\not\to f(k)$ as $x\to k^-$ and $f(x)\not\to f(k)$ as $x\to k^+$, irrespective of the value chosen for $f(k)$.

(3) Figure 1.34 is a sketch of the graph $\{(x, y)|y=\lceil x \rceil\}$, which has breaks at infinitely many points. For example, for any value of x between zero and one, no matter how close to one, the value of y is one but for any value of x between one and two, no matter how close to one, the value of y is two. Since $f(1)\equiv\lceil 1 \rceil=1, f(x)\to f(1)$ as $x\to 1^-$ but $f(x)\not\to f(1)$ as $x\to 1^+$.

From the above examples, it can be seen that in order for f to be continuous at $x=c, f(c)$ must exist and $f(x)$ must tend to $f(c)$ as x approaches c through *any* values in the domain of f.

Definition 3.3.2 Continuity at a Point
A real function $f: X\to \mathbb{R}, X\subset\mathbb{R}$, is *continuous* at $c\in X$ if, given any number $\varepsilon>0$, no matter how small, a number $\delta>0$ can be found such that

$$|f(x)-f(c)|<\varepsilon \text{ for all } x\in X \text{ satisfying } |x-c|<\delta$$

Note that the value of δ depends on ε *and* on the value c.

Example 3.3.2

Let $f(x)=1+4x$; then $f(2)=9$. If $\varepsilon=10^{-6}$, then

$$|f(x)-f(2)|<\varepsilon\supset -10^{-6}<4x-8<10^{-6}$$

$$\supset -\frac{1}{4}\times 10^{-6}<x-2<\frac{1}{4}\times 10^{-6}$$

$$\supset |x-2|<\delta$$

where $\delta=2.5\times 10^{-7}$.

Definition 3.3.3 Continuity on an Interval

(a) f is continuous on the open interval (a, b) if it is continuous at each point in the interval;

(b) f is continuous on the closed interval $[a, b]$ if

(i) it is continuous on (a, b),

(ii) $f(x)\to f(a)$ as $x\to a^+$ and $f(x)\to f(b)$ as $x\to b^-$.

The set of functions continuous on $[a, b]$ is denoted by $C[a, b]$.

Example 3.3.3

If $f(x)\equiv\lceil x \rceil$ then $f\in C[1/2, 1]$ but $f\notin C[-1, 1]$.

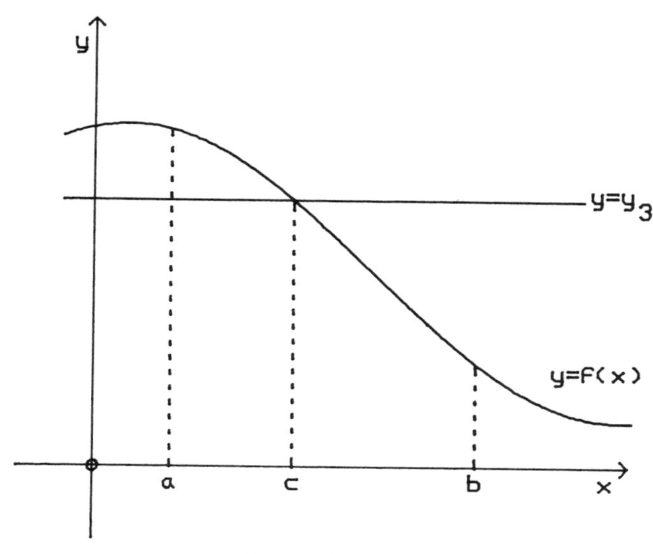

Figure 3.6

Properties of Continuous Functions

Theorem 3.3.1
If f and g are both continuous at $x=c$, then $f \pm g$ and $f \times g$ are continuous at $x=c$, and, if $g(c) \neq 0$, f/g is also continuous at $x=c$.

Proof These results may be proved using arguments analogous to those used to prove the corresponding facts for sequences in section 3.1. In particular, in order to establish the continuity of f/g, it is sufficient to show that $1/g$ is continuous at c, the more general result then following from the result for products. Since the rule for products of sequences is not actually proved in the text, but left as an exercise for the reader, the proof of the corresponding result for functions is now given.

The continuity of $f \times g$ may be established by noting that

$$f(x)g(x) - f(c)g(c) = [f(x) - f(c)][g(x) - g(c)] + [f(x) - f(c)]g(c) + [g(x) - g(c)]f(c)$$

and at least one factor in each of the terms in the expression on the right-hand side tends to zero as x approaches c.

Theorem 3.3.2 Intermediate-value Property
Suppose $f \in C[a, b]$ and $f(a) = y_1$, $f(b) = y_2$, where $y_1 \neq y_2$. If y_3 is any value between y_1 and y_2, there is a value $c \in (a, b)$ for which $f(c) = y_3$.

Figure 3.6 serves to illustrate the truth of this theorem. Assuming for the sake of argument that $y_1 > y_3 > y_2$, the intermediate-value property

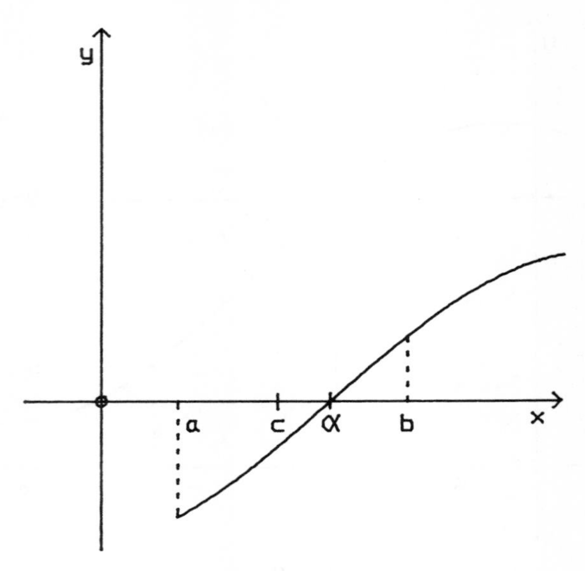

Figure 3.7

says that the straight line $y = y_3$ cuts the curve $y = f(x)$ at least once between $x = a$ and $x = b$.

The intermediate-value property may be used to determine whether an interval $[a, b]$ contains a root of the equation $f(x) = 0$. Suppose that $f(a) < 0$ and $f(b) > 0$, that is $f(a) < 0 < f(b)$. Then, by the intermediate-value property, there must be at least one value α between a and b such that $f(\alpha) = 0$ (see figure 3.7). In general, if $f \in C[a, b]$ and $f(a)f(b) < 0$, then at least one root of the equation $f(x) = 0$ lies between $x = a$ and $x = b$.

Algorithm 3.3.1 describes how the width of an interval bracketing a root α of $f(x) = 0$ may be successively halved, or *bisected*, by the repeated application of the intermediate-value theorem. The algorithm stops when the width of the interval is less than a specified tolerance, δ.

Algorithm 3.3.1 Bisection Method

[Given an interval $[a, b]$ over which f changes sign, that is, $f(a)f(b) < 0$.]

```
c←(a+b)/2;
while b−a>δ do
    if f(a)*f(c)⩽0 then
        b←c
```

else

$a \leftarrow c$

endif;

$c \leftarrow (a+b)/2$

endwhile;

accept c as the root

If $[a, b]$ is the initial interval enclosing α and if c_n is used to denote the nth value of c in the algorithm then

$$|\alpha - c_n| \leqslant (b-a)/2^n$$

and, since $(b-a)/2^n \to 0$ as $n \to \infty$, the bisection method is convergent. Unfortunately, its rate of convergence is generally slow and, in practice, it is only used to provide an initial approximation, correct to perhaps two significant digits, for use in a faster method.

The computer science student may already be familiar with the principle of bisection, since it is also the basis of a technique for searching for a particular record in a sorted list of records, known as the *binary-search* method.

Theorem 3.3.3
If f is continuous on a closed interval $[a, b]$, then it is bounded on $[a, b]$ and attains its least upper bound and its greatest lower bound.

This theorem is merely saying that, if $f \in C[a, b]$, then $f(x)$ has a greatest and a least value over $[a, b]$.

Note that a function continuous on an open interval (a, b) is not necessarily bounded; this is illustrated in figure 3.8. The function given by $f(x) = 1/(x-2)$ is continuous on $(2, 3)$ but it is not bounded on this interval. In fact, $f(x) \to \infty$ as $x \to 2^+$.

Polynomial Functions

As defined in section 1.4, a real function f is a *polynomial of degree n* if $f(x)$ may be expressed in the form

$$f(x) \equiv a_0 + a_1 x + a_2 x^2 + \ldots + a_n x^n$$

where the *coefficients* $a_0, a_1, \ldots, a_n$ are real constants.

A polynomial may be evaluated for any value of x using a finite number of arithmetic operations. This means that such functions may be evaluated directly on a computer. The majority of functions, however, are defined by

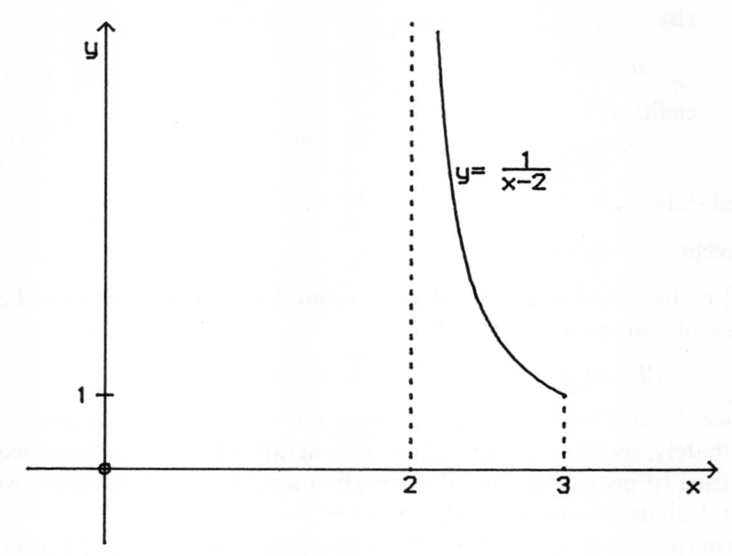

Figure 3.8

infinite processes and cannot, in general, be evaluated exactly. A rather surprising fact is that *any* continuous function may be approximated arbitrarily closely by a polynomial of sufficiently high degree. This result is usually referred to as *Weierstrass' theorem* and is the theoretical basis for the common practice of using polynomials to approximate more general functions. As a consequence of the use of polynomials in this fashion, the evaluation of a polynomial is a frequently occurring computational task. It is therefore important that this task should be performed as efficiently as possible.

Algorithm 3.3.2 evaluates the polynomial $a_0+a_1x+\ldots+a_nx^n$ in a straightforward manner, using the previously computed value of x^i in the calculation of x^{i+1}. This algorithm involves n additions and $2n$ multiplications.

Algorithm 3.3.2

```
sum←a₀;
y←1;
for i from 1 in steps of 1 to n do
    y←y * x;
    sum←sum + aᵢ * y
endfor
```

Algorithm 3.3.3 evaluates the equivalent *nested form* of the polynomial

$$a_0 + x(a_1 + x(a_2 + \ldots + x(a_{n-1} + xa_n))\ldots)$$

which requires only n additions and n multiplications, a saving of n multiplications over algorithm 3.3.2.

Algorithm 3.3.3 Nested Multiplication

sum $\leftarrow a_n$;

for i **from** 1 **in steps of** 1 **to** n **do**

sum $\leftarrow$ sum $* \; x + a_{n-i}$

endfor

Example 3.3.4

(1) Evaluate $p(3.1)$, where $p(x) = x^3 - 2x^2 + 2x + 3$.

Initialisation: sum $\leftarrow 1$

$i=1$: sum $\leftarrow 1 \times 3.1 - 2 = 1.1$

$i=2$: sum $\leftarrow 1.1 \times 3.1 + 2 = 5.41$

$i=3$: sum $\leftarrow 5.41 \times 3.1 + 3 = 19.771$

(2) A special case of algorithm 3.3.3 was used in example 1.3.4 to convert a hexadecimal number into its decimal equivalent. More generally, if $(a_n a_{n-1} \ldots a_0)_b$ represents a base-b integer, then its decimal equivalent is given by

$$a_n b^n + a_{n-1} b^{n-1} + \ldots + a_0 b^0$$

which may be evaluated using algorithm 3.3.3 with x replaced by b. Thus, the decimal equivalent of 110011_2 may be computed as follows

initialisation: sum $\leftarrow 1$

$i=1$: sum $\leftarrow 1 \times 2 + 1 = 3$

$i=2$: sum $\leftarrow 3 \times 2 + 0 = 6$

$i=3$: sum $\leftarrow 6 \times 2 + 0 = 12$

$i=4$: sum $\leftarrow 12 \times 2 + 1 = 25$

$i=5$: sum $\leftarrow 25 \times 2 + 1 = 51$

The Binomial Theorem

The reader will be familiar with the *expansion* of $(a+x)^2$ to give the quadratic polynomial $a^2+2ax+x^2$, and with the expansion of $(a+x)^3$ to give the cubic polynomial $a^3+3a^2x+3ax^2+x^3$. In general, $(a+x)^p$, where $a\in\mathbb{R}$ and $p\in N$, may be expanded as a polynomial of degree p. The particular form that this polynomial takes is given by the *binomial theorem.*

Theorem 3.3.4
For any $p\in N$

$$(a+x)^p=\sum_{n=0}^{p} {}^pC_n a^{p-n}x^n$$

where

$${}^pC_n\equiv\frac{p!}{n!(p-n)!}$$

(Here ${}^pC_n$ denotes the number of combinations of p objects, taking n at a time; see definition 4.1.8.)

Proof The proof is by induction. Since $(a+x)^1=a^1x^0+a^0x^1$, the result is true for $p=1$. Assume that it is true for $p=m$, and consider the case $p=m+1$

$$(a+x)^{m+1}=(a+x)\sum_{n=0}^{m} {}^mC_n a^{m-n}x^n$$

$$=a^{m+1}+a\sum_{n=1}^{m} {}^mC_n a^{m-n}x^n+x\sum_{n=0}^{m-1} {}^mC_n a^{m-n}x^n+x^{m+1}$$

Clearly

$$\sum_{n=0}^{m-1} {}^mC_n a^{m-n}x^n=\sum_{n=1}^{m} {}^mC_{n-1}a^{m+1-n}x^{n-1}$$

and thus

$$(a+x)^{m+1}=a^{m+1}+\sum_{n=1}^{m}({}^mC_n+{}^mC_{n-1})a^{m+1-n}x^n+x^{m+1}$$

Now

$${}^mC_n+{}^mC_{n-1}=\frac{m!}{n!(m-n)!}+\frac{m!}{(n-1)!(m+1-n)!}$$

$$= \frac{m!}{(n-1)!(m-n)!}\left[\frac{1}{n} + \frac{1}{(m+1-n)}\right]$$

$$= \frac{m!}{(n-1)!(m-n)!}\left[\frac{m+1}{n(m+1-n)}\right]$$

$$= \frac{(m+1)!}{n!(m+1-n)!}$$

$$= {}^{m+1}C_n$$

Thus

$$(a+x)^{m+1} = a^{m+1} + \sum_{n=1}^{m} {}^{m+1}C_n a^{m+1-n} x^n + x^{m+1}$$

$$= \sum_{n=0}^{m+1} {}^{m+1}C_n a^{m+1-n} x^n$$

and so, if the result is true for $p=m$, it is true for $p=m+1$. Since it is true for $p=1$, by induction it is true for all $p \in N$.

Example 3.3.5

Evaluate $(1.01)^9$ correct to six decimal places.

By the binomial theorem

$$(1+x)^9 = 1 + 9x + 36x^2 + 84x^3 + 126x^4 + 126x^5 + 84x^6 + 36x^7 + 9x^8 + x^9$$

Putting $x=0.01$ gives

$$(1.01)^9 \approx 1 + 0.09 + 0.0036 + 0.000084 + 0.00000126 + 0.0000000126$$

$$\approx 1.093685$$

which is correct to six decimal places.

Power Series

As stated above, many functions are defined by infinite processes. A simple way in which this may be done is by means of a power series.

Definition 3.3.4

An infinite series of the form

$$\sum_{n=0}^{\infty} c_n x^n = c_0 + c_1 x + c_2 x^2 + \ldots$$

in which x is a real variable and the *coefficients* $c_0, c_1, c_2, \ldots$ are real constants, is called a *power series*. An interval $(-R, R)$ of the real line is said to be an *interval of convergence* if the power series is absolutely convergent for all $x \in (-R, R)$. The largest value of R for which this is true is called the *radius of convergence* of the power series.

Example 3.3.6

(1) The power series

$$1+cx+\frac{c(c-1)}{2!}x^2+\frac{c(c-1)(c-2)}{3!}x^3+\ldots$$

where $c \in \mathbb{R}$, is known as the *binomial series*. Note that if $c \in N$, the binomial series has only $c+1$ non-zero terms, that is, it is a finite series and is simply the expansion of $(1+x)^c$ as given by the binomial theorem. The behaviour of the infinite series may be investigated by applying the ratio test. The nth term of the series is

$$a_n=\frac{c(c-1)\ldots(c-n+2)}{(n-1)!}x^{n-1}$$

Hence

$$\frac{a_{n+1}}{a_n}=\frac{(c-n+1)}{n}x=[-1+(c+1)/n]x$$

$$\to -x \text{ as } n\to\infty$$

since c is a fixed value; that is

$$\lim_{n\to\infty}\left|\frac{a_{n+1}}{a_n}\right|=|x|$$

The binomial series is thus convergent provided $|x|<1$. Furthermore, if c is a rational number, the series converges to $(1+x)^c$. When $c=-1$, the binomial series is an infinite geometric series with first term 1 and common ratio $-x$.

(2) The power series $x-x^2/2+x^3/3-x^4/4+\ldots$ is known as the *logarithmic series*. Applying the ratio test

$$\left|\frac{a_{n+1}}{a_n}\right|=\left|\frac{nx}{n+1}\right|=\frac{|x|}{1+1/n}\to|x| \text{ as } n\to\infty$$

The logarithmic series is thus absolutely convergent for $|x|<1$. It is also convergent if $x=1$, since $1-1/2+1/3-1/4+\ldots$ is an alternating and decreasing sequence. If $x=-1$, however, the series is the harmonic series, with the signs of the terms reversed, and is therefore divergent. It is shown in example 3.4.12 that the sum of the logarithmic series is $\log_e(1+x)$ for $|x|<1$.

Definition 3.3.5

The *product series* of the two power series

$$\sum_{n=0}^{\infty} c_n x^n$$

and

$$\sum_{n=0}^{\infty} d_n y^n$$

is defined by

$$\sum_{n=0}^{\infty} c_n x^n \sum_{n=0}^{\infty} d_n y^n = \sum_{n=0}^{\infty} (c_n d_0 x^n + c_{n-1} d_1 x^{n-1} y + \ldots c_0 d_n y^n)$$

In particular

$$\sum_{n=0}^{\infty} c_n x^n \sum_{n=0}^{\infty} d_n x^n = \sum_{n=0}^{\infty} (c_n d_0 + c_{n-1} d_1 + \ldots + c_0 d_n) x^n$$

This definition is merely the extension to power series of the rule for multiplying two polynomials.

Clearly, the product series will be absolutely convergent only when *both* of the series on the left-hand side are absolutely convergent. Hence, if

$$\sum_{n=0}^{\infty} c_n x^n$$

and

$$\sum_{n=0}^{\infty} d_n x^n$$

have radii of convergence R_1 and R_2, respectively, then the radius of convergence of their product is the minimum of R_1 and R_2, min (R_1, R_2). Furthermore, it follows from theorem 3.2.4 and its corollary that the sum of the product series, when it exists, is the product of the sums of the two series being multiplied.

Example 3.3.7

(1) $\displaystyle\sum_{n=0}^{\infty} x^n \sum_{n=0}^{\infty} x^n = \sum_{n=0}^{\infty} (n+1)x^n$

Since

$$\sum_{n=0}^{\infty} x^n$$

is absolutely convergent whenever $|x| < 1$, with sum $1/(1-x)$, the product series has unit radius of convergence and sum $1/(1-x)^2$. Hence

$$\frac{1}{(1-x)^2} = 1 + 2x + 3x^2 + \ldots, \text{ provided } |x| < 1$$

(2)

$$\sum_{n=0}^{\infty} (2x)^n \sum_{n=0}^{\infty} (n+1)x^n = \sum_{n=0}^{\infty} a_n x^n$$

where

$$a_n = 2^n \times 1 + 2^{n-1} \times 2 + 2^{n-2} \times 3 + \ldots + 2^0 \times (n+1)$$
$$= 2^n\left[1 + 2\left(\frac{1}{2}\right) + 3\left(\frac{1}{2}\right)^2 + \ldots + (n+1)\left(\frac{1}{2}\right)^n\right]$$

From example 3.2.6

$$1 + 2x + 3x^2 + \ldots + (n+1)x^n = (1 - x^{n+1})/(1-x)^2 - (n+1)x^{n+1}/(1-x)$$

Thus

$$1 + 2\left(\frac{1}{2}\right) + 3\left(\frac{1}{2}\right)^2 + \ldots + (n+1)\left(\frac{1}{2}\right)^n = 4(1 - 1/2^{n+1}) - (n+1)/2^n$$
$$= 2^2 - 1/2^{n-1} - (n+1)/2^n$$

Hence

$$a_n = 2^{n+2} - 2 - (n+1) = 2^{n+2} - 3 - n$$

The geometric series

$$\sum_{n=0}^{\infty} (2x)^n$$

has sum $1/(1-2x)$, provided $|x| < \frac{1}{2}$, and, from above

$$\sum_{n=0}^{\infty} (n+1)x^n$$

has sum $1/(1-x)^2$, provided $|x| < 1$. Thus, provided $|x| < \frac{1}{2}$, the series

$$\sum_{n=0}^{\infty} (2^{n+2} - 3 - n)x^n$$

is convergent, with sum $1/[(1-2x)(1-x)^2]$.

Solution of Linear Recurrence Relations by Generating Functions

Definition 3.3.6
A power series

$$\sum_{n=0}^{\infty} c_n x^n$$

is called a *recurring series of order* r if, for all n sufficiently large, the coefficients c_n satisfy a linear recurrence relation of the form

$$c_n + \alpha_1 c_{n-1} + \alpha_2 c_{n-2} + \ldots + \alpha_r c_{n-r} = 0$$

where $\alpha_1, \alpha_2, \ldots, \alpha_r$ are constants and r is a fixed integer.
The associated polynomial

$$1 + \alpha_1 x + \alpha_2 x^2 + \ldots + \alpha_r x^r$$

is known as the *scale of relation* of the recurring series.

Example 3.3.8

The power series $1 + 3x + 12x^2 + 54x^3 + \ldots + \frac{1}{2}((3 + \sqrt{3})^n + (3 - \sqrt{3})^n)x^n + \ldots$, is a recurring series of order 2. The coefficients satisfy the recurrence relation $c_n - 6c_{n-1} + 6c_{n-2} = 0$, $n \geqslant 2$. The scale of relation is $1 - 6x + 6x^2$.

Definition 3.3.7
The sum to infinity of a convergent recurring series is called the *generating function* for the series.

The generating function for a recurring series is a rational function, the denominator of which is the scale of relation of the series. If the scale of relation is known, the numerator in the generating function may be determined by equating coefficients.

Example 3.3.9

Obtain the generating function for the series

$$1 + 3x + 12x^2 + 54x^3 + \ldots$$

Assume

$$\frac{P(x)}{1 - 6x + 6x^2} = 1 + 3x + 12x^2 + 54x^3 + \ldots$$

Multiplying both sides by $1 - 6x + 6x^2$ gives

$$P(x) = 1 - 3x$$

where all other terms on the right-hand side vanish because of the scale of relation. The generating function is thus $(1-3x)/(1-6x+6x^2)$.

Generating functions may be used to solve linear recurrence relations. The method is illustrated in the following example.

Example 3.3.10

Solve the linear recurrence relation

$$a_n - 4a_{n-1} + 5a_{n-2} - 2a_{n-3} = 0$$

given $a_1 = 1$, $a_2 = 0$, $a_3 = -5$.

With $c_0 = a_1$, $c_1 = a_2$, $c_2 = a_3$, ..., $c_n = a_{n+1}$, ..., the resulting power series

$$\sum_{n=0}^{\infty} c_n x^n$$

is a recurring series of order 3, with scale of relation $1-4x+5x^2-2x^3$. Its generating function is given by

$$\begin{aligned}\frac{P(x)}{1-4x+5x^2-2x^3} &= c_0 + c_1 x + c_2 x^2 + c_3 x^3 + \ldots \\ &= a_1 + a_2 x + a_3 x^2 + a_4 x^3 + \ldots \\ &= 1 + 0x - 5x^2 + a_4 x^3 + \ldots\end{aligned}$$

Multiplying through by the scale of relation gives $P(x) = 1-4x$. Hence the generating function is

$$\frac{1-4x}{1-4x+5x^2-2x^3} = \frac{1-4x}{(1-2x)(1-x)^2}$$

From example 3.3.7(2)

$$1/[(1-2x)(1-x)^2] = \sum_{n=0}^{\infty} (2^{n+2} - 3 - n)x^n$$

and hence

$$(1-4x)/[(1-2x)(1-x)^2] = \sum_{n=0}^{\infty} (2^{n+2} - 3 - n)x^n - 4\sum_{n=0}^{\infty} (2^{n+2} - 3 - n)x^{n+1}$$

Now

$$\begin{aligned}\sum_{n=0}^{\infty} (2^{n+2} - 3 - n)x^{n+1} &= \sum_{n=1}^{\infty} [2^{n+1} - 3 - (n-1)]x^n \\ &= \sum_{n=0}^{\infty} [2^{n+1} - 3 - (n-1)]x^n\end{aligned}$$

since $(2^{n+1}-2-n)=0$ for $n=0$. Thus

$$(1-4x)/[(1-2x)(1-x)^2]=\sum_{n=0}^{\infty}(2^{n+2}-3-n-2^{n+3}+8+4n)x^n$$

$$=\sum_{n=0}^{\infty}(3n+5-2^{n+2})x^n$$

Since $a_n=c_{n-1}$, an explicit formula for the nth term of the sequence (a_n) is

$$a_n=3(n-1)+5-2^{(n-1)+2}=3n+2-2^{n+1}$$

The Exponential and Logarithmic Functions

Definition 3.3.8
The power series

$$1+x+\frac{x^2}{2!}+\frac{x^3}{3!}+\ldots+\frac{x^n}{n!}+\ldots$$

is known as the *exponential function* and is denoted by $\exp(x)$.

If $a_n=x^{n-1}/(n-1)!$, then

$$\left|\frac{a_{n+1}}{a_n}\right|=\frac{|x|}{n}\to 0 \text{ as } n\to\infty$$

whatever the value of x. By the ratio test, the exponential function is thus absolutely convergent for all $x\in\mathbb{R}$. Clearly, $\exp(x)>0$ for any $x\geqslant 0$.

Theorem 3.3.5

$$\exp(x)\times\exp(y)=\exp(x+y)$$

Proof From definition 3.3.5

$$\exp(x)\times\exp(y)=\sum_{n=0}^{\infty}\left(\frac{x^n}{n!}+\frac{x^{n-1}y}{(n-1)!1!}+\ldots+\frac{y^n}{n!}\right)$$

By theorem 3.3.4

$$(x+y)^n=\sum_{r=0}^{n}{}^nC_r x^{n-r}y^r$$

$$=x^n+nx^{n-1}y+\ldots+\frac{n(n-1)\ldots(n-r+1)x^{n-r}y^r}{r!}+\ldots+y^n$$

$$=n!\left[\frac{x^n}{n!}+\frac{x^{n-1}y}{(n-1)!1!}+\ldots+\frac{x^{n-r}y^r}{(n-r)!r!}+\ldots+\frac{y^n}{n!}\right]$$

Hence

$$\exp(x) \times exp(y) = \sum_{n=0}^{\infty} \frac{(x+y)^n}{n!}$$

$$= \exp(x+y)$$

It follows immediately from this theorem that $\exp(0)=1$. It then follows that $\exp(-x)=1/\exp(x)$. Hence $\exp(x)>0$ for *all* $x \in \mathbb{R}$.

Corollary

$$\exp(x_1) \times \exp(x_2) \times \ldots \times \exp(x_q) = \exp(x_1 + x_2 + \ldots + x_q)$$

Definition 3.3.9

$$e = 1 + 1 + \frac{1}{2!} + \frac{1}{3!} + \ldots, \text{ that is, } e = \exp(1)$$

The irrational number e has already made a number of informal appearances in this book. Like π, it is one of the basic mathematical constants. Its approximate value is 2.718.

Using theorem 3.3.5, it is easy to show that if x is a rational number, then $\exp(x) = e^x$. This prompts the following definition.

Definition 3.3.10

$$e^x = \exp(x) \text{ for } all\ x \in \mathbb{R}$$

This definition gives meaning to e^x even when x is irrational.

Logarithms to base e were considered in chapter 1. It is now possible to give a formal definition of the logarithmic function.

Definition 3.3.11

The *logarithmic function* is the inverse function of the exponential function.

Thus, $y = \log_e x$ if $x = e^y$ and it follows that $\log_e x$ is only defined for x positive. Sketches of the graphs of $y = e^x$ and $y = \log_e x$ are given in example 1.5.6.

It is easy to show that for any $k \in N$, $e^x/x^k \to \infty$ as $x \to \infty$ but $\log_e x/x^k \to 0$ as $x \to \infty$ (see exercise 3.3). Thus, for x sufficiently large, e^x is bigger than any given polynomial, while $\log_e x$ is smaller than any given polynomial.

If x is a rational number then

$$2^x = e^{x \log_e 2}$$

and if x is irrational, this is taken to be the definition of 2^x.

Clearly, for sufficiently large x, 2^x is also bigger than any given polynomial. Similarly, for sufficiently large x, $\log_2 x$ is smaller than any given polynomial, since $\log_2 x = \log_e x / \log_e 2 \approx 1.44 \log_e x$.

Thus, in computer science, an algorithm with execution time proportional to $n \log_2 n$, for n sufficiently large, is generally preferable to an algorithm having execution time proportional to n^2. Here n is a measure of the 'size' of the problem being solved, for example, it might be the number of records in a file to be sorted, or the number of equations in a system to be solved. The *complexity measure* (for example, $n \log_2 n$ and n^2, above) is usually determined by counting the number of basic operations performed by the algorithm: for example, the number of arithmetic operations in a numerical algorithm, or the number of key comparisons in a search algorithm.

Definition 3.3.12
An algorithm is *order* $f(n)$, or $O(f(n))$, where n is an integer indicating the size of the problem to be solved, if, for all n sufficiently large, the complexity measure of the algorithm is less than $cf(n)$, where c is a positive real constant.

Example 3.3.11

The method of Gaussian elimination (see chapter 2) for the solution of an $n \times n$ system of equations is an $O(n^3)$ algorithm.

Exercise 3.3

1. Show that the function defined by

$$f(x) = \begin{cases} (x-1)/(x^2-1), & x \neq 1 \\ \frac{1}{2}, & x = 1 \end{cases}$$

is continuous.

2. Prove that $f+g$ and $f-g$ are continuous at $x=c$ if f and g are continuous at $x=c$.

3. Rewrite the following polynomial in nested form

$$p_3(x) \equiv x^3 - 0.75x^2 - 4.5x + 4.75$$

Using algorithm 3.3.1 and a calculator, determine to an accuracy of five decimal places the root of $p_3(x)=0$ that lies in the interval [1.5, 2.0]. All evaluations of $p_3(x)$ should be performed using algorithm 3.3.3.

4. Using the binomial theorem, expand $(2+3x)^6$ as a polynomial of degree six.

5. Determine the coefficient of x^7 in the expansion of $(1+2x)^4(1-2x)^6$.

6. Find the radius of convergence for each of the following power series.

(i) $x+\frac{x^2}{2}+\frac{x^3}{3}+\ldots+\frac{x^n}{n}+\ldots$

(ii) $1+\frac{1}{2}x+\frac{2}{3}x^2+\frac{3}{4}x^3+\ldots+\frac{n-1}{n}x^{n-1}+\ldots$

(iii) $1-\frac{x^2}{2!}+\frac{x^4}{4!}-\frac{x^6}{6!}+\ldots+\frac{(-1)^n x^{2n}}{(2n)!}+\ldots$

(iv) $x-\frac{x^3}{3!}+\frac{x^5}{5!}-\frac{x^7}{7!}+\ldots+\frac{(-1)^n x^{2n+1}}{(2n+1)!}+\ldots$

7. Obtain a power series expansion for $1/(1-x)^3$ by multiplying the power series for $1/(1-x)$ and $1/(1-x)^2$.

8. A sequence (a_n) is defined by the following linear recurrence relation

$$a_1=\frac{2}{3},\ a_2=-\frac{7}{9},\ 3a_n+5a_{n-1}+2a_{n-2}=0,\ n\geqslant 3$$

Obtain an explicit formula of the form $a_n=f(n)$ by the generating function method.

9. Show that:

(a) if x is a rational number, then $\exp(x)=e^x$. [Hint: Do first for $x \in N$; then for $x=p/q$, write $p=\sum_{i=1}^{q} x$ and use the corollary to theorem 3.3.5.]

(b) $e^x/x^k \to \infty$ as $x\to\infty$, for all $k \in N$;

(c) $(\log_e x)/x \to 0$ as $x\to\infty$. [Hint: If $y=\log_e x$ then $x=e^y$; hence $(\log_e x)/x = y/e^y$.]

10. Determine a value of n for which $2^n \approx n^3$.

3.4 DIFFERENTIATION

In general, the value of a continuous, real function f: $X\to\mathbb{R}$ changes when the value of its argument, $x \in X$, changes and then it is important to be able to determine the *rate of change of* $f(x)$ *with respect to* x. It is with this fundamental problem that differentiation is concerned.

As an illustration of the principle involved, consider the motion of a car. At any instant the car has a velocity, as recorded on the speedometer. If the car is moving at constant velocity then the acceleration, that is, the rate of change of velocity with respect to time, is equal to zero. However, if the driver of the car starts increasing the pressure on the accelerator pedal, the velocity changes from one instant to the next and the acceleration is thus no longer equal to zero. If $v(t)$ denotes the velocity of the car at time t, the average value of the acceleration during the time interval t_1 to t_2 is given by $[v(t_2)-v(t_1)]/(t_2-t_1)$. If $t_2=t_1+\delta t$, where δt denotes a very small interval of time then, intuitively, the acceleration at time t_1 is approximately equal to $[v(t_1+\delta t)-v(t_1)]/\delta t$. The limit of this expression as $\delta t \to 0$ gives the acceleration of the car at time t_1.

Definition 3.4.1
A real function $f: X \to \mathbb{R}$, $X \subseteq \mathbb{R}$, is *differentiable* at $x \in X$ if the expression

$$\frac{f(x+\delta x)-f(x)}{\delta x}$$

tends to a finite limit as $\delta x \to 0$. In such a case, the limit is called the *derivative* or *differential coefficient* of $f(x)$ and is denoted by $f'(x)$ or $(\mathrm{d}/\mathrm{d}x)f(x)$. The process of determining $f'(x)$, when it exists, is called *differentiation.*

If $y=f(x)$, it is usual to write $\mathrm{d}y/\mathrm{d}x=f'(x)$. Acceleration is thus the derivative of velocity, when both are regarded as functions of time.

If f is differentiable at x then it must be continuous at x. However, it is demonstrated in example 3.4.1(5) that the converse of this statement is not necessarily true.

Definition 3.4.2
If $f'(x)$ exists for each $x \in (a, b) \subseteq X$ then f is said to be *continuously differentiable* in the interval (a, b). The function $f': x \mapsto f'(x)$ is called the *derived function*, or simply the *derivative*, of f in (a, b).

In the following examples, a number of commonly occurring functions are *differentiated from first principles*, that is, by using the definition of derivative given above.

Example 3.4.1

(1) $f: x \mapsto c$ where $c \in \mathbb{R}$.
For any $x \in \mathbb{R}$, $[f(x+\delta x)-f(x)]/\delta x=(c-c)/\delta x=0$. The derivative of a constant is thus equal to zero and the derived function of a constant function is defined by $f': x \mapsto 0$.

(2) $f: x \mapsto x^n$ where $n \in N$.
By the binomial theorem (see theorem 3.3.4)

$$\frac{(x+\delta x)^n - x^n}{\delta x} = \frac{nx^{n-1}\delta x + \frac{1}{2}n(n-1)x^{n-2}(\delta x)^2 + \ldots + (\delta x)^n}{\delta x}$$

$$= nx^{n-1} + \delta x\left[\frac{1}{2}n(n-1)x^{n-2} + \ldots + (\delta x)^{n-2}\right]$$

which tends to nx^{n-1} as $\delta x \to 0$. The derived function of f is thus $f': x \mapsto nx^{n-1}$.

(3) $f: x \mapsto \sin x$.
Replacing y by $x+\delta x$ in the trigonometric identity

$$\sin y - \sin x = 2\cos\left[\frac{1}{2}(y+x)\right]\sin\left[\frac{1}{2}(y-x)\right]$$

and dividing through by δx, gives

$$\frac{\sin(x+\delta x) - \sin x}{\delta x} = \frac{2\cos\left(x+\frac{1}{2}\delta x\right)\sin\frac{1}{2}\delta x}{\delta x}$$

Using the definitions of the trigonometric ratios, it may be shown that $\sin\theta < \theta < \tan\theta$, for $0 < \theta < \pi/2$. Hence

$$\frac{1}{\sin\theta} > \frac{1}{\theta} > \frac{\cos\theta}{\sin\theta}$$

that is

$$1 > \frac{\sin\theta}{\theta} > \cos\theta$$

Since $\cos 0 = 1$

$$\lim_{\theta\to 0}\frac{\sin\theta}{\theta} = 1$$

Hence

$$\cos\left(x+\frac{1}{2}\delta x\right)\frac{\sin\frac{1}{2}\delta x}{\frac{1}{2}\delta x} \to \cos x, \text{ as } \delta x \to 0$$

The derivative of $\sin x$ is thus equal to $\cos x$. It may similarly be shown (see exercise 3.4) that the derivative of $\cos x$ is $-\sin x$.

(4) $f\colon x \mapsto e^x$.

$$\frac{e^{x+\delta x}-e^x}{\delta x}=\frac{e^x(e^{\delta x}-1)}{\delta x}$$

It is easy to show that $(e^{\delta x}-1)/\delta x \to 1$ as $\delta x \to 0$. The derivative of e^x is thus e^x, that is, $f' \equiv f$.

(5) $f\colon x \mapsto |x|$.
For $x>0$ and δx sufficiently small in magnitude

$$[f(x+\delta x)-f(x)]/\delta x=\delta x/\delta x=1$$

that is, $f'(x)=1$. Similarly, $f'(x)=-1$ for $x<0$. Hence, the derived function of f is given by

$$f'\colon x \mapsto \begin{cases} -1, & x<0 \\ 1, & x>0 \end{cases}$$

f' is not defined for $x=0$, that is, f is continuous but not differentiable at $x=0$.

Geometrical Interpretation of Differentiation

Let P and Q be two neighbouring points on the graph of $y=f(x)$ (see figure 3.9a) with coordinates (x_1, y_1) and (x_2, y_2), respectively, where $x_2=x_1+\delta x$ and $y_2=y_1+\delta y$. Then

$$\frac{f(x_1+\delta x)-f(x_1)}{\delta x}=\frac{\delta y}{\delta x}$$

where the ratio $\delta y/\delta x$ is the gradient of the secant PQ. As $\delta x \to 0$, Q moves along the curve towards P and in the limit coincides with P. The secant has now become a tangent (see figure 3.9b) and the derivative of f at $x=x_1$ may thus be interpreted as the *gradient of the tangent to the curve* $y=f(x)$ *at P*, that is, as the slope of the curve at P.

Differentiation and Arithmetic Operations

Differentiating complicated expressions from first principles can be very tedious. Fortunately, the process can be simplified by applying the following general rules for differentiating functions made up of sums, products or quotients of simpler functions.

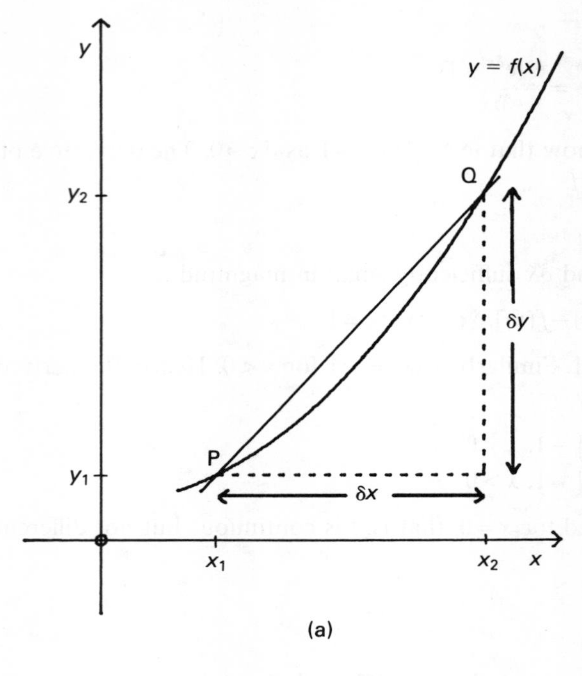

(a)

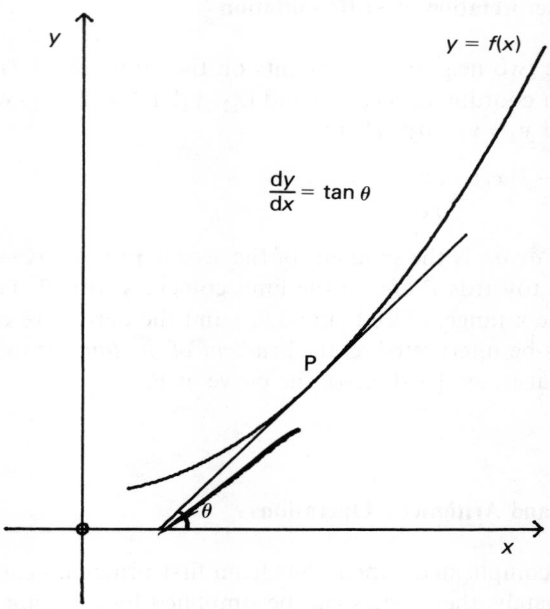

(b)

Figure 3.9

Theorem 3.4.1

(i) $(af+bg)'(x)=af'(x)+bg'(x), \forall a,b\in\mathbb{R}$.

(ii) $(fg)'(x)=f(x)g'(x)+g(x)f'(x)$.

(iii) $(f/g)'(x)=[g(x)f'(x)-f(x)g'(x)]/[g(x)]^2$, provided $g(x)\neq 0$.

Proof (i) This is established by differentiating $af+bg$ from first principles

$$(af+bg)(x+\delta x)-(af+bg)(x)=af(x+\delta x)+bg(x+\delta x)-af(x)-bg(x)$$
$$=a[f(x+\delta x)-f(x)]+b[g(x+\delta x)-g(x)]$$

Dividing throughout by δx and letting $\delta x\to 0$, gives the required result.

(ii) This is usually called the *product rule* for derivatives. It is obtained by noting that

$$\frac{f(x+\delta x)g(x+\delta x)-f(x)g(x)}{\delta x}=f(x+\delta x)\frac{g(x+\delta x)-g(x)}{\delta x}+g(x)\frac{f(x+\delta x)-f(x)}{\delta x}$$

(iii) This is called the *quotient rule* for derivatives. By differentiating from first principles, it is easy to show that

$$(1/g)'(x)=-g'(x)/[g(x)]^2, \text{ provided } g(x)\neq 0$$

The general result then follows from the product rule.

Example 3.4.2

(1) $y=5x^3+16x^2-11x+7$.
Since the derivative of x^n is nx^{n-1}

$$\frac{dy}{dx}=15x^2+32x-11$$

(2) $y=e^x \cos x$.
By the product rule

$$\frac{dy}{dx}=-e^x \sin x+e^x \cos x$$

since

$$\frac{d}{dx}e^x=e^x \text{ and } \frac{d}{dx}\cos x=-\sin x$$

(3) $y = \tan x$.
By the quotient rule

$$\frac{dy}{dx} = \frac{d}{dx}\frac{\sin x}{\cos x} = (\cos^2 x + \sin^2 x)/\cos^2 x = \sec^2 x$$

(4) $y = x^k$, where $k = -n$, $n \in N$.
By the quotient rule

$$\frac{dy}{dx} = \frac{d}{dx}(1/x^n) = -nx^{n-1}/x^{2n}$$

$$= -nx^{-n-1} = kx^{k-1}$$

Hence, the derivative of x^k is kx^{k-1}, $\forall k \in Z$. This is a generalisation of the result given in example 3.4.1(2). In fact, the result holds $\forall k \in Q$ (see exercise 3.4). For example, if $y = x^{1/2}$ then $dy/dx = \frac{1}{2}x^{-1/2}$ and if $y = x^{-1/2}$ then $dy/dx = -\frac{1}{2}x^{-3/2}$.

Differentiation of Composite Functions

If the function $g: X \to U$ is differentiable at $x \in X$ and if the function $f: U \to \mathbb{R}$ is differentiable at $y = g(x)$ then the composite function $f_0 g: X \to \mathbb{R}$ is also differentiable at x.

Let $\delta y = g(x + \delta x) - g(x)$. Then $g(x + \delta x) = y + \delta y$. Furthermore, since g is continuous at x, $\delta y \to 0$ as $\delta x \to 0$. Thus

$$\frac{f[g(x+\delta x)] - f[g(x)]}{\delta x} = \frac{f(y+\delta y) - f(y)}{\delta x}$$

$$= \frac{f(y+\delta y) - f(y)}{\delta y}\frac{\delta y}{\delta x}$$

$$= \frac{f(y+\delta y) - f(y)}{\delta y}\frac{g(x+\delta x) - g(x)}{\delta x}$$

Hence, $(f_0 g)'(x) = (f'_0 g)(x)g'(x)$, which is sometimes called the rule for *differentiating a function of a function.* If $y = g(x)$ and $z = f(y)$ then it is usual to write

$$\frac{dz}{dx} = \frac{dz}{dy}\frac{dy}{dx}$$

Example 3.4.3

(1) $h: x \mapsto e^{ax}$ where $a \in \mathbb{R}$.
Let $y = ax$, $z = e^y$. Then

$$h'(x) = \frac{dz}{dy}\frac{dy}{dx} = e^y a = ae^{ax}$$

since $y = ax$.

(2) $h: x \mapsto \sin^2 x$.
Let $y = \sin x$, $z = y^2$. Then

$$h'(x) = \frac{dz}{dy}\frac{dy}{dx} = 2y \cos x = 2 \sin x \cos x$$

$$= \sin 2x$$

A common problem requiring the rule for differentiating a function of a function is: given $f(x, y) = 0$, where $y = g(x)$, determine dy/dx.

Example 3.4.4

Determine dy/dx given $xy^3 - 3x^2 - xy - 5 = 0$.

Using the product rule, together with the rule for differentiating a function of a function gives

$$y^3 + 3xy^2\frac{dy}{dx} - 6x - y - x\frac{dy}{dx} = 0$$

that is

$$\frac{dy}{dx} = (6x - y^3 + y)/(3xy^2 - x)$$

[Note that if $z = y^3$ then $dz/dx = 3y^2(dy/dx)$.]

The most intuitively obvious rate-of-change problems are those occurring with respect to time. Such problems may be solved by applying the rule for differentiating a function of a function. The following example illustrates the general approach.

Example 3.4.5

Determine the rate at which the volume of a spherical balloon is decreasing when the radius is 10 cm, given that the radius is decreasing at a rate of 0.5 cm/sec.

If v is the function given by $v: t \mapsto 4\pi[r(t)]^3/3$, then $v(t)$ denotes the volume of the sphere at time t.

Let $y = r(t)$, $z = 4\pi y^3/3$. Then

$$v'(t)=\frac{dz}{dy}\frac{dy}{dt}=4\pi y^2 r'(t)=4\pi[r(t)]^2 r'(t)$$

But $r'(t) = -0.5$ cm/s and $r(t)=10$ cm. Hence

$$v'(t) = -200\pi \text{ cm}^3/\text{s}$$

If the derivative of a differentiable function f is known, a special form of the rule for differentiating a function of a function may be used to determine the derivative of f^{-1}, if it exists.

Let $y=f^{-1}(x)$ and $z=f(y)$. Then

$$\frac{dz}{dx}=\frac{dz}{dy}\frac{dy}{dx}$$

But $dz/dx=(f_o f^{-1})'(x)=(d/dx)x=1$. Hence

$$\frac{dy}{dx}=1\bigg/\frac{dz}{dy}$$

that is, $(d/dx)f^{-1}(x)=1/(d/dy)f(y)$, where $y=f^{-1}(x)$. Since $x=f(y)$ if $y=f^{-1}(x)$, the above rule is often written as $dy/dx=1/(dx/dy)$.

Example 3.4.6

Let f be defined by $x \mapsto e^x$. Then f^{-1} is defined by $x \mapsto \log_e x$, $x>0$. Hence

$$\frac{d}{dx}\log_e x=\frac{1}{e^y}$$

where $y=\log_e x$, that is

$$\frac{d}{dx}\log_e x=\frac{1}{x}$$

since $e^{\log_e x}=x$.

The reader should verify that the derivatives of $\sin^{-1} x$, $\cos^{-1} x$ and $\tan^{-1} x$ are $1/\surd(1-x^2)$, $-1/\surd(1-x^2)$ and $1/(1+x^2)$, respectively.

Numerical Differentiation

Example 3.4.7

The value of the derivative of $\sin x$ at $x=\pi/6$ is the limit as $\delta x \to 0$ of the expression

$$[\sin(\pi/6+\delta x)-\sin(\pi/6)]/\delta x$$

If this expression is evaluated on a calculator that displays up to eight digits but performs its arithmetic using a further three guard digits, then with $\delta x = 10^{-n}$, $n = 1, 2, \ldots$, the following results are obtained

n	
1	0.8396036
2	0.8635110
3	0.8657753
4	0.8660000
5	0.8660150
6	0.8660100

The correct value to seven significant digits is 0.8660254. On the calculator being used, only four correct significant digits can be obtained. The problem arises as a result of the subtraction of two nearly equal floating-point numbers, with many of the leading digits of the two mantissae consequently cancelling each other out. (The reader is referred to example 3.2.17 for another illustration of this behaviour.) Division of the resulting floating-point number by a very small number further compounds the loss of accuracy.

Of course, since the derivative of sin x is known to be cos x, there is no need to attempt to evaluate it for a particular value of x in the above fashion. However, in practice, an explicit expression for a particular function may not be known. All that may be available is a table of approximate values, $f_n \approx f(x_n)$, $n = 1, 2, \ldots, p$. In such a case, if the value of the derivative at one of the tabular points x_n is required, there may be no alternative but to approximate the value of $f'(x_n)$ by the value of a *difference approximation*, for example

$$f'(x_n) \approx \frac{\Delta f_n}{\Delta x_n} \equiv \frac{f_{n+1} - f_n}{x_{n+1} - x_n}$$

As illustrated in the example above, $\Delta f_n / \Delta x_n$ will often be substantially less accurate as an estimate to f' than f_n is as an estimate to f.

Repeated Differentiation

If the derivative f' of a real function f is itself differentiable at a point x in its domain, f is said to be *twice differentiable* at x. The derivative of $f'(x)$ is variously denoted by $f''(x)$, $f^{(2)}(x)$ and $(\mathrm{d}^2/\mathrm{d}x^2)f(x)$ and is called the *second derivative* of $f(x)$. This notion extends in the obvious way to derivatives higher than 2.

Definition 3.4.3
If it exists, the nth *derivative* of $f(x)$ is denoted by $f^{(n)}(x)$ or $(\mathrm{d}^n/\mathrm{d}x^n)f(x)$ and is obtained by differentiating the $(n-1)$th derivative of $f(x)$, that is

$$f^{(n)}(x)=\frac{\mathrm{d}}{\mathrm{d}x}f^{(n-1)}(x)$$

Another name for $f^{(n)}(x)$ is the *derivative of order n* for $f(x)$.

If $y=f(x)$, it is usual to write

$$\frac{\mathrm{d}^n y}{\mathrm{d}x^n}=f^{(n)}(x)$$

Example 3.4.8

1. $y=\sin x$.

$$\frac{\mathrm{d}^2 y}{\mathrm{d}x^2}=\frac{\mathrm{d}}{\mathrm{d}x}\left(\frac{\mathrm{d}y}{\mathrm{d}x}\right)=\frac{\mathrm{d}}{\mathrm{d}x}\cos x=-\sin x$$

2. $y=\mathrm{e}^x$.

$$\frac{\mathrm{d}^n y}{\mathrm{d}x^n}=\mathrm{e}^x,\ \forall n\in N$$

3. $y=4x^3-2x^2+3x+4$
$\mathrm{d}y/\mathrm{d}x=12x^2-4x+3$ and $\mathrm{d}^2y/\mathrm{d}x^2=24x-4$, while $\mathrm{d}^3y/\mathrm{d}x^3=24$ and $\mathrm{d}^4y/\mathrm{d}x^4=0$. The $(n+1)$th derivative of a polynomial of degree n is always identically zero.

Turning Points

Definition 3.4.4
If, for every point c in $[a, b]$, $f(x)<f(c)\ \forall x\in[a, c)$ and $f(x)>f(c)\ \forall x\in(c, b]$ then f is said to be *increasing* on $[a, b]$.

If, for every point c in $[a, b]$, $f(x)>f(c)\ \forall x\in[a, c)$ and $f(x)<f(c)\ \forall x\in(c, b]$ then f is said to be *decreasing* on $[a, b]$.

The slope of an increasing function is positive, while that of a decreasing function is negative (see figure 3.10). Since $f'(x)$ is equal to the slope of the curve $y=f(x)$, it follows that $f'(x)>0$ when f is increasing and $f'(x)<0$ when f is decreasing.

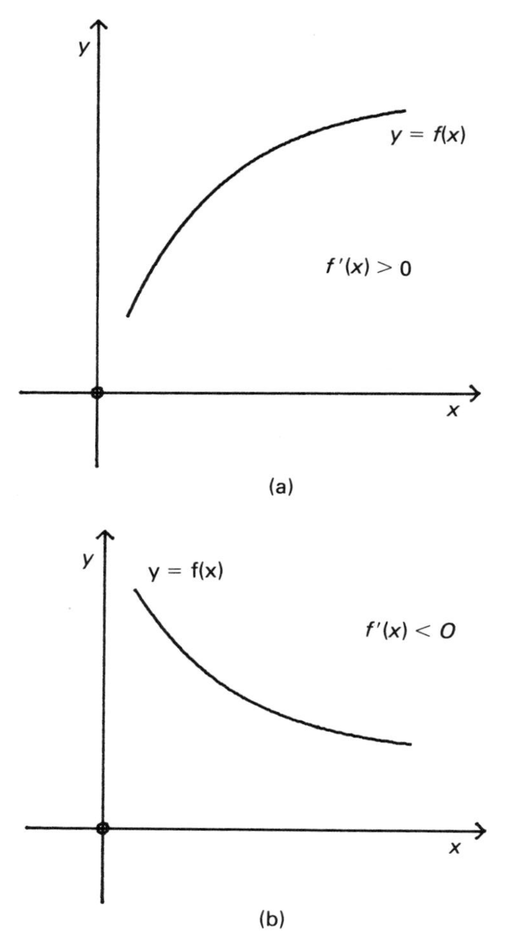

Figure 3.10 (a) The function f is increasing; (b) f is decreasing

Definition 3.4.5
The point $c \in X$ is a *maximum turning point* of a real function f: $X \to \mathbb{R}$ if there exists a number $\delta > 0$ such that

$$f(c) > f(x), \ \forall x \in [c-\delta, c+\delta], \ x \neq c$$

while c is a *minimum turning point* of f if there exists $\delta > 0$ such that

$$f(c) < f(x), \ \forall x \in [c-\delta, c+\delta], \ x \neq c$$

Suppose $x = c$ is a maximum turning point of f (see figure 3.11a). Then f is increasing for values of x close to but less than c, that is, $f'(x) > 0$ immediately to the left of c. At $x = c$ itself, the tangent to the curve $y = f(x)$ is horizontal and its gradient is thus equal to zero, that is, $f'(c) = 0$. For

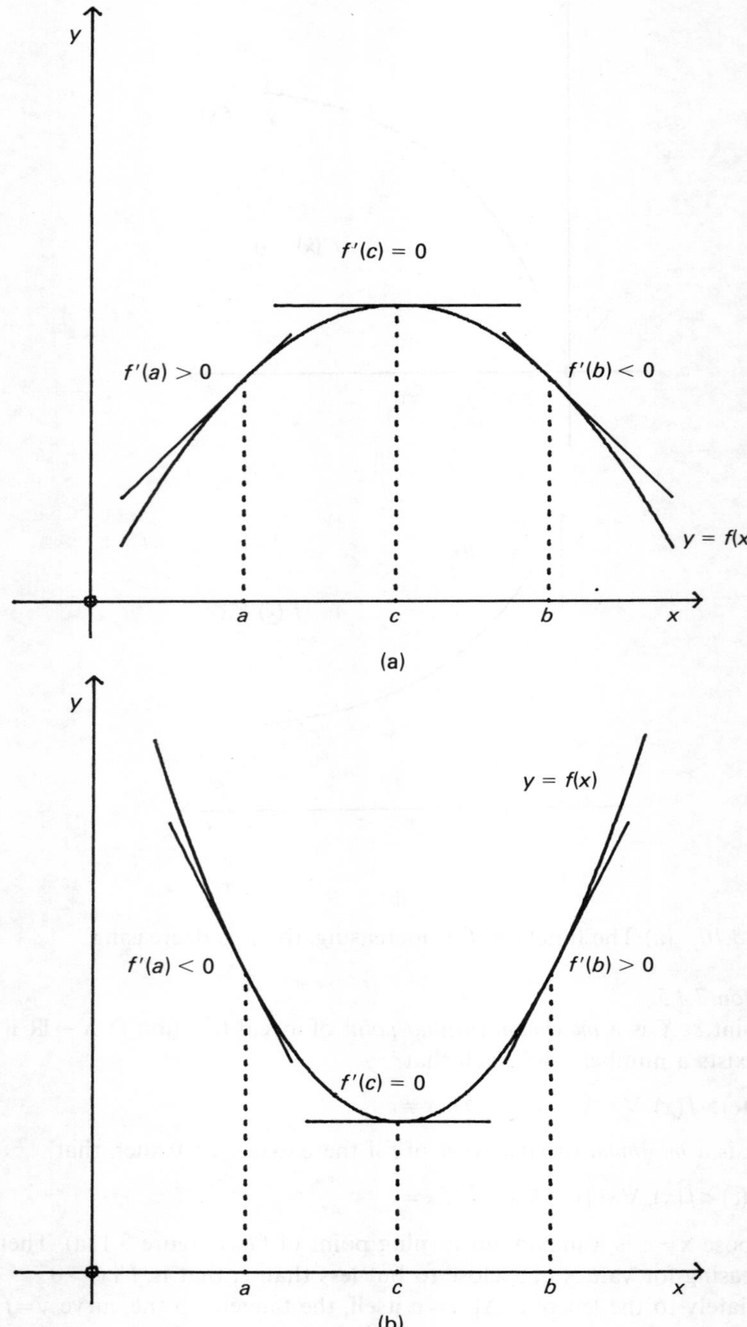

Figure 3.11 (a) Maximum turning point; (b) minimum turning point

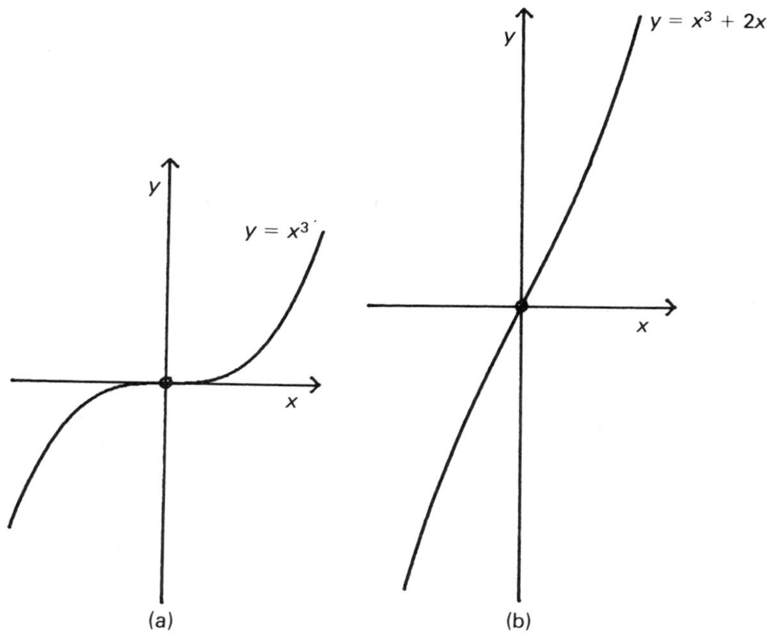

Figure 3.12 Points of inflexion

values of x close to but greater than c, f is decreasing, that is, $f'(x)<0$ for values of x immediately to the right of the turning point.

Since $f'(c^-)>0$, $f'(c)=0$ and $f'(c^+)<0$, f' is decreasing in a neighbourhood of the maximum turning point, c. Hence, $f''(c)<0$. Therefore, $x=c$ is a maximum turning point of f if

(i) $f'(c)=0$

(ii) $f''(c)<0$

Similarly, it may be shown that $x=c$ is a minimum turning point if

(i) $f'(c)=0$;

(ii) $f''(c)>0$.

In order to locate the maximum and minimum turning points of a function f it is thus first necessary to solve the non-linear equation $f'(x)=0$. Each root of this equation is then substituted into the expression for $f''(x)$ and the sign of the resulting value determined. If $f'(c)=f''(c)=0$ then c is *not* a turning point but a *point of inflexion*. At such a point, $f''(x)$ is always equal to zero but $f'(x)$ is not necessarily equal to zero. For example, if $y=x^3$ (figure 3.12a) then $dy/dx=d^2y/dx^2=0$ at $x=0$ but if $y=x^3+2x$ (figure 3.12b) then $dy/dx=2$ and $d^2y/dx^2=0$ at $x=0$. In both cases, $x=0$ is a point of inflexion. Note that the curve $y=f(x)$ does not change direction when it passes through a point of inflexion. At such a point, the *curvature*, given by $y''(x)/\{1+[y'(x)]^2\}^{3/2}$, is zero.

Example 3.4.9

The speed at which information can be transmitted along a copper cable is proportional to $v(x)=-x^2 \log_e x$, where $x>0$ denotes the ratio of the radius of the core to the thickness of the insulation. It is required to determine the value of x which maximises the transmission speed.

$$v'(x)=-2x \log_e x - x, \quad v''(x)=-2 \log_e x - 3$$

Now $v'(x)=0$ when $x=e^{-1/2}$, and for $x=e^{-1/2}$, $v''(x)=-2$, which is less than zero. Thus $x=e^{-1/2}$ is a maximum turning point and the maximum transmission speed is attained when x has this value.

Curve Sketching

It is possible to sketch the general shape of a function f provided the following pieces of information can be determined:

(a) the location of any zeros of f;

(b) the location of any turning points and points of inflexion of f;

(c) any values of x for which f is undefined, in particular, any values of x for which $f(x)$ is infinite;

(d) the behaviour of f as $x \to \pm\infty$.

Example 3.4.10

Sketch the general shape of the curve $y=x^2/(x+1)$.

(a) $y=0$ when $x=0$.

(b) $dy/dx=(x^2+2x)/(x+1)^2$; hence, $dy/dx=0$ when $x=0$ or $x=-2$.

$$\frac{d^2y}{dx^2}=[(2x+2)(x+1)^2-2(x^2+2x)(x+1)]/(x+1)^4$$

Hence, $d^2y/dx^2=2$ when $x=0$, that is, $x=0$ is a minimum turning point; and $d^2y/dx^2=-2$ when $x=-2$, that is, $x=-2$ is a maximum turning point. When $x=0$, $y=0$; when $x=-2$, $y=-4$.

(c) As x approaches -1 through values to the left of -1, y becomes large negative, that is, as $x \to -1^-$, $y \to -\infty$. Similarly, $y \to \infty$ as $x \to -1^+$.

(d) $y \to \infty$ as $x \to \infty$, $y \to -\infty$ as $x \to -\infty$.
Figure 3.13 is the required sketch.

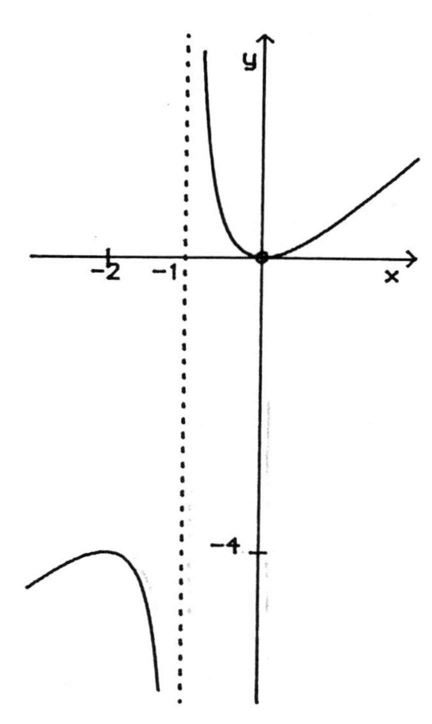

Figure 3.13 Example 3.4.10: sketch of $y = x^2/(x+1)$

Except in simple cases, solving the equations $f(x) = 0$ and $f'(x) = 0$ for the zeros and turning points of f, respectively, would require a relatively large amount of computational effort. This being the case, many computer graph-plotting packages, such as the one used to generate most of the diagrams in this book, simply draw curves passing through a given set of points in the x,y plane.

Some Important Theorems

Theorem 3.4.2 Rolle's theorem
If a function f is continuously differentiable in the open interval (a, b) and if $f(a) = f(b) = 0$ then there is a value $c \in (a, b)$ such that $f'(c) = 0$.

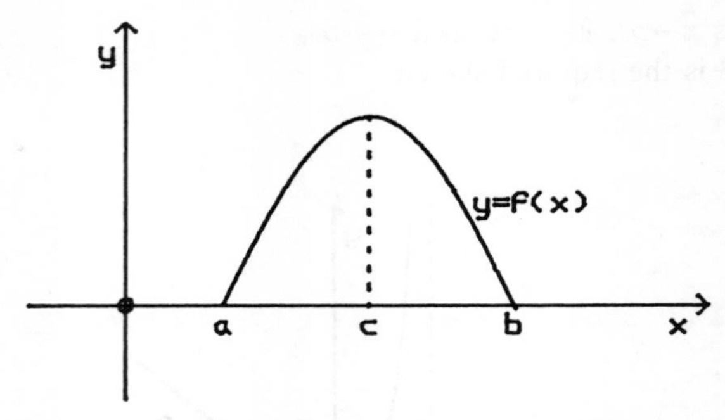

Figure 3.14 Theorem 3.4.2: $f(a)=f(b)=0$; at some point $c\in(a, b)$, the gradient of the curve must be zero

Figure 3.14 serves to illustrate the truth of this theorem. For a formal proof, the interested reader is referred to Burkill (1962).

Theorem 3.4.3 The Mean Value Theorem
If f is continuously differentiable in (a, b) then there is a value $c\in(a, b)$ such that

$$\frac{f(b)-f(a)}{b-a}=f'(c)$$

Proof Let

$$g(x)\equiv f(b)-f(x)-\frac{(b-x)}{(b-a)}[f(b)-f(a)]$$

Then $g(a)=g(b)=0$ and so, by Rolle's theorem, $g'(c)=0$ for some $c\in(a, b)$, that is

$$-f'(c)+\frac{1}{(b-a)}[f(b)-f(a)]=0$$

as required.

Example 3.4.11

Let f be defined by $x\mapsto x(x-1)(x-2)$ and consider $x\in[0, 1/2]$. Determine the value(s) of c in the mean value theorem.

$$\frac{f(b)-f(a)}{b-a}=\frac{(1/2)(-1/2)(-3/2)-0}{(1/2)-0}=\frac{3}{4}$$

$$f'(x)=3x^2-6x+2$$

The required value c satisfies

$$3c^2-6c+2=3/4$$

that is

$$3c^2-6c+\frac{5}{4}=0$$

Hence

$$c=1\pm\sqrt{(21)}/6=1.7637626 \text{ or } 0.2362374$$

Theorem 3.4.4 Taylor's Theorem
If the nth derivative of f is continuously differentiable in the interval $(a, a+h)$ then

$$f(a+h)=f(a)+hf'(a)+h^2\,\frac{f''(a)}{2!}+\ldots+h^n\,\frac{f^{(n)}(a)}{n!}+h^{n+1}\,\frac{f^{(n+1)}(\theta)}{(n+1)!}$$

where $\theta\in(a, a+h)$.

Proof The proof is given for the case $n=1$. The approach for a general value of n is similar and is left as an exercise for the reader. For $n=1$, it is required to prove that

$$f(a+h)=f(a)+hf'(a)+h^2\,\frac{f''(\theta)}{2!},\ \theta\in(a, a+h)$$

Let

$$g(x)\equiv f(a+h)-f(x)-(a+h-x)f'(x)$$

and

$$h(x)\equiv g(x)-\frac{(a+h-x)^2}{h^2}\,g(a)$$

Then $g(a+h)=0$ and $h(a)=h(a+h)=0$. By Rolle's theorem, $h'(\theta)=0$ for some value of $\theta\in(a, a+h)$, that is

$$g'(\theta)+\frac{2(a+h-\theta)}{h^2}g(a)=0$$

Thus

$$g(a)=\frac{-h^2g'(\theta)}{2(a+h-\theta)}=\frac{h^2f''(\theta)}{2}$$

since $g'(x)=-f'(x)-(a+h-x)f''(x)+f'(x)$. Hence

$$f(a+h)-f(a)-hf'(a)=h^2\,\frac{f''(\theta)}{2},\ \theta\in(a, a+h)$$

as required.

Putting $h=x-a$ in Taylor's theorem gives

$$f(x)=f(a)+(x-a)f'(a)+(x-a)^2\ \frac{f''(a)}{2!}+\ldots+(x-a)^n\ \frac{f^{(n)}(a)}{n!}+R_n(x)$$

where $R_n(x)=(x-a)^{n+1}f^{(n+1)}(\theta)/(n+1)!$, and θ lies between x and a.

Definition 3.4.6
The polynomial $c_0+c_1(x-a)+c_2(x-a)^2+\ldots+c_n(x-a)^n$, where $c_k=f^{(k)}(a)/k!$, $k=0, 1, \ldots, n$, is called the *Taylor polynomial approximation of degree n to f*; the polynomial is said to be *expanded about* $x=a$.

If $R_n(x)\to 0$ as $n\to\infty$ then the infinite series

$$\sum_{k=0}^{\infty} c_k(x-a)^k$$

converges to $f(x)$ and is called the *Taylor series* for $f(x)$, expanded about $x=a$.

In particular, if $a\equiv 0$

$$f(x)=f(0)+xf'(0)+x^2\ \frac{f''(0)}{2!}+\ldots+x^n\ \frac{f^{(n)}(0)}{n!}+R_n(x)$$

Example 3.4.12

(1) $f: x\mapsto e^x$, $f^{(k)}(x)=e^x$, $\forall k\in N$, and so $f^{(k)}(0)=1$, $k=0, 1, 2, \ldots$.
If $x\in(-r, r)$, for any *fixed value* $r\geqslant 0$, then

$$R_n(x)\leqslant r^{n+1}e^r/(n+1)!,$$

which tends to 0 as $n\to\infty$. Hence

$$e^x=1+x+\frac{x^2}{2!}+\ldots$$

which is the power series used to define the exponential function in definition 3.3.8.

(2) $f: x\mapsto \sin x$, $f(0)=\sin 0=0$, $f'(0)=\cos 0=1$, $f''(0)=-\sin 0=0$, $f'''(0)=-\cos 0=-1$, etc. The Taylor polynomial approximation of degree $2n+1$ to $\sin x$ is thus

$$\sin x\approx x-\frac{x^3}{3!}+\frac{x^5}{5!}-\ldots\frac{(-1)^n x^{2n+1}}{(2n+1)!}$$

The corresponding Taylor series converges for all values of x, that is, it is *uniformly convergent* $\forall x\in\mathbb{R}$. The Taylor series for $\cos x$ is

$$\cos x = 1 - \frac{x^2}{2!} + \frac{x^4}{4!} - \ldots$$

which is also uniformly convergent $\forall x \in \mathbb{R}$.

(3) $f: x \mapsto \log_e(1+x)$, $f'(x) = 1/(1+x)$, $f^{(k)}(x) = (-1)^{k-1}(k-1)!/(1+x)^k$, $k \geqslant 2$. Hence, $f(0) = 0$, $f'(0) = 1$, $f''(0) = -1$, $f'''(0) = 2!$, $f^{iv}(0) = -3!$, etc. The Taylor polynomial approximation of degree n for $\log_e(1+x)$ is thus

$$\log_e(1+x) \approx x - \frac{x^2}{2} + \frac{x^3}{3} - \frac{x^4}{4} + \ldots + \frac{(-1)^{n-1}x^n}{n}$$

Since $|R_n(x)| \leqslant 1/(n+1)$ provided $x \in (-1, 1]$, the corresponding Taylor series is uniformly convergent on this interval.

Exercise 3.4

1. Differentiate from first principles each of the following functions.

(a) $x \mapsto \dfrac{3+x}{3-x}$;

(b) $x \mapsto \sqrt{x}$ [hint: $(\sqrt{(x+\delta x)} - \sqrt{x})(\sqrt{(x+\delta x)} + \sqrt{x}) = \delta x$];

(c) $x \mapsto \cos x$ {hint: use the trigonometric identity $\cos y - \cos x = -2 \sin[\frac{1}{2}(y+x)] \sin[\frac{1}{2}(y-x)]$}.

2. Differentiate each of the following functions.

(a) $x \mapsto 5x^4 + 7x^3 + 2x - 3$;

(b) $x \mapsto (x-2)(x+3)(x^2-1)$;

(c) $x \mapsto (x \sin x)/(1-x^2)$;

(d) $x \mapsto \tan^2 x$;

(e) $x \mapsto \sqrt{(1-x^2)}$;

(f) $x \mapsto \tan^{-1} x$, $x \in [-1, 1]$.

3. Determine dy/dx given $e^{xy} + y \log_e x = \cos 2x$.

4. Show that

$$\frac{d}{dx} x^{1/q} = \frac{1}{q} x^{(1-q)/q}$$

[Hint: if $y=x^{1/q}$ then $y^q=x$.] Hence, show that

$$\frac{d}{dx}x^{p/q}=\frac{p}{q}x^{(p-q)/q},\ \forall p,\ q\in N,\ q\neq 0$$

5. Determine the turning points of each of the following functions. Establish whether each turning point is a maximum or a minimum and determine the corresponding maximum or minimum value.

(a) $x\mapsto x^3-6x^2+11x-6$;

(b) $x\mapsto x/(1+x^2)$;

(c) $x\mapsto xe^{-x}$.

6. A spherical balloon is being inflated at the rate of 100 cm^3 of air per second. Determine the rate at which the radius is increasing when the radius is 5 cm.

7. The cost c in pounds per mile of an electric cable is given by $c=(120/x)+600x$, where x is the cross-section, in square inches, of the cable. Determine

(a) the cross-section which minimises the cost of the cable;

(b) the least-cost per mile.

8. Using a scientific calculator, evaluate the expression

$$\frac{\tan\left(\frac{\pi}{6}+\delta x\right)-\tan\frac{\pi}{6}}{\delta x}$$

for $\delta x=10^{-n}$, $n=1, 2, \ldots$ Compare the values obtained with $\sec^2(\pi/6)$.

9. Sketch the general shape of each of the curves given by the following equations.

(a) $y=(x-1)(x-2)(x-3)$;

(b) $y=x^2/(x+1)$;

(c) $y=(2x-5)(x-3)/(x-2)(x-4)$.

10. Determine the Taylor polynomial approximation of degree $2n-1$ to $\tan^{-1} x$. Show that the corresponding Taylor series is uniformly convergent $\forall x\in[-1, 1]$.

3.5 INTEGRATION

The Indefinite Integral

The fundamental problem of integration is to determine a function given its derivative.

Definition 3.5.1
An *indefinite integral* of $f(x)$ with respect to x is denoted by $\int f(x)\mathrm{d}x$. If $F'(x)=f(x)$, then

$$\int f(x)\mathrm{d}x = F(x) + c$$

where the *constant of integration* c can be any real number. The process of determining $\int f(x)\mathrm{d}x$ for a given function f is called *integration* and $f(x)$ is called the *integrand*.

Clearly, the derivative of $\int f(x)\mathrm{d}x$ is $f(x)$, since $F'(x)=f(x)$ and the derivative of a constant is zero. The integral is 'indefinite' in the sense that it is only determined up to the addition of an arbitrary constant.

While it is usually a relatively straightforward process to differentiate a given function, it is generally difficult, and sometimes impossible, to express a given indefinite integral in terms of known elementary functions. Indeed, integration can lead to the definition of new functions.

Certain indefinite integrals, however, may be expressed in terms of elementary functions by recognising the integrands as the derivatives of particular functions. Such integrands are known as *standard forms*. Each of the results in the following examples may be checked by differentiating both sides. In every example of an indefinite integral given in this section, an arbitrary constant of integration may be added to the expression on the right-hand side.

Example 3.5.1

(1) $\int x^n \mathrm{d}x = x^{n+1}/(n+1),\ n \neq -1$

(2) $\int \frac{1}{x}\,\mathrm{d}x = \log_e x$

(3) $\int \cos x\,\mathrm{d}x = \sin x$

(4) $\int \sin x\,\mathrm{d}x = -\cos x$

(5) $\int \mathrm{e}^x\,\mathrm{d}x = \mathrm{e}^x$

(6) $\int \sec^2 x\,\mathrm{d}x = \tan x$

(7) $\int \frac{1}{\sqrt{(1-x^2)}}\,\mathrm{d}x = \sin^{-1} x$

(8) $\int \frac{1}{1+x^2}\,\mathrm{d}x = \tan^{-1} x$

Integration and Arithmetic Operations

Theorem 3.5.1

(i) $\int [af(x)+bg(x)]\mathrm{d}x = a\int f(x)\,\mathrm{d}x + b\int g(x)\,\mathrm{d}x,\ \forall a, b \in \mathbb{R}$;

(ii) $\int f'(x)g(x)\,\mathrm{d}x = f(x)g(x) - \int f(x)g'(x)\,\mathrm{d}x$.

Proof (i) Suppose $F'(x)=f(x)$ and $G'(x)=g(x)$. Then

$$\begin{aligned}\int [af(x)+bg(x)]\mathrm{d}x &= \int [aF'(x)+bG'(x)]\mathrm{d}x\\ &= \int (aF+bG)'(x)\,\mathrm{d}x\end{aligned}$$

by theorem 3.4.1. Hence

$$\begin{aligned}&= aF(x)+bG(x)\\ &= a\int f(x)\,\mathrm{d}x + b\int g(x)\,\mathrm{d}x\end{aligned}$$

(ii) This is the rule for *integration by parts*. By the product rule for derivatives (see theorem 3.4.1)

$$(fg)'(x) = f(x)g'(x) + g(x)f'(x)$$

Integrating both sides gives

$$f(x)g(x) = \int f(x)g'(x)\,\mathrm{d}x + \int g(x)f'(x)\,\mathrm{d}x$$

from which the required result follows.

Example 3.5.2

(1)

$$\begin{aligned}\int (5x^3+4x^2+7)\mathrm{d}x &= 5\int x^3\,\mathrm{d}x + 4\int x^2\,\mathrm{d}x + 7\int 1\,\mathrm{d}x\\ &= 5x^4/4 + 4x^3/3 + 7x\end{aligned}$$

(2) Integrating xe^x by parts, with $f'(x)=e^x$ and $g(x)=x$, gives

$$\int xe^x\,\mathrm{d}x = xe^x - \int e^x\,\mathrm{d}x = xe^x - e^x$$

(3) Integrating $\log_e x$ by parts, with $f'(x)=1$ and $g(x)=\log_e x$, gives

$$\int \log_e x \, dx = x \log_e x - \int x \frac{1}{x} \, dx = x \log_e x - x$$

(4) Let $I_n = \int \sin^n x \, dx$, $n \in N$ and $n > 1$. Integrating $\sin^n x$ by parts, with $f'(x) = \sin x$ and $g(x) = \sin^{n-1} x$, gives

$$\begin{aligned} I_n &= -\cos x \sin^{n-1} x + \int \cos x \times (n-1) \sin^{n-2} x \cos x \, dx \\ &= -\cos x \sin^{n-1} x + \int (1 - \sin^2 x)(n-1) \sin^{n-2} x \, dx \\ &= -\cos x \sin^{n-1} x + (n-1) I_{n-2} - (n-1) I_n \end{aligned}$$

Hence

$$I_n = (-\cos x \sin^{n-1} x + (n-1) I_{n-2})/n$$

Such a recursive formula is known as a *reduction formula*. By applying it successively it is possible to express I_n in terms of $I_1 = \int \sin x \, dx = -\cos x$, if n is odd and in terms of $I_0 = \int 1 \, dx = x$, if n is even.

When attempting to integrate a given function by parts, the factor $f'(x)$ should be 'easy' to integrate and the derivative of the factor $g(x)$ should be 'simpler' than $g(x)$ itself. It is hoped that the above examples clarify this principle.

Integrating Fractional Integrands

The result given in example 3.5.1(2) may be generalised to give a rule which can sometimes be applied to integrate fractional integrands. If $z = \log_e y$ and $y = f(x)$ then

$$\frac{dz}{dx} = \frac{dz}{dy}\frac{dy}{dx} = \frac{1}{y} \; f'(x) = \frac{f'(x)}{f(x)}$$

Hence

$$\int \frac{f'(x)}{f(x)} \, dx = \log_e f(x)$$

Example 3.5.3

(1) $\displaystyle \int \frac{x}{x^2+3} \, dx = \frac{1}{2} \int \frac{2x}{x^2+3} \, dx = \frac{1}{2} \log_e (x^2+3)$

(2) $\displaystyle \int \tan x \, dx = \int \frac{\sin x}{\cos x} \, dx = -\int \frac{-\sin x}{\cos x} \, dx = -\log_e \cos x = \log_e (\sec x)$

When the integrand is a rational function (see definition 3.1.4), with the degree of the denominator greater than that of the numerator, the *method of partial fractions* may be used.

Example 3.5.4

(1) Integrate $1/(x^2-a^2)$.

$$\frac{1}{x^2-a^2}=\frac{1}{(x+a)(x-a)}=\frac{A}{(x-a)}+\frac{B}{(x+a)}$$

Thus $1=A(x+a)+B(x-a)$, where A and B are constants to be determined. Putting $x=a$ gives $A=1/2a$, putting $x=-a$ gives $B=-1/2a$. $1/(x^2-a^2)$ may thus be *resolved into the partial fractions*

$$\frac{1}{2a(x-a)} \text{ and } \frac{-1}{2a(x+a)}$$

Hence

$$\int\frac{1}{x^2-a^2}\,dx=\frac{1}{2a}\left[\int\frac{1}{(x-a)}\,dx-\int\frac{1}{x+a}\,dx\right]=\frac{1}{2a}\log_e\frac{(x-a)}{(x+a)}$$

(2) Integrate $(3x+1)/(x+1)^3$.

$$\frac{3x+1}{(x+1)^3}=\frac{A}{(x+1)}+\frac{B}{(x+1)^2}+\frac{C}{(x+1)^3}$$

Thus $3x+1=A(x+1)^2+B(x+1)+C=Ax^2+(2A+B)x+(A+B+C)$. Comparing coefficients of powers of x on each side of this identity gives

$$A+B+C=1,\ 2A+B=3 \text{ and } A=0$$

from which $A=0$, $B=3$, $C=-2$. Hence

$$\int\frac{3x+1}{(x+1)^3}\,dx=\int\frac{3}{(x+1)^2}\,dx+\int\frac{-2}{(x+1)^3}\,dx=\frac{-3}{x+1}+\frac{1}{(x+1)^2}$$

In general, a factor of the form $(x-a)^r$ in the denominator of a rational function gives rise to a sum of partial fractions of the form

$$\frac{A_1}{(x-a)}+\frac{A_2}{(x-a)^2}+\ldots+\frac{A_r}{(x-a)^r}$$

(3) Integrate $(x-1)/[(x+1)(x^2+1)]$.

$$\frac{x-1}{(x+1)(x^2+1)}=\frac{A}{(x+1)}+\frac{Bx+C}{(x^2+1)}$$

Thus, $x-1=(A+B)x^2+(B+C)x+(A+C)$. Comparing coefficients of powers of x gives

$$A+C=-1,\ B+C=1,\ A+B=0$$

from which $A=-1$, $B=1$ and $C=0$. Hence

$$\int\frac{x-1}{(x+1)(x^2+1)}\,\mathrm{d}x=\int\frac{-1}{(x+1)}\,\mathrm{d}x+\int\frac{x}{x^2+1}\,\mathrm{d}x$$
$$=-\log_e(x+1)+\tfrac{1}{2}\log_e(x^2+1)$$

In general, if a quadratic factor ax^2+bx+c in the denominator of a rational function cannot be factorised into real factors, then it gives rise to a partial fraction of the form $(Bx+C)/(ax^2+bx+c)$.

Integration by Substitution

If $\int f(x)\mathrm{d}x$ cannot be expressed immediately in terms of known elementary functions, it may be possible to transform the integrand into a more amenable form by a *change of variable*, that is, by putting $x=g(t)$, where g is some function.

Theorem 3.5.2
If f: $X\to\mathbb{R}$ is continuous $\forall x\in X$ and if g: $T\to X$ is differentiable $\forall t\in T$ and is a bijection, then

$$\int f(x)\,\mathrm{d}x=\int(f_0g)(t)g'(t)\,\mathrm{d}t$$

where $t=g^{-1}(x)$.

Proof Suppose $F'(x)=f(x)$, $\forall x\in X$. By the rule for differentiating a function of a function

$$(F_0g)'(t)=(F'_0g)(t)g'(t)=(f_0g)(t)g'(t)$$

Integrating both sides with respect to t

$$(F_0g)(t)=\int(f_0g)(t)g'(t)\,\mathrm{d}t$$

But $(F_0g)(t)=F(x)$, since $x=g(t)$. Hence

$$\int f(x)\,\mathrm{d}x=F(x)=\int(f_0g)(t)g'(t)\,\mathrm{d}t$$

Since $\mathrm{d}x/\mathrm{d}t=g'(t)$, the rule for integration by substitution may also be written in the form

$$\int f(x)\,\mathrm{d}x=\int f[g(t)]\frac{\mathrm{d}x}{\mathrm{d}t}\,\mathrm{d}t$$

Example 3.5.5

(1) Determine $\int x \sin x^2 \, dx$. Let $x = t^{1/2}$, then $dx/dt = \frac{1}{2}t^{-1/2}$. Hence

$$\int x \sin x^2 \, dx = \frac{1}{2}\int \sin t \, dt = -\frac{1}{2} \cos t = -\frac{1}{2} \cos x^2$$

(2) Determine $\int \operatorname{cosec} x \, dx$. Let $x = 2 \tan^{-1} t$. Then

$$\operatorname{cosec} x = \frac{1 + \tan^2 \frac{1}{2}x}{2 \tan \frac{1}{2}x} = \frac{1+t^2}{2t}$$

and $dx/dt = 2/(1+t^2)$. Hence

$$\int \operatorname{cosec} x \, dx = \int \frac{1}{t} \, dt = \log_e t = \log_e \tan \frac{1}{2}x$$

The integral of any rational function of $\cos x$ and $\sin x$ can always be transformed into the integral of a simple rational function of $t = \tan x/2$.

The Definite Integral

Definition 3.5.2
If it exists, the *definite integral*

$$\int_a^b f(x) \, dx$$

of a real function f between $x = a$ and $x = b$ $(b \geqslant a)$ is a real number satisfying

$$\int_a^b f(x) \, dx = [F(x)]_a^b \equiv F(b) - F(a)$$

where F is any indefinite integral of f. a and b are called the *lower limit* and the *upper limit*, respectively, of the *range of integration*, $[a, b]$.

While this definition has the virtue of being very straightforward, it is a little unsatisfactory in that it defines the definite integral in terms of its most fundamental property. A second, more formal definition, from which this fundamental property may be deduced, is given in definition 3.5.4.

Example 3.5.6

(1) $$\int_0^1 x\,dx=\left[\frac{1}{2}x^2\right]_0^1=\frac{1}{2}-0=\frac{1}{2}.$$

(2) $$\int_{-1}^1 e^x\,dx=[e^x]_{-1}^1=e-e^{-1}.$$

(3) $$\int_0^{\pi/2} \sin x\,dx=[-\cos x]_0^{\pi/2}=-\cos\frac{1}{2}\pi+\cos 0=1.$$

Properties of the Definite Integral

Theorem 3.5.3
If $f\in C[a, b]$ then

(i) $$\int_a^a f(x)\,dx=0,\ \forall a\in\mathbb{R};$$

(ii) $$\int_a^b f(x)\,dx=-\int_b^a f(x)\,dx;$$

(iii) $$\int_a^b f(x)\,dx=\int_a^c f(x)\,dx+\int_c^b f(x)\,dx,\ \forall c\in[a, b];$$

(iv) $$\int_a^b kf(x)\,dx=k\int_a^b f(x)\,dx,\ \forall k\in\mathbb{R};$$

(v) $$\int_a^b f(x)\,dx=(b-a)f(\zeta),\ \text{for some } \zeta\in(a, b);$$

(vi) if $f(x)\geqslant 0,\ \forall x\in[a, b]$, then $\int_a^b f(x)\,dx\geqslant 0;$

(vii) if g is continuous and one-signed in $[a, b]$ then

$$\int_a^b f(x)g(x)\,dx = f(\zeta)\int_a^b g(x)\,dx$$

for some $\zeta \in (a, b)$.

Proof Parts (i) to (iv) of the theorem follow immediately from definition 3.5.2.

(v) This is known as the *integral mean value theorem*. Suppose that $F'(x) = f(x)$, $\forall x \in [a, b]$. By the (differential) mean value theorem (theorem 3.4.3)

$$F(b) - F(a) = (b - a)F'(\zeta)$$

that is

$$\int_a^b f(x)\,dx = (b - a)f(\zeta)$$

for some $\zeta \in (a, b)$.

(vi) Let m denote the least value of $f(x)$ in the interval $[a, b]$. Since $f(x) \geqslant 0$ $\forall x \in [a, b]$, $m \geqslant 0$. Thus

$$\int_a^b f(x)\,dx = (b - a)f(\zeta) \geqslant m(b - a) \geqslant 0$$

Furthermore, since

$$\int_a^b f(x)\,dx = -\int_a^b [-f(x)]\,dx$$

if $f(x) \leqslant 0$ $\forall x \in [a, b]$ then

$$\int_a^b f(x)\,dx \leqslant 0$$

More generally, if $f(x)$ R $g(x)$, $\forall x \in [a, b]$, where R denotes $<$, $\leqslant$, $>$ or $\geqslant$, then it is straightforward to show that

$$\int_a^b f(x)\,dx \ \mathrm{R} \int_a^b g(x)\,dx$$

(vii) This is known as the *generalised integral mean value theorem.* Without loss of generality, assume $g(x) \geqslant 0, \forall x \in [a, b]$. Define

$$w(t) \equiv f(t) \int_a^b g(x)\,\mathrm{d}x - \int_a^b f(x)g(x)\,\mathrm{d}x$$

$$= \int_a^b [f(t) - f(x)]g(x)\,\mathrm{d}x$$

where $t \in [a, b]$. The function w: $t \mapsto w(t)$ is of the form $t \mapsto (\alpha f(t) + \beta)$, where α and β are constants, and is thus continuous in $[a, b]$. Let $t = t_m$ be such that $f(t_m) = m \leqslant f(t), \forall t \in [a, b]$. Then, from (vi)

$$w(t_m) = \int_a^b [m - f(x)]g(x)\,\mathrm{d}x \leqslant 0$$

since $[m - f(x)]g(x) \leqslant 0, \forall x \in [a, b]$. Similarly, if $t = t_M$ such that $f(t_M) = M \geqslant f(t), \forall t \in [a, b]$, then

$$w(t_M) = \int_a^b [M - f(x)]g(x)\,\mathrm{d}x \geqslant 0,$$

since $[M - f(x)]g(x) \geqslant 0, \forall x \in [a, b]$. Thus w changes sign in $[a, b]$ and, by the intermediate value property (theorem 3.3.2), must have a zero in (a, b): that is, there exists $\zeta \in (a, b)$ such that

$$w(\zeta) = f(\zeta) \int_a^b g(x)\,\mathrm{d}x - \int_a^b f(x)g(x)\,\mathrm{d}x = 0$$

The Definite Integral as an Area

Historically, the definite integral was developed in an attempt to formalise the intuitive but surprisingly difficult concept of 'area'. Consider the region ABCD in figure 3.15.

Let P be an arbitrary point (x, y) on the curve $y = f(x)$, lying between D and C, and let P′ be a neighbouring point on the curve, with coordinates $(x + \delta x, y + \delta y)$. The area $\mathscr{A}$ of the region ARPD depends on the position of the vertical line RP, which is itself determined by the value of x. Thus $\mathscr{A}$ is a function of x. Suppose

$$\mathscr{A} = G(x)$$

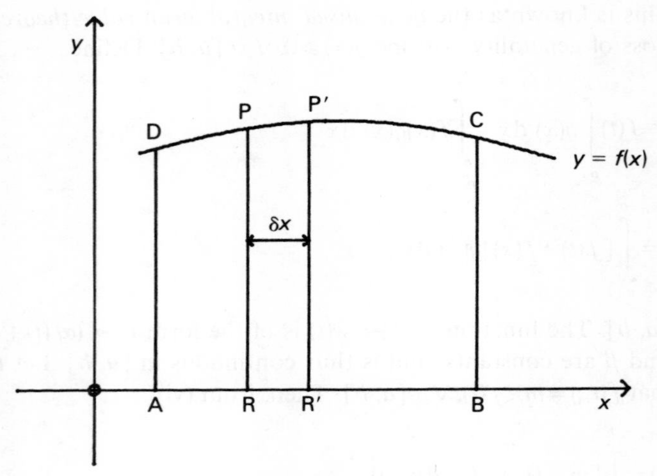

Figure 3.15

then the area $\delta\mathscr{A}$ of the region RR′P′P is given by

$$\delta\mathscr{A}=G(x+\delta x)-G(x)$$

If m and M denote the least value and the greatest value, respectively, of $f(x)$ in $[x, x+\delta x]$, then

$$m\ \delta x\leqslant\delta\mathscr{A}\leqslant M\ \delta x$$

that is

$$m\ \delta x\leqslant G(x+\delta x)-G(x)\leqslant M\ \delta x$$

Assuming f is continuous, $\delta y\to 0$ as $\delta x\to 0$. Both m and M therefore tend to y as $\delta x\to 0$ and thus

$$\frac{G(x+\delta x)-G(x)}{\delta x}\to y \text{ as } \delta x\to 0$$

that is

$$G'(x)=f(x)$$

Thus, $G(x)$ is an indefinite integral of $f(x)$. Suppose $F(x)$ is any other indefinite integral of $f(x)$. Then $F(x)$ and $G(x)$ are related by

$$G(x)=F(x)+c$$

where c is some constant. If A is the point $(a, 0)$ then $G(a)=0$, since the area of ARPD is zero when RP coincides with AD. Hence

$$0=F(a)+c$$

and so

$$G(x)=F(x)-F(a)$$

If B is the point $(b, 0)$, the area of ABCD is given by

$$G(b)=F(b)-F(a)$$

But

$$F(b)-F(a)=\int_a^b f(x)\,dx$$

and the definite integral of $f(x)$ between $x=a$ and $x=b$ may thus be interpreted as the area between the curve $y=f(x)$, the x axis and the lines $x=a$, $x=b$. This interpretation needs to be qualified slightly if f has a zero in the interval (a, b); see example 3.5.7(2).

Example 3.5.7

(1) Determine the area between the curve $y=x^2/2$, the x axis and the lines $x=0$, $x=2$ (figure 3.16a).

$$\text{Area AOB}=\int_0^2 \frac{1}{2}x^2\,dx=\left[\frac{x^3}{6}\right]_0^2=\frac{4}{3}\text{ square units}$$

(2) Determine the area enclosed between the curve $y=x^2-3x+2$, the x axis and the lines $x=0$, $x=2$.

The general shape of the curve may be sketched by applying the curve-sketching procedure described in section 3.4. Part of the required area lies beneath the x axis (figure 3.16b). By theorem 3.5.3(vi)

$$\int_0^1 (x^2-3x+2)dx\geqslant 0$$

and

$$\int_1^2 (x^2-3x+2)dx\leqslant 0$$

Since an area cannot be negative, the required area is obtained by changing the sign of the negative definite integral and adding it to the positive definite integral. The area is thus given by

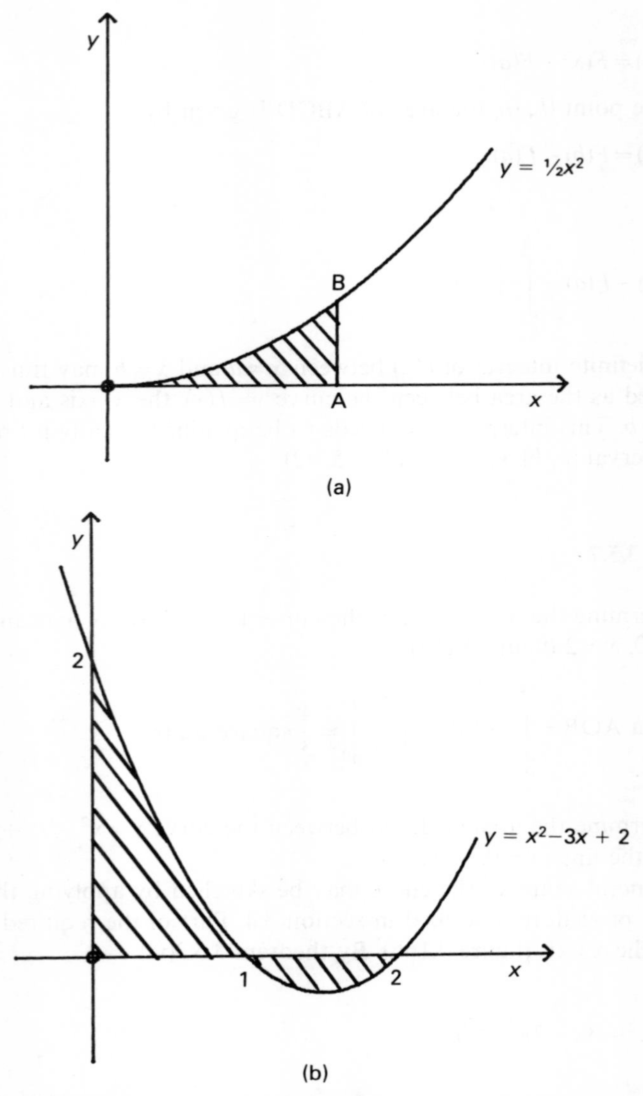

Figure 3.16 (a) Example 3.5.7(1); (b) example 3.5.7(2): the shaded regions represent the desired areas

$$\int_0^1 (x^2-3x+2)\mathrm{d}x + \left[-\int_1^2 (x^2-3x+2)\mathrm{d}x\right] = \frac{5}{6}+\frac{1}{6} = 1 \text{ square unit}$$

Note, however, that

$$\int_0^2 (x^2 - 3x + 2)\mathrm{d}x = \frac{2}{3}$$

(that is, $5/6 - 1/6$).

(3) Using calculus, show that the area of a circle of radius r is πr^2.

The equation of a circle of radius r is $x^2 + y^2 = r^2$. Its area is thus equal to

$$4\int_0^r \sqrt{(r^2 - x^2)}\mathrm{d}x$$

Let $x = r \sin \theta$. Then $\sqrt{(r^2 - x^2)} = r \cos \theta$ and $\mathrm{d}x/\mathrm{d}\theta = r \cos \theta$. Furthermore, when $\theta = 0$, $x = r \sin \theta = 0$ and when $\theta = \pi/2$, $x = r \sin \theta = r$. Hence

$$\int_0^r (\sqrt{r^2 - x^2})\mathrm{d}x = r^2 \int_0^{\pi/2} \cos^2 \theta \, \mathrm{d}\theta = \frac{1}{2} r^2 \int_0^{\pi/2} (\cos 2\theta + 1)\mathrm{d}\theta$$

$$= r^2 \left[\frac{1}{4} \sin 2\theta + \frac{1}{2}\theta\right]_0^{\pi/2} = \frac{1}{4}\pi r^2$$

The area of a circle of radius r is thus πr^2 square units.

The Definite Integral as the Limit of a Sum

Definition 3.5.3

Let $x_1, x_2, \ldots, x_{n-1}$ be *any* $n-1$ points satisfying

$$a < x_1 < x_2 < \ldots < x_{n-1} < b$$

With $x_0 = a$ and $x_n = b$, the set $\{x_i | i = 0, 1, \ldots, n\}$ forms a *partition* of the interval $[a, b]$. Let ζ_i be an arbitrary point in the subinterval $[x_{i-1}, x_i]$. Then

$$\sum_{i=1}^{n} f(\zeta_i)(x_i - x_{i-1})$$

is called a *Riemann sum.*

For a function such as the one depicted in figure 3.17, a Riemann sum represents an approximation to the area between the curve $y = f(x)$, the x axis and the lines $x = a$, $x = b$. If $n \to \infty$ in such a way that the width of each subinterval in the partition tends to 0, the limit of the corresponding

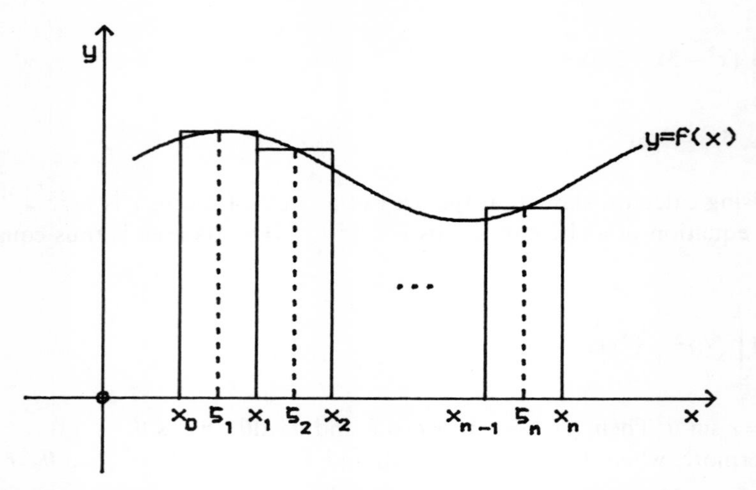

Figure 3.17 The Riemann sum represents the total area of all the rectangles

sequence of Riemann sums will be equal to the area under the curve between $x=a$ and $x=b$.

More generally, the limit of a sequence of Riemann sums of a function f exists provided f is bounded and *continuous almost everywhere* in $[a, b]$, that is, provided f is bounded and continuous except perhaps for a countable set of points of discontinuity.

Definition 3.5.4

If it exists, the limit as both $n \to \infty$ and $\max_{1 \leqslant i \leqslant n} |x_i - x_{i-1}| \to 0$ of a sequence of Riemann sums of f on $[a, b]$ is called the *Riemann integral* of f and is denoted by

$$\int_a^b f(x)\,\mathrm{d}x$$

In such a case, f is said to be (*Riemann*) *integrable* on $[a, b]$.

The property used in definition 3.5.2 to define the definite integral of f, that is

$$\int_a^b f(x)\,\mathrm{d}x = F(b) - F(a)$$

where $F'(x) = f(x)$, $\forall x \in [a, b]$, is usually called the *fundamental theorem of integral calculus*, and may be derived from the definition of the Riemann integral (Burkill, 1962, p. 129).

Approximate Integration

From above

$$\int_a^b f(x)\, \mathrm{d}x = F(b) - F(a)$$

where $F(x)$ is an indefinite integral of $f(x)$. If $F(x)$ is readily available and is not too complicated, this equation provides the obvious means of evaluating the definite integral. As mentioned at the start of this section, however, integration often leads to new functions that cannot be expressed as finite combinations of known elementary functions. The above technique for evaluating definite integrals is useless in such a case. Even if $F(x)$ can be obtained, its expression in terms of elementary functions may be extremely complicated. The evaluation of $F(a)$ and $F(b)$ then involves considerable computational effort and almost certainly a degree of approximation. This being so, it may be more efficient to replace the definite integral by an approximation right from the start.

Approximate integration by numerical techniques is thus often the most appropriate and sometimes the only available way of evaluating a given definite integral. Most methods of approximate integration (known as quadrature rules) approximate the integral by a weighted sum of function values

$$\int_a^b f(x)\mathrm{d}x \approx \sum_{j=0}^{n} w_j f(x_j)$$

Definition 3.5.5

A *quadrature rule* is a prescription for determining the *weights* w_j and the *abscissae*, or *quadrature points*, x_j in a sum of the form

$$\sum_{j=0}^{n} w_j f(x_j)$$

Some simple quadrature rules may be derived by approximating the region under the curve $y=f(x)$ between the lines $x=a$ and $x=b$ by a region whose area is easy to determine.

Example 3.5.8

(1) The *trapezium rule* corresponds to approximating the area under the curve $y=f(x)$ by the area of a trapezium (see figure 3.18a). Thus

$$\int_a^b f(x)\,\mathrm{d}x \approx \frac{1}{2}(b-a)[f(a)+f(b)]$$

In this quadrature rule, $n=1$, $w_0 = w_1 = \frac{1}{2}(b-a)$, $x_0=a$, $x_1=b$.

(2) *Simpson's rule* corresponds to approximating the area under the curve $y=f(x)$ by the area under the quadratic curve $y=p_2(x)$, say, that cuts $y=f(x)$ at $x=a$, $x=\frac{1}{2}(a+b)$ and $x=b$ (see figure 3.18b). This gives rise to the approximation

$$\int_a^b f(x)\,\mathrm{d}x \approx \frac{b-a}{6}\{f(a)+4f[\tfrac{1}{2}(a+b)]+f(b)\}$$

In this quadrature rule, $n=2$, $w_0=(b-a)/6$, $w_1=4(b-a)/6$, $w_2=(b-a)/6$, $x_0=a$, $x_1=\frac{1}{2}(a+b)$, $x_2=b$.

If f is a sufficiently 'smooth' function (that is, has a number of continuous derivatives) and if the interval $[a, b]$ is sufficiently small then, intuitively, both the trapezium rule and Simpson's rule should afford reasonably good approximations to the definite integral. More generally, it is necessary to apply such rules over each of a number of subintervals of the given range of integration.

Let $\{x_i | i=0, 1, \ldots, n\}$ be a partition of $[a, b]$ in which each subinterval has the same width, h. Then, by the trapezium rule

$$\int_a^b f(x)\,\mathrm{d}x = \sum_{i=1}^{n} \int_{x_{i-1}}^{x_i} f(x)\,\mathrm{d}x \approx \sum_{i=1}^{n} \frac{1}{2}h(f_{i-1}+f_i)$$

where $f_i \equiv f(x_i)$. Hence

$$\int_b^a f(x)\,\mathrm{d}x \approx h\left(\frac{1}{2}f_0+f_1+f_2+\ldots+f_{n-1}+\frac{1}{2}f_n\right)$$

This is known as the *repeated trapezium rule.* The area under the curve $y=f(x)$ is approximated by the area under a 'broken-line' approximation to the curve.

Similarly, if $\{x_i | i=0, 1, \ldots, 2n\}$ is a partition of $[a, b]$, with $x_i=x_0+ih$, then by Simpson's rule

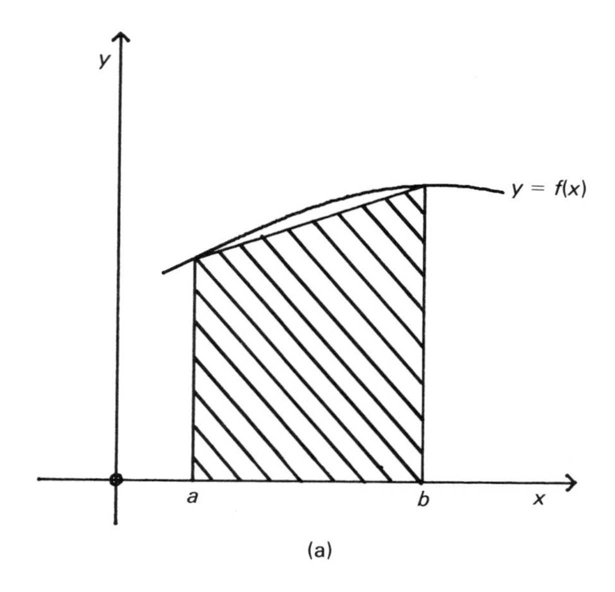

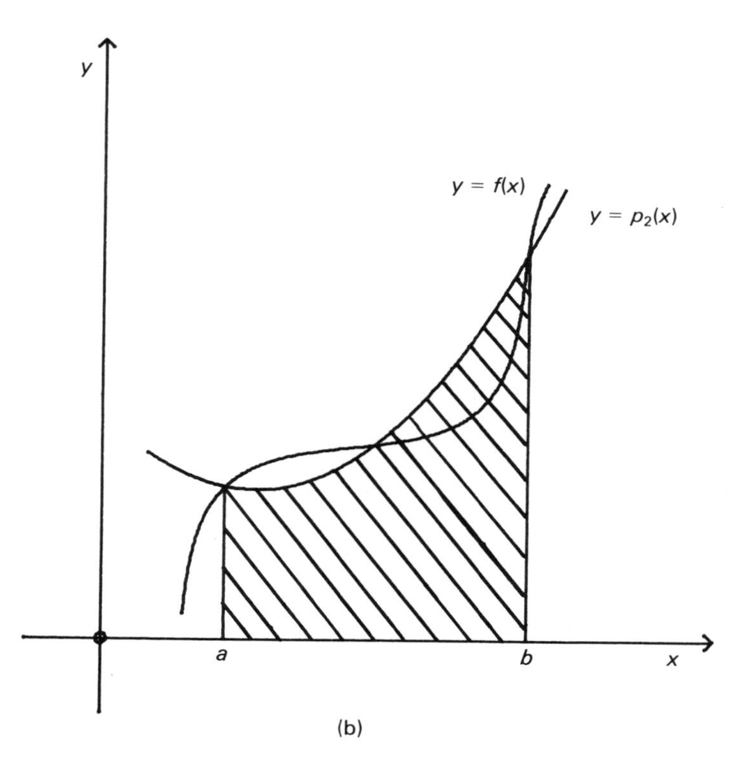

Figure 3.18 (a) The trapezium rule; (b) Simpson's rule

$$\int_a^b f(x)\,dx = \sum_{i=1}^{n} \int_{x_{2i-2}}^{x_{2i}} f(x)\,dx \approx \sum_{i=1}^{n} \frac{2h}{6}(f_{2i-2}+4f_{2i-1}+f_{2i})$$

that is

$$\int_a^b f(x)\,dx \approx \frac{h}{3}[f_0+4(f_1+f_3+\ldots+f_{2n-1})+2(f_2+f_4+\ldots+f_{2n-2})+f_{2n}]$$

This is called the *repeated Simpson's rule*. Repeated rules are also referred to as *composite* or *compound rules*.

Algorithm 3.5.1 describes how the repeated Simpson's rule might be used in practice. Successive approximations to the value of the integral are obtained by successively doubling the number of subintervals. The computation ceases when two successive approximations to the value of the integral differ by less than a prescribed value, tol. On doubling the number of subintervals, the function need only be evaluated at the new quadrature points $x_1, x_3, \ldots, x_{2n-1}$, since it has already been evaluated at $x_0, x_2, .., x_{2n}$ while computing the previous approximation.

Algorithm 3.5.1

```
h←(b−a)/4;
n←2;
s1←f(a)+f(b);
s2←f(a+h)+f(a+3*h);
s3←f(a+2*h);
simp0←(b−a) * (s1+4 * s3)/6;
simp1←h * (s1+4 * s2+2 * s3)/3;
while (abs(simp1−simp0)>tol) do
    n←2 * n;
    h←0.5 * h;
    s3←s2+s3;
    s2←0;
    for i from 1 in steps of 2 to (2n−1) do
        s2←s2+f(a+i * h)
```

```
        endfor;
        simp0←simp1;
        simp1←h * (s1 + 4 * s2 + 2 * s3)/3
    endwhile;
    accept simp1 as the value of the integral.
```

The repeated trapezium rule and the repeated Simpson's rule are just two of the many commonly used quadrature rules. For a comprehensive treatment of approximate integration, the reader is referred to Davis and Rabinowitz (1975).

Improper Integrals

An *improper integral* is one in which either the range of integration is not finite or the integrand is unbounded at one or more points in the range. Such integrals are defined as the limits, if they exist, of certain proper integrals.

Definition 3.5.6

(i) $$\int_0^\infty f(x)\,\mathrm{d}x = \lim_{M\to\infty}\int_0^M f(x)\,\mathrm{d}x;$$

(ii) $$\int_{-\infty}^\infty f(x)\,\mathrm{d}x = \lim_{M\to\infty}\int_{-M}^0 f(x)\,\mathrm{d}x + \lim_{M\to\infty}\int_0^M f(x)\,\mathrm{d}x;$$

(iii) if f is unbounded in the neighbourhood of $x=a$ then

$$\int_a^b f(x)\,\mathrm{d}x = \lim_{\varepsilon\to 0^+}\int_{a+\varepsilon}^b f(x)\,\mathrm{d}x;$$

similarly, if f is unbounded in the neighbourhood of $x=b$ then

$$\int_a^b f(x)\,\mathrm{d}x = \lim_{\varepsilon\to 0^+}\int_a^{b-\varepsilon} f(x)\,\mathrm{d}x$$

Example 3.5.9

(1) $$\int_0^M \frac{1}{1+x^2}\,dx=[\tan^{-1} x]_0^M=\tan^{-1} M$$

Hence

$$\int_0^\infty \frac{1}{1+x^2}\,dx=\lim_{M\to\infty}\int_0^M \frac{1}{1+x^2}\,dx=\lim_{M\to\infty} \tan^{-1} M=\frac{\pi}{2}$$

(2) $$\int_\varepsilon^1 x^{-1/2}\,dx=[2x^{1/2}]_\varepsilon^1=2-2\varepsilon^{1/2}$$

Hence

$$\int_0^1 x^{-1/2}\,dx=\lim_{\varepsilon\to 0^+}\int_\varepsilon^1 x^{-1/2}\,dx=\lim_{\varepsilon\to 0^+}[2-2\varepsilon^{1/2}]=2$$

Double Integration

Let R denote the rectangle consisting of the set of points $\{(x, y)|a\leqslant x\leqslant b, c\leqslant y\leqslant d\}$ and let $\{x_i|i=0, 1, \ldots, n\}$ and $\{y_j|j=0, 1, \ldots, m\}$ be partitions of $[a, b]$ and $[c, d]$, respectively. For each $i\in\{1, 2, \ldots, n\}$ and $j\in\{1, 2, \ldots, m\}$ let R_{ij} denote the subrectangle $\{(x, y)|x_{i-1}\leqslant x\leqslant x_i, y_{j-1}\leqslant y\leqslant y_j\}$ and let (ζ_i, η_i) denote an arbitrary point in R_{ij}.

Definition 3.5.7
A sum of the form

$$\sum_{i=1}^{n}\sum_{j=1}^{m} f(\zeta_i, \eta_j)(x_i-x_{i-1})(y_j-y_{j-1})$$

is called a *two-dimensional Riemann sum.* Let n and m tend to infinity in such a way that $\max_{1\leqslant i\leqslant n}|x_i-x_{i-1}|$ and $\max_{1\leqslant j\leqslant m}|y_j-y_{j-1}|$ both tend to zero. If the resulting sequence of Riemann sums tends to a limit, f is said to be *(Riemann) integrable* over R. The limit is denoted by

$$\iint_R f(x, y)\,dx\,dy$$

and is called a *double integral.*

Suppose B denotes a non-rectangular bounded region. Let R denote *any* rectangle containing B and define $f(x, y)=0$, $\forall (x, y)\notin B$. The double integral of f over B is then defined by

$$\iint_B f(x, y)\, dx\, dy = \iint_R f(x, y)\, dx\, dy$$

provided the latter integral exists. The area of the region B is given by

$$\iint_B dx\, dy$$

The Iterated Integral

Let φ_1 and φ_2 be continuous functions in $[a, b]$, with $\varphi_1(x)\leqslant\varphi_2(x)$, $\forall x\in[a, b]$. Let B denote the region bounded by $x=a$, $x=b$, $y=\varphi_1(x)$ and $y=\varphi_2(x)$. Then

$$\iint_B f(x, y)\, dx\, dy = \int_a^b \int_{\varphi_1(x)}^{\varphi_2(x)} f(x, y)\, dy\, dx$$

$$= \int_a^b g(x)\, dx$$

where

$$g(x) = \int_{\varphi_1(x)}^{\varphi_2(x)} f(x, y)\, dy$$

Definition 3.5.8

The two single integrals

$$g(x) \equiv \int_{\varphi_1(x)}^{\varphi_2(x)} f(x, y)\, dy$$

and

$$\int_a^b g(x)\, dx$$

are called *iterated integrals.*

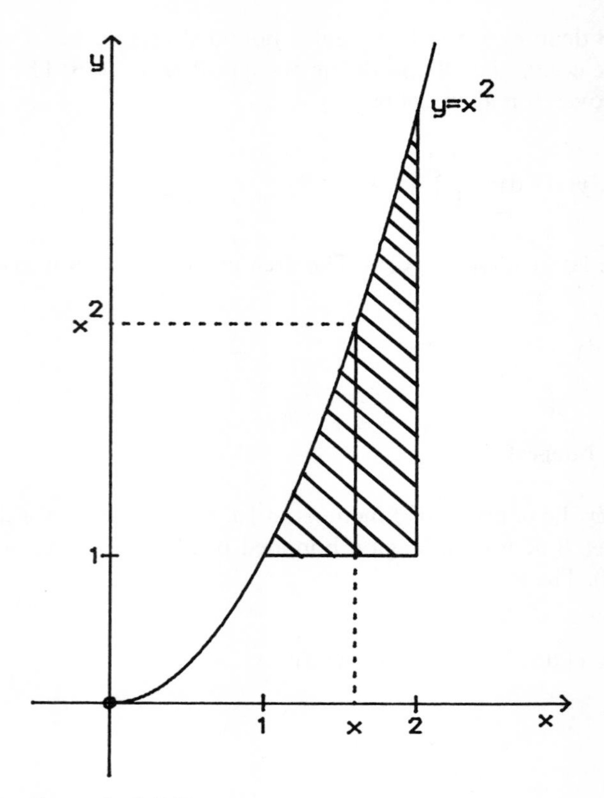

Figure 3.19 Example 3.5.10

Example 3.5.10

Evaluate the double integral

$$\iint_B (x^2 + y^2)\, \mathrm{d}x\, \mathrm{d}y$$

where B is the region in the cartesian plane consisting of the set of points $\{(x, y) | y \leqslant x^2, x \leqslant 2, y \geqslant 1\}$, and is represented by the shaded area in figure 3.19.

For each value of x in the interval $[1, 2]$, y varies continuously between $y = 1$ and $y = x^2$. The value of the given double integral may thus be obtained by evaluating the iterated integrals

$$g(x) \equiv \int_1^{x^2} (x^2 + y^2)\, \mathrm{d}y$$

and

$$\int_1^2 g(x)\,\mathrm{d}x$$

Now, $g(x) = [x^2y + y^3/3]_1^{x^2} = x^4 + x^6/3 - x^2 - 1/3$. Hence

$$\int_1^2 g(x)\,\mathrm{d}x = \left[\frac{x^5}{5} + \frac{x^7}{21} - \frac{x^3}{3} - \frac{x}{3}\right]_1^2 = \frac{1006}{105}$$

Exercise 3.5

1. Integrate each of the following

(a) $5x^4 + 7x$;

(b) $x^{1/3}$;

(c) xe^{-x^2}.

2. Using the rule for integration by parts, determine each of the following integrals

(a) $\int xe^{cx}\mathrm{d}x,\ c \in \mathbb{R}$;

(b) $\int e^x \cos x\,\mathrm{d}x$;

(c) $\int x \sec^2 x\,\mathrm{d}x$.

3. If $I_n = \int x^n e^x\,\mathrm{d}x$, show that

$$I_n = x^n e^x - nI_{n-1}$$

Hence determine $\int x^3 e^x\,\mathrm{d}x$.

4. Using the method of partial fractions, determine each of the following integrals

(a) $\int \frac{23 - 2x}{2x^2 + 9x - 5}\mathrm{d}x$

(b) $\int \frac{2x + 5}{(x + 2)^2}\mathrm{d}x$

(c) $\int \frac{x}{(x + 1)(x^2 + 4)}\mathrm{d}x$

5. By making an appropriate change of variable, integrate each of the following

(a) $1/[x(x^2+4)^{1/2}]$;

(b) $2x/(1+x^4)$;

(c) $(\log_e x)/x$;

(d) $1/(3+2\sin^2 x)$.

6. Evaluate each of the following definite integrals:

(a) $$\int_1^2 4x(x-1)(x-2)\,dx;$$

(b) $$\int_0^\pi x\cos 3x\,dx;$$

(c) $$\int_0^1 xe^{-x^2}\,dx.$$

7. Determine the area between the curve $y=1+\cos 2x$, the x axis and the lines $x=0$, $x=\pi$.

8. Calculate the area between the curve $y=5x-2x^2$ and the line $y=x$.

9. Implement algorithm 3.5.1 as a computer program. Use your program to evaluate each of the following integrals

(a) $$\int_0^\pi e^x\cos x\,dx;$$

(b) $$\int_0^5 \frac{1}{1+x^4}\,dx;$$

(c) $$\int_0^\pi e^x\cos 4x\,dx.$$

10. Evaluate each of the following improper integrals

(a) $$\int_0^\infty x^2 e^{-3x}\, dx;$$

(b) $$\int_0^4 \frac{1}{\sqrt{(4-x)}} dx;$$

(c) $$\int_0^\infty \frac{1}{(1+x^3)}\, dx.$$

11. Evaluate the double integral

$$\iint_B (x+y) dx\, dy$$

where B is the region in the x–y plane consisting of the set of points $\{(x, y) | 1 \leqslant x \leqslant 2,\ y \leqslant 4 - x^2,\ y \geqslant 0\}$.

12. Using a programming language that allows recursion, write a recursive function to evaluate

$$I_n = \int_0^{\pi/2} \sin^n x\, dx$$

[Hint: See example 3.5.2(4).]

4 Probability

4.1 INTRODUCTION

A *stochastic event* is a *non-deterministic* or *random* experiment, that is, its outcome cannot be determined in advance. Probability is the study of such experiments. The theory of probability discussed below relates to a mathematical model in which each possible outcome may be assigned a real number in the interval [0, 1], called the probability of that outcome.

Probability theory is applied in many areas of computer science. The use of probability models in the design and analysis of algorithms and in analysing the performance of data base systems are two particularly important applications and the increasingly popular technique of computer simulation is a direct application of probability theory. In addition, the fields of coding and information theory, which are highly relevant to computer science, depend heavily on the use of probability. Familiarity with the basic ideas of probability theory is thus very important for the computer scientist. It is hoped that the survey of the topic given in this chapter is sufficient to provide the student with a reasonable basis for further study. The reader is referred in particular to Allen (1978).

Sample Spaces and Events

As an aid to the mathematical theory of probability, use is made of set theory (see section 1.2).

Definition 4.1.1
Each occurrence of a stochastic event is called a *trial*. A *sample point* is a possible outcome of a trial and the set of all such points is called the *sample space*, $\mathscr{S}$, for the experiment. If $\mathscr{S}$ is a finite or countably infinite set, it is said to be a *discrete sample space*; otherwise it is said to be a *continuous sample space*.

Since $\mathbb{R}$ is not countable (see theorem 1.3.4), a sample space consisting of an interval of the real line is a continuous sample space. Detailed discussion of such spaces is deferred until section 4.4.

Example 4.1.1

1. If h denotes the outcome 'heads' and t denotes the outcome 'tails', then the sample space for the experiment of tossing two coins is $\{(h, h), (h, t), (t, h), (t, t)\}$.

2. If a die is thrown and the number appearing on the top noted, then the sample space for this experiment is $\{1, 2, 3, 4, 5, 6\}$.

Definition 4.1.2
Let $\mathscr{S}$ be the sample space corresponding to a random experiment. An *event* is a subset of $\mathscr{S}$. If the subset has only one element, it is called a *simple event*. An event $E \subset \mathscr{S}$ occurs if the outcome of a trial of the experiment is an element of E.

The set $\mathscr{S}$ is the *certain event*, since all possible outcomes belong to $\mathscr{S}$, while the empty set, $\emptyset$, is the *impossible event*, since each trial must have an outcome.

Example 4.1.2

(1) The event: 'at least one coin shows heads', when two coins are tossed, is $\{(h, h), (h, t), (t, h)\}$.

(2) When a die is thrown and the number appearing on the top noted, the event: 'the number noted is even' is $\{2, 4, 6\}$.

The fundamental set operations of union and intersection have special significance when applied to events. If E_1 and E_2 are two events from the same sample space, then

$E_1 \cup E_2$ is the event that *at least one of* E_1 *or* E_2 occurs;

$E_1 \cap E_2$ is the event that *both* E_1 *and* E_2 occur.

Definition 4.1.3
If $E_1 \cap E_2 = \emptyset$ then E_1 and E_2 are said to be *mutually exclusive events.*

Mutually exclusive events have no sample points in common. Distinct simple events are thus mutually exclusive.

The complement of a set also has special significance when the set is an event from a sample space, $\mathscr{S}$: the complement E' of an event $E \subset \mathscr{S}$ is the event that occurs if and only if E does not occur.

Example 4.1.3

(1) For the experiment of tossing two coins, let

$$E_1 = \{(h, h)\},\ E_2 = \{(t, t)\}$$

Then

$$E_1 \cup E_2 = \{(h, h), (t, t)\}$$

and is the event: 'both of the coins show the same face', while

$$E_1 \cap E_2 = \emptyset$$

that is E_1 and E_2 are mutually exclusive. This is because they are distinct simple events. Finally

$$E_1' = \{(h, t), (t, h), (t, t)\}$$

and is thus the event: 'one head at most shows'.

(2) For the experiment of throwing a die and noting the top number, let

$$E_1 = \{1, 2, 3\},\ E_2 = \{2, 4, 6\}$$

Then

$$E_1 \cup E_2 = \{1, 2, 3, 4, 6\}$$

which is the event: 'the number noted is not 5' while

$$E_1 \cap E_2 = \{2\}$$

and is the event: 'the number noted is 2'. Finally

$$E_2' = \{1, 3, 5\}$$

which is the event: 'the number noted is odd'.

If $\mathscr{S} = \{s_1, s_2, s_3, \ldots\}$ then $\mathscr{S}$ may be partitioned by

$$\mathscr{S} = E_1 \cup E_2 \cup E_3 \cup \ldots$$

where $E_i = \{s_i\}$ and $E_i \cap E_j = \emptyset$, $i \neq j$. A discrete sample space is thus the union of a sequence of mutually exclusive events.

Probabilities in Discrete Sample Spaces

Suppose that the certain event is assigned a probability equal to one and the impossible event a probability equal to zero. Then any event may be assigned a probability between zero and one such that the sum of the probabilities of all possible distinct simple events is equal to one.

Sometimes probabilities may be assigned *a priori* to the possible outcomes of an experiment. For example, if a fair coin is tossed, it is equally likely to show heads as it is tails, that is,

$$\text{probability of heads} = \text{probability of tails} = \frac{1}{2}$$

Similarly, for a fair die, the probability of throwing a given number in the sample space $\{1, 2, 3, 4, 5, 6\}$ is equal to $1/6$.

Definition 4.1.4
An *equiprobable sample space* is one in which each sample point is assigned the same probability.

Let $\mathscr{S}$ be a finite, equiprobable space, having n sample points. If the event $E \subset \mathscr{S}$ contains n_E sample points, then the probability, $\mathscr{P}(E)$, that the event E will occur is given by

$$\mathscr{P}(E) = \frac{\text{number of ways that } E \text{ can occur}}{\text{number of ways that } \mathscr{S} \text{ can occur}} = \frac{n_E}{n}$$

Example 4.1.4

If two dice are thrown and the number appearing on the top of each die is noted, then the sample space for this experiment is

$$\mathscr{S} = \{(i, j) | i, j \in N, 1 \leqslant i \leqslant 6, 1 \leqslant j \leqslant 6\}$$

which has 36 elements. Let E be the event: 'the sum of the two noted numbers equals 8'. Then

$$E = \{(2, 6), (3, 5), (4, 4), (5, 3), (6, 2)\}$$

which has five elements. Hence

$$\mathscr{P}(E) = \frac{5}{36}$$

The assignment of probabilities in the above examples may be justified by appealing to symmetry. Alternatively, a probability may be assigned to an event by taking the limiting value of the relative frequency of that event, as the number of trials becomes very large.

Definition 4.1.5
If a random experiment is performed M times, a *sample of size M* is said to be taken. If an event E_i occurs m_i times in a sample of size M, then its *relative frequency* with respect to this sample is given by $f_i = m_i/M$. If n

Table 4.1

Number of compilation errors	Number of jobs (m)	Relative frequency ($m/600$)
0	55	0.09167
1	65	0.10833
2 or 3	151	0.25167
4 or 5	120	0.20000
6 to 10	148	0.24667
10 to 15	61	0.10167

distinct events occur in the sample, then f_i, $i=1, 2, \ldots, n$, is the *frequency distribution* for the sample of size M.

Example 4.1.5

Table 4.1 summarises the results of a survey carried out by a university computing centre. The number of compilation errors in each of 600 jobs submitted by first-year computer science students were noted and the given relative frequencies computed. Frequency distributions are further discussed in section 4.3.

Probability Functions

As far as the mathematical theory of probability is concerned, it does not matter how the probabilities are actually assigned to the events in a sample space: in certain applications they may even be assigned subjectively. It is only necessary that the assigned probabilities form the image set of a probability function.

Definition 4.1.6
A *probability function*, $\mathscr{P}$, for a sample space, $\mathscr{S}$, assigns to each event $E \subset \mathscr{S}$ a real number, $\mathscr{P}(E)$, in the interval $[0, 1]$. Here $\mathscr{P}(E)$ is called the *probability* of E and satisfies the following properties:

(1) $\mathscr{P}(\mathscr{S})=1$;

(2) if G and H are mutually exclusive events, then

$$\mathscr{P}(G \cup H)= \mathscr{P}(G)+ \mathscr{P}(H)$$

Property (2) is often called the *addition rule for two mutually exclusive events.*

From this definition, each of the statements in the following theorem may be proved.

Theorem 4.1.1

(i) $\mathscr{P}(\emptyset)=0$;

(ii) if $E_i \cap E_j=\emptyset$, $i \neq j$, then

$$\mathscr{P}(E_1 \cup E_2 \cup E_3 \cup \ldots)=\mathscr{P}(E_1)+\mathscr{P}(E_2)+\mathscr{P}(E_3)+\ldots$$

(iii) $\mathscr{P}(E)=\mathscr{P}(E \cap F)+\mathscr{P}(E \cap F')$;

(iv) $\mathscr{P}(F)+\mathscr{P}(F')=1$;

(v) $\mathscr{P}(E \cup F)=\mathscr{P}(E)+\mathscr{P}(F)-\mathscr{P}(E \cap F)$.

Proof (i)

$$\begin{aligned}\mathscr{P}(\mathscr{S}) &= \mathscr{P}(\mathscr{S} \cup \emptyset)\\ &= \mathscr{P}(\mathscr{S})+\mathscr{P}(\emptyset)\end{aligned}$$

since $\mathscr{S}$ and $\emptyset$ are mutually exclusive. Hence, $\mathscr{P}(\emptyset)=0$.

(ii) This is a generalisation of property (2) in definition 4.1.6 and may be proved by mathematical induction (see exercise 4.1).

(iii) If $G=E \cap F'$ and $H=E \cap F$, then G and H are disjoint and $E=G \cup H$. The required result thus follows from property (2) in definition 4.1.6.

(iv) Putting $E=\mathscr{S}$ in statement (iii), above, gives

$$\mathscr{P}(\mathscr{S})=\mathscr{P}(\mathscr{S} \cap F)+\mathscr{P}(\mathscr{S} \cap F')$$

that is, $1=\mathscr{P}(F)+\mathscr{P}(F')$, as required.

Note for some problems, it is easier to compute the probability that an event will *not* occur, rather than the probability that it will occur. In such a case, $\mathscr{P}(F)$ may be calculated by first determining $\mathscr{P}(F')$ and then subtracting this value from 1.

(v) If $G=E \cap F'$ and $H=F$, then G and H are mutually exclusive events (see figure 4.1). Furthermore, $G \cup H=E \cup F$. Thus

$$\mathscr{P}(E \cup F)=\mathscr{P}(E \cap F')+\mathscr{P}(F)$$

from property (2) in definition 4.1.6. But

$$\mathscr{P}(E \cap F')=\mathscr{P}(E)-\mathscr{P}(E \cap F)$$

from statement (iii) above. Hence

$$\mathscr{P}(E \cup F)=\mathscr{P}(E)+\mathscr{P}(F)-\mathscr{P}(E \cap F)$$

Note here E and F are *any* pair of events from $\mathscr{S}$. If they happen to be mutually exclusive, that is, $E \cap F=\emptyset$, then $\mathscr{P}(E \cap F)=0$ and the above result reduces to property (2) in definition 4.1.6.

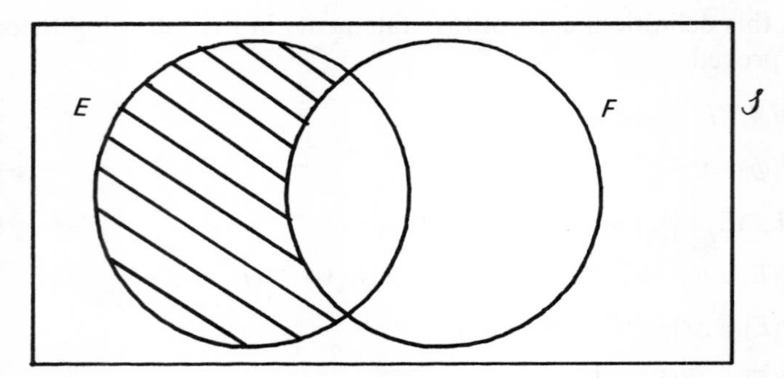

Figure 4.1 The region corresponding to $E \cap F'$ is shaded

Example 4.1.6

Of 100 students, 30 are taking computer science and 20 are taking mathematics. Of these 50 students, 10 are taking a joint degree in computer science, and mathematics. Find the probability, *p*, that a student selected at random is taking just computer science or just mathematics.

Since half of the student sample are taking computer science and/or mathematics and one-fifth of these are taking the joint degree, the required probability is

$$p = \frac{4}{5} \times \frac{1}{2} = \frac{2}{5}$$

This result, which is intuitively correct, is also predicted by the probability theory so far developed.

Let C be the event: 'the student is taking computer science', M the event: 'the student is taking mathematics' and J the event: 'the student is taking the joint degree'. Since the sample space is equiprobable

$$\mathscr{P}(C) = 30/100 = 3/10$$

$$\mathscr{P}(M) = 20/100 = 1/5$$

$$\mathscr{P}(J) = \mathscr{P}(C \cap M) = 10/100 = 1/10$$

Now

$$p = \mathscr{P}(C \cup M)$$

and by statement (v) in theorem 4.1.1.

$$\mathscr{P}(C \cup M) = \mathscr{P}(C) + \mathscr{P}(M) - \mathscr{P}(C \cap M)$$

Hence

$$p=3/10+1/5-1/10$$
$$=2/5$$

Permutations and Combinations

The development of the mathematical theory of probability was begun by two French mathematicians, Fermat and Pascal, in the 17th century and was continued by another French mathematician, Laplace, in the 18th and early 19th centuries. The latter developed the theory for equiprobable spaces. As stated above, to determine the probability of an event E in such a sample space, $\mathscr{S}$, the number of sample points in E is divided by the number of sample points in $\mathscr{S}$. If the number of sample points in $\mathscr{S}$ or a subset of $\mathscr{S}$ is very large, explicit enumeration will, in general, be out of the question. Fortunately, the cardinality of $\mathscr{S}$, $\#(\mathscr{S})$, and of specific subsets of $\mathscr{S}$, may often be determined by applying certain formulae for permutations and combinations.

Definition 4.1.7
A *permutation* of $m \leqslant n$ elements of $\mathscr{S}$, where $\#(\mathscr{S})=n$, is any arrangement of m distinct elements of $\mathscr{S}$. The number of all such possible arrangements is called the *number of permutations of n elements taking m at a time* and is denoted by ${}^nP_m$.

(*Note* in a permutation, the order of the elements is important.)

Example 4.1.7

The permutations of the three letters *abc* taking two at a time are: *ab*, *ba*, *ac*, *ca*, *bc*, *cb*.

Thus, *ab* and *ba*, for example, are different permutations since the order in which the letters are taken is important.

Definition 4.1.8
A *combination* of $m \leqslant n$ elements of a set $\mathscr{S}$, where $\#(\mathscr{S})=n$, is any selection of m distinct elements of $\mathscr{S}$. The number of all possible selections is called the *number of combinations of n elements taking m at a time* and is denoted by ${}^nC_m$ or $\binom{n}{m}$.

(*Note*, unlike in a permutation, the order of the elements is unimportant in a combination.)

Example 4.1.8

(1) The combinations of the three letters *abc* taking two at a time are: *ab*, *ac*, *bc*

(2) *abc*, *acb*, *bac*, *bca*, *cab*, *cba* are different permutations of the three letters *abc*, but the same combination.

The study of permutations and combinations belongs to a branch of mathematics called *combinatorial theory*, which is concerned with counting possibilities and to which probability theory is closely related.

Combinatorial theory is based upon the following fundamental principle, called the *multiplication rule of counting:*

> if the ith operation in a sequence of n successive operations can be performed in r_i different ways, then the number of ways in which the n operations can be performed together is
>
> $$r_1 \times r_2 \times \ldots \times r_n$$

Theorem 4.1.2

(i) The number of ways of arranging m elements from n when repetitions are allowed is equal to n^m.

(ii) The number of permutations of n elements taking m at a time is given by

$$^nP_m = \frac{n!}{(n-m)!}$$

(iii) The number of permutations of n elements taking n at a time, when r of the elements are the same is equal to $n!/r!$.

(iv) The number of combinations of n elements taking m at a time is given by

$$^nC_m = \frac{n!}{m!(n-m)!}$$

Proof (i) This follows directly from the multiplication rule, since there are n ways in which each successive element in the arrangement can be chosen.

(ii) The first element may be any one of n, the second any one of $n-1$, and so on through to the mth, which may be any one of the $n-m+1$ remaining elements. The multiplication rule thus gives

$$^nP_m = n(n-1)\ldots(n-m+1) = \frac{n!}{(n-m)!}$$

In particular

$$^{m}P_{m}=m!$$

(iii) This result follows immediately from the fact that $^{n}P_{n}=n!$ and $^{r}P_{r}=r!$ and may be generalised to the case where r_1 elements are alike of a first kind, r_2 alike of a second kind, ..., r_t alike of a tth kind. In this case, the number of permutations is given by

$$\frac{n!}{r_1!r_2!\ldots r_t!}$$

(iv) From above, $^{m}P_{m}=m!$, that is, m elements may be ordered in $m!$ different ways. Since order is not important in a combination, it follows that

$$^{n}C_{m}=\frac{^{n}P_{m}}{m!}=\frac{n!}{m!(n-m)!}$$

Note, $^{n}C_{m}$ is the coefficient which appears in the *binomial expansion*

$$(a+x)^n=\sum_{m=0}^{n} {}^{n}C_{m}a^{n-m}x^{m}$$

(see theorem 3.3.4).

Theorem 4.1.3
(i) $^{n+1}C_{m}={}^{n}C_{m}+{}^{n}C_{m-1}$

(ii) The number of subsets of a set with n elements is 2^n.

Proof (i) This property is established during the proof of theorem 3.3.4.
(ii) $^{n}C_{r}$ denotes the number of subsets with cardinality equal to r. The total number of subsets is thus

$$\sum_{r=0}^{n} {}^{n}C_{r}$$

and the required result follows on substituting $a=x=1$ in the binomial expansion.

Example 4.1.9

(1) Each member of a university who wishes to use the central computing facilities is assigned a unique user-identifier consisting of two letters followed by two digits. Since each letter may be any one of 26 and each digit any one of 10, it follows that the total number of possible identifiers is

$$26\times 26\times 10\times 10=67\,600$$

(2) Two cards are picked at random from a normal deck of 52 playing cards. Find the probability, p, that: (a) both are diamonds, (b) one is a diamond and the other a club.

(a) There are ${}^{52}C_2$ ways of picking two cards from the deck and ${}^{13}C_2$ ways of picking two diamonds. Thus

$$p = {}^{13}C_2/{}^{52}C_2 = 78/1326 = 3/51$$

(b) Since there are 13 diamonds and 13 clubs, there are $13 \times 13 = 169$ ways of picking one diamond and one club. Hence

$$p = 169/1326 = 13/102$$

(3) How many people must be chosen at random so that the probability, p, that at least two of them will have the same birthday is more than 0.6?

Suppose that there are m people and that each person's birthday may fall on any of the 365 days with equal probability. Then, from statement (i) in theorem 4.1.2, there are 365^m ways in which the m people could have their birthdays. The number of ways in which they can have *distinct* birthdays is equal to ${}^{365}P_m$.

Since the sample space is equiprobable, the probability q that they will have distinct birthdays is given by

$$q = {}^{365}P_m/365^m$$

$$= \frac{365 \times 364 \times \ldots \times (365 - m + 1)}{365^m}$$

Since $p = 1 - q$ (from statement (iv) in theorem 4.1.1), q must be less than 0.4 in order for p to be greater than 0.6. This will be the case if $m \geqslant 27$.

If asked to guess the answer to the above question, most people give a value for m far bigger than that just determined and are surprised to discover just how few people need to be chosen at random for there to be a good chance that two of them will have the same birthday.

Apart from being fascinating in its own right, this result is important in computing since a variation of it occurs when analysing *hashing* techniques. The latter are an important class of methods for searching a file of records stored in a computer. (See Knuth, 1973.)

Exercise 4.1

1. Two dice are thrown and the number appearing on the top of each die is noted. Let E be the event: 'the sum of the two numbers is even' and let F be the event: 'at least one of the numbers is a 2'.

List the sample points which belong to the events $E \cup F$, $E \cap F$ and $E \cap F'$, respectively, and determine the probability of each of these events, assuming that the sample space for the experiment is an equiprobable space.

2. Prove theorem 4.1.1(ii).

3. Each one of 50 computer programmers can program in at least one of the three languages, Fortran, Cobol and Pascal

25 can program in Fortran

35 can program in Cobol

10 can program in Pascal

None of the programmers can program in both Cobol and Pascal but 15 can program in both Fortran and Cobol.

Compute the probability that a programmer selected at random can program in

(a) Fortran only;

(b) Fortran or Cobol.

4. A lecture is attended by 20 male and 10 female students. Half of the males and half of the females have brown eyes. Compute the probability that a student selected at random is male or has brown eyes.

5. Five married couples are standing in a room. Compute the probability that two people chosen at random are married to each other.

6. Three cards are dealt from a normal pack of 52 playing cards. Compute the probability that the cards are three of a kind (that is, three 'kings' or three 'fives', etc.).

4.2 CONDITIONAL PROBABILITY. MULTI-STEP EXPERIMENTS

Definition 4.2.1
If the outcome of a trial is known to be a sample point in an event $E \subset \mathscr{S}$, where $\mathscr{S}$ is the sample space for the experiment, then the probability that the outcome is also in an event $F \subset \mathscr{S}$ is called the *conditional probability* of F given that E has occurred and is denoted by $\mathscr{P}(F|E)$.

Example 4.2.1

The sample space for the experiment of tossing two coins is $\{(h, h), (h, t), (t, h), (t, t)\}$. If the event 'at least one of the coins shows heads' is known to

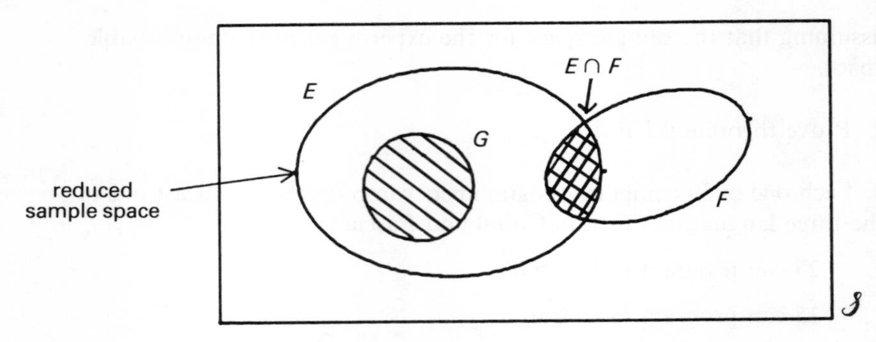

Figure 4.2

have occurred, then the only possible outcomes are those in the *reduced sample space*: $\{(h, h), (h, t), (t, h)\}$. In this case, the probability of the event 'both coins show heads' is equal to 1/3.

In general, if an event E is known to have occurred, then the only way in which F can also have occurred is if the outcome is in $E \cap F$; see figure 4.2. The conditional probability of F given that E has occurred is thus equal to the probability of the event $E \cap F$ in the reduced sample space, E. The probabilities of all distinct events in the reduced sample space must sum to unity. The probability that an event $G \subset E$ occurs when E occurs is thus $\mathscr{P}(G)/\mathscr{P}(E)$, where $\mathscr{P}(G)$ and $\mathscr{P}(E)$ are the probabilities of G and E in the original sample space, $\mathscr{S}$. Putting $G = E \cap F$ gives the following theorem.

Theorem 4.2.1

$$\mathscr{P}(F|E) = \frac{\mathscr{P}(E \cap F)}{\mathscr{P}(E)}$$

Corollary 4.2.1(1)
If $\mathscr{S}$ is an equiprobable space, then

$$\mathscr{P}(F|E) = \frac{\text{number of sample points in } E \cap F}{\text{number of sample points in } E}$$

Corollary 4.2.1(2)

$$\mathscr{P}(E \cap F) = \mathscr{P}(E)\mathscr{P}(F|E)$$

The latter formula is known as the *multiplication rule for conditional probability*. Using mathematical induction, the multiplication rule may be generalised to give

$$\mathscr{P}(F_1 \cap F_2 \cap \ldots \cap F_r) = \mathscr{P}(F_1)\mathscr{P}(F_2|F_1)\mathscr{P}(F_3|F_1 \cap F_2)\ldots\mathscr{P}(F_r|F_1 \cap F_2 \cap \ldots \cap F_{r-1})$$

(see exercise 4.2).

Because of the reduction in the size of the sample space, it is often easier to compute $\mathscr{P}(E \cap F)$ using corollary 4.2.1(2) rather than calculating it directly. Once $\mathscr{P}(E \cap F)$ has been computed, $\mathscr{P}(E \cup F)$ may be calculated using statement (v) in theorem 4.1.1.

Theorem 4.2.2 Bayes' Theorem

$$\mathscr{P}(E|F) = \frac{\mathscr{P}(E)\mathscr{P}(F|E)}{\mathscr{P}(F)}$$

Proof From theorem 4.2.1

$$\mathscr{P}(E|F) = \frac{\mathscr{P}(F \cap E)}{\mathscr{P}(F)}$$

$$= \frac{\mathscr{P}(E \cap F)}{\mathscr{P}(F)}$$

$$= \frac{\mathscr{P}(E)\mathscr{P}(F|E)}{\mathscr{P}(F)}$$

from the multiplication rule.

Bayes' theorem may be generalised to the following case. If $\mathscr{S} = E_1 \cup E_2 \cup \ldots \cup E_n$, where $E_i \cap E_j = \emptyset$, $i \neq j$, and $F \subset \mathscr{S}$, then, for each i

$$\mathscr{P}(E_i|F) = \frac{\mathscr{P}(E_i)\mathscr{P}(F|E_i)}{\sum_{j=1}^{n} \mathscr{P}(E_j)\mathscr{P}(F|E_j)}$$

Example 4.2.2

(1) Two cards are drawn from a normal deck of 52 playing cards. Compute the probability that they will both be aces.

Let E be the event: 'the first card is an ace' and F be the event: 'the second card is an ace'; $E \cap F$ is then the event: 'both cards are aces'. Clearly, $\mathscr{P}(E) = 4/52$. If E has occurred, then there are 3 aces left in 51 cards. Hence, $\mathscr{P}(F|E) = 3/51$ and by the multiplication rule

$$\mathscr{P}(E \cap F) = \mathscr{P}(E)\mathscr{P}(F|E)$$

$$= \frac{4}{52} \times \frac{3}{51} = \frac{1}{221}$$

(2) A lecture is attended by 25 students. Given that 16 of the students are male, calculate the probability that 3 students selected at random will all be male.

Let F_1 be the event: 'the first student selected is male', F_2 the event: 'the second student selected is male' and F_3 the event: 'the third student selected is male'. $F_1 \cap F_2 \cap F_3$ is then the event: 'each of the three students is male'. Now $\mathscr{P}(F_1)=16/25$, $\mathscr{P}(F_2|F_1)=15/24$ and $\mathscr{P}(F_3|F_1 \cap F_2)=14/23$. Hence

$$\mathscr{P}(F_1 \cap F_2 \cap F_3)=\mathscr{P}(F_1)\mathscr{P}(F_2|F_1)\mathscr{P}(F_3|F_1 \cap F_2)$$
$$=\frac{16}{25}\times\frac{15}{24}\times\frac{14}{23}=\frac{28}{115}$$

(3) A rack contains the output from the line-printer for 25 computer programs submitted by first-year undergraduates. 10 of the programs failed to compile. If the output for 2 jobs is picked at random from the rack, what is the probability that: (a) both programs compiled? (b) one program compiled and one failed to compile?

(a) Let E be the event: 'the first program compiled' and F the event: 'the second program compiled'; $E \cap F$ is then the event: 'both programs compiled'.

$$\mathscr{P}(E \cap F)=\mathscr{P}(E)\mathscr{P}(F|E)$$
$$=\frac{15}{25}\times\frac{14}{24}=\frac{7}{20}$$

(b) Let G be the event: 'the first program failed to compile', and H the event: 'the second program failed to compile'; $G \cap F$ is then the event: 'the first program failed to compile and the second program compiled' and $E \cap H$ is the event: 'the first program compiled but the second failed to compile'.

$$\mathscr{P}(G \cap F)=\mathscr{P}(G)\mathscr{P}(F|G)=\frac{10}{25}\times\frac{15}{24}=\frac{1}{4}$$

and

$$\mathscr{P}(E \cap H)=\mathscr{P}(E)\mathscr{P}(H|E)=\frac{15}{25}\times\frac{10}{24}=\frac{1}{4}$$

$(G \cap F) \cup (E \cap H)$ is the event: 'one program compiled and one program failed to compile'. Since $(G \cap F)$ and $(E \cap H)$ are mutually exclusive events

$$\mathscr{P}((G \cap F) \cup (E \cap H))=\mathscr{P}(G \cap F)+\mathscr{P}(E \cap H)=\frac{1}{2}$$

(4) 60% of the students in a computer science department are male. 15% of the males and 5% of the females are taller than 6 feet. In an experiment, a student is selected at random. Given that the chosen student is taller than 6 feet, compute the probability that the student is male.

Let M be the event: 'the student is male' and F the event: 'the student is female'. If $\mathscr{S}$ is the sample space for the experiment then

$$\mathscr{S}=M\cup F$$

Let T be the event: 'the student is taller than 6 feet'. The required probability is thus $\mathscr{P}(M|T)$. By Bayes' theorem

$$\mathscr{P}(M|T)=\frac{\mathscr{P}(M)\mathscr{P}(T|M)}{\mathscr{P}(M)\mathscr{P}(T|M)+\mathscr{P}(F)\mathscr{P}(T|F)}$$

$$=\frac{(0.6)(0.15)}{(0.6)(0.15)+(0.4)(0.05)}=\frac{9}{11}$$

Independent Events

Definition 4.2.2
If $\mathscr{P}(F|E)=\mathscr{P}(F)$, then F is said to be *independent* of E.

It follows from Bayes' theorem that, if F is independent of E, E is independent of F. Two events which are not independent are said to be *dependent events*.

Since either F or F' must occur in any trial, $\mathscr{P}(F'|E)=1-\mathscr{P}(F|E)$. Thus, if E and F are independent events then $\mathscr{P}(F'|E)=1-\mathscr{P}(F)=\mathscr{P}(F')$, that is, E and F' are also independent events.

Corollary 4.2.1(3)
If E and F are independent events then

$$\mathscr{P}(E\cap F)=\mathscr{P}(F)\mathscr{P}(E)$$

The latter formula is called the *multiplication rule for independent events* and is analogous to the addition rule for two mutually exclusive events (see definition 4.1.6).

It should, perhaps, be emphasised at this point that mutually exclusive events are not, in general, independent. For, if G and H are mutually exclusive events then

$$G\cap H=\emptyset \text{ and so } \mathscr{P}(G\cap H)=0$$

If they are also independent events, then

$$\mathscr{P}(G)\mathscr{P}(H)=\mathscr{P}(G\cap H)=0$$

from above. Thus, either G or H, or both, is the impossible event. In general, therefore, mutually exclusive events are not independent.

Example 4.2.3

(1) A die is thrown and the number appearing on the top is noted. Let E be the event $\{1, 3, 5\}$ and F the event $\{2, 4, 6\}$. Then $\mathscr{P}(E) = \mathscr{P}(F) = 1/2$. But $\mathscr{P}(F|E) = \mathscr{P}(E \cap F)/\mathscr{P}(E) = 0$, that is, $\mathscr{P}(F|E) \neq \mathscr{P}(F)$. E and F are thus dependent events. They are, however, mutually exclusive, since they have no sample points in common.

(2) In an experiment, two coins are tossed. Let E be the event $\{(h, h), (h, t)\}$ and F the event $\{(h, h), (t, h)\}$. E is thus the event: 'the first coin shows heads' and F is the event: 'the second coin shows heads'. Clearly

$$\mathscr{P}(E) = \mathscr{P}(F) = \frac{1}{2}$$

Now

$$\mathscr{P}(F|E) = \mathscr{P}(E \cap F)/\mathscr{P}(E) = \frac{1}{4}\Big/\frac{1}{2} = \frac{1}{2}$$

that is

$$\mathscr{P}(F|E) = \mathscr{P}(F)$$

E and F are thus independent events. They are not mutually exclusive since the sample point (h, h) belongs to both events.

In practice, if two events, E and F, are clearly independent, then corollary 4.2.1(3) may be used to compute $\mathscr{P}(E \cap F)$. Once this has been done, $\mathscr{P}(E \cup F)$ may be calculated using statement (v) in theorem 4.1.1.

Example 4.2.4

A and B each fire one shot at a target. The probabilities of each hitting the target are 0.3 and 0.25, respectively. What is the probability that: (a) the target is hit? (b) if one bullet hits, it is fired by A?

(a) Let E be the event: 'A's shot hits the target' and F the event 'B's shot hits the target'. $E \cap F$ is then the event: 'both shots hit the target' and $E \cup F$ is the event: 'the target is hit'. Since E and F are clearly independent events

$$\mathscr{P}(E \cap F) = \mathscr{P}(E)\mathscr{P}(F) = (0.3)(0.25) = 0.075$$

From statement (v) in theorem 4.1.1

$$\begin{aligned}\mathscr{P}(E \cup F) &= \mathscr{P}(E) + \mathscr{P}(F) - \mathscr{P}(E \cap F)\\ &= 0.3 + 0.25 - 0.075\\ &= 0.475\end{aligned}$$

(b) Let G be the event: 'precisely one bullet hits the target'. The required probability is thus $\mathscr{P}(E|G)$. The sample space for this experiment is $\{(h, h), (h, m), (m, h), (m, m)\}$, where h denotes a hit and m denotes a miss. Thus, $G=\{(h, m), (m, h)\}$, $E=\{(h, h), (h, m)\}$ and $F=\{(h, h), (m, h)\}$.

Since E and F are independent events, E and F' are also independent events (see above), and so are E' and F. Hence, if $H=\{(h, m)\}$ and $K=\{(m, h)\}$, then

$$\begin{aligned}\mathscr{P}(H) &= \mathscr{P}(E \cap F') \\ &= \mathscr{P}(E)\ P(F') \\ &= (0.3)(0.75) = 0.225\end{aligned}$$

and

$$\begin{aligned}\mathscr{P}(K) &= \mathscr{P}(E' \cap F) \\ &= \mathscr{P}(E')\ P(F) \\ &= (0.7)(0.25) = 0.175\end{aligned}$$

Since H and K are mutually exclusive events

$$\begin{aligned}\mathscr{P}(G) &= \mathscr{P}(H \cup K) \\ &= \mathscr{P}(H) + \mathscr{P}(K) \\ &= 0.225 + 0.175 = 0.4\end{aligned}$$

Now, $G \cap E = H$ and so

$$\mathscr{P}(G \cap E) = \mathscr{P}(H) = 0.225$$

from above. Finally

$$\mathscr{P}(E|G) = \frac{\mathscr{P}(G \cap E)}{\mathscr{P}(G)} = (0.225)/(0.4) = 0.5625$$

Multi-step Random Experiments

Definition 4.2.3

A *multi-step random experiment* or *process* consists of two or more steps, each of which has a number of possible outcomes. Such an experiment is also called a *stochastic process.*

Suppose that an experiment has n steps and that the outcome of the ith step is a sample point from a discrete sample space, $\mathscr{S}_i$. The sample space, $\mathscr{T}$, for the multi-step experiment is then given by the cartesian product (see definition 1.5.8) of $\mathscr{S}_1, \mathscr{S}_2, \ldots, \mathscr{S}_n$, that is

$$\mathscr{T} = \mathscr{S}_1 \times \mathscr{S}_2 \times \ldots \times \mathscr{S}_n$$

Definition 4.2.4
An event $E \subset \mathscr{T}$, where $\mathscr{T}$ is the sample space for a stochastic process, is called a *compound* or *composite* event.

Example 4.2.5

(1) A first-year computer science 'problem analysis' class is attended by both male (m) and female (f) students. A student is chosen at random to present his or her solution to a particular problem. The student's solution may be either correct (c) or incorrect (i). This is a two-step random experiment in which

$$\mathscr{S}_1 = \{m, f\}, \ \mathscr{S}_2 = \{c, i\}$$

and

$$\mathscr{T} = \mathscr{S}_1 \times \mathscr{S}_2 = \{(m, c), (m, i), (f, c), (f, i)\}$$

(2) Each of three boxes, a, b and c, contains a number of machine parts; some of the parts in each box are defective (d) and the remainder are non-defective (n). Selecting a box at random and picking a part at random from the chosen box is a two-step random experiment, in which

$$\mathscr{S}_1 = \{a, b, c\}, \ \mathscr{S}_2 = \{d, n\}$$

and

$$\mathscr{T} = \mathscr{S}_1 \times \mathscr{S}_2 = \{(a, d), (a, n), (b, d), (b, n), (c, d), (c, n)\}$$

The following example illustrates how the concept of conditional probability may be used to determine the probability of an event in a multi-step random experiment.

Example 4.2.6

Boxes a, b and c contain 25, 15 and 12 parts, respectively, of which 8, 3 and 4, respectively are defective. Compute the probability, p, that a part selected at random will be defective.

Let A denote the event: 'the part is taken from box a', B the event: 'the part is taken from box b', C the event: 'the part is taken from box c', D the event: 'the part is defective' and N the event: 'the part is non-defective'. Then

$$\mathscr{P}(A) = \mathscr{P}(B) = \mathscr{P}(C) = 1/3$$

and

$$\mathscr{P}(D|A) = 8/25, \ \mathscr{P}(D|B) = 3/15, \ \mathscr{P}(D|C) = 4/12$$

By the multiplication rule for conditional probability

$$\mathscr{P}(A \cap D) = \mathscr{P}(A)\mathscr{P}(D|A) = (1/3) \times (8/25) = 8/75$$

$$\mathscr{P}(B \cap D) = \mathscr{P}(B)\mathscr{P}(D|B) = (1/3) \times (3/15) = 1/15$$

and

$$\mathscr{P}(C \cap D) = \mathscr{P}(C)\mathscr{P}(D|C) = (1/3) \times (4/12) = 1/9$$

Thus, since $A \cap D$, $B \cap D$ and $C \cap D$ are mutually exclusive events

$$p = 8/75 + 1/15 + 1/9 = 64/225$$

Probability Trees

In general, tree structures are useful for describing hierarchical systems: family trees and the Dewey decimal system for classifying library books are familiar examples, and a tree is also an important type of computer data structure.

If each step of a stochastic process has a finite number of possible outcomes, then the process may be described by a tree.

One way of defining a tree is given below. It may also be defined as a special type of digraph; see section 5.2. A recursive definition is also possible; see, for example, Page and Wilson (1978).

Definition 4.2.5
A *tree* consists of a finite set of elements, or *nodes*, possessing the following properties

(a) each node may be assigned a non-negative integer, called the *level* of the node;
(b) there is precisely one node at level zero and this is called the *root* of the tree;
(c) a node at level i may be connected to a number of nodes at level $i+1$ and each such connection is called a *branch* of the tree;
(d) two nodes at level i may not be connected to the same node at level $i+1$.

A *leaf*, or *terminal node*, is a node which is not connected to a node at the next level.

The diagrammatic representation of a typical tree is illustrated in figure 4.3; every node apart from the root is connected to the root by a unique *path* composed of one or more branches.

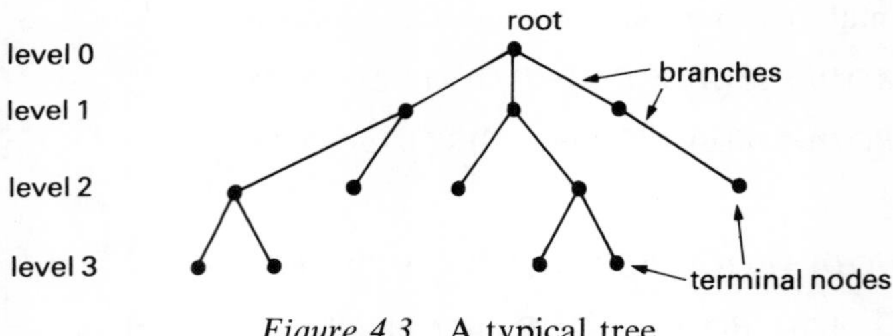

Figure 4.3 A typical tree

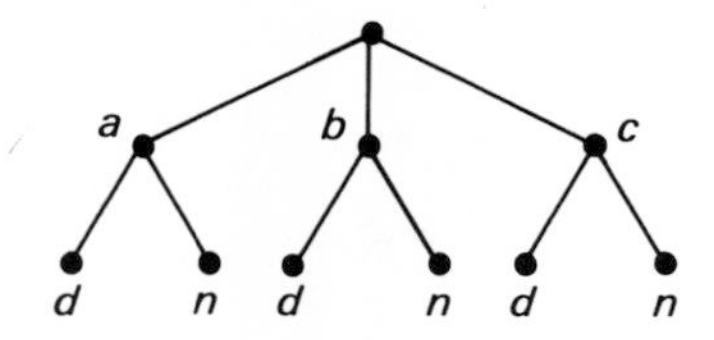

Figure 4.4 Example 4.2.7

Example 4.2.7

A part is selected at random from one of three boxes, *a*, *b* or *c*; each box contains both defective (*d*) and non-defective (*n*) parts. This two-step random experiment may be represented by the tree in figure 4.4.

The path leading to each of the six terminal nodes represents a point in the sample space for the two-step process.

In general, if an *r*-step process has sample space $\mathscr{T} = \mathscr{S}_1 \times \mathscr{S}_2 \times \ldots \times \mathscr{S}_r$, where each $\mathscr{S}_i$ is finite, then any node at level *i* in the corresponding tree represents a point in $\mathscr{S}_i$ and may be labelled accordingly. Furthermore, a branch connecting a particular node at level *i* to a particular node at level $i+1$ may be labelled with the conditional probability that this branch will be followed, given that the path to the node at level *i* has been followed.

Definition 4.2.6
If each branch in the tree corresponding to a finite stochastic process is labelled with the appropriate conditional probability, then the tree is called a *probability tree.*

Example 4.2.8

The probability tree for the experiment described in example 4.2.6 is given in figure 4.5.

Each path leading to a terminal node is unique and the probability, *p*,

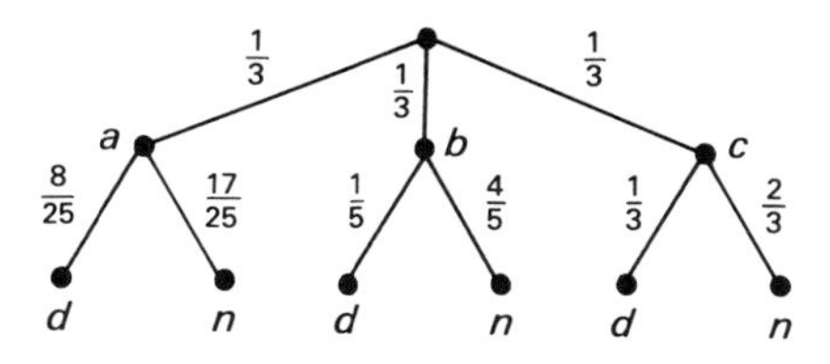

Figure 4.5 Example 4.2.8

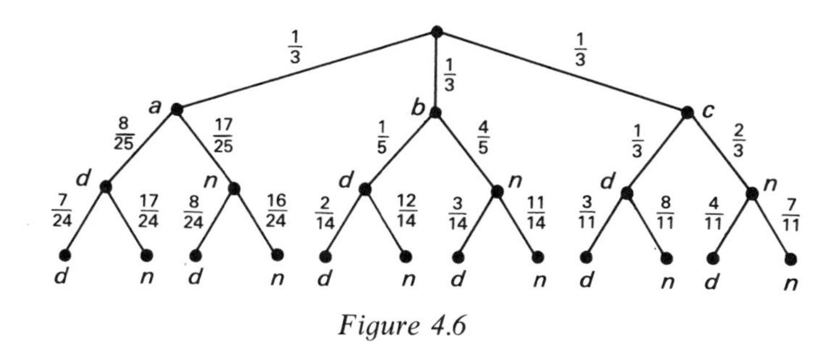

Figure 4.6

that a part selected at random is non-defective is thus given by

$$\begin{aligned} p &= (1/3) \times (17/25) + (1/3) \times (4/5) + (1/3) \times (2/3) \\ &= 161/225 \end{aligned}$$

If a second part is now picked at random from the same box as was the first part, then the probability tree for the resulting three-step process is given in figure 4.6. The probability, q, that both parts will be defective is given by

$$\begin{aligned} q &= (1/3) \times (8/25) \times (7/24) + (1/3) \times (1/5) \times (1/7) \\ &\quad + (1/3) \times (1/3) \times (3/11) \\ &= (7/225) + (1/105) + (1/33) \\ &\approx 0.071 \end{aligned}$$

Markov Chains

Definition 4.2.7
A *Markov chain* is a stochastic process having the following properties

(i) the outcome at each step is a sample point from the same discrete sample space, that is, $\mathscr{T} = \mathscr{S} \times \mathscr{S} \times \mathscr{S} \times \ldots$;

(ii) the outcome at any step in the process depends on the outcome of the immediately preceding step but does not depend upon any further preceding events.

Definition 4.2.8
The sample space, $\mathscr{S}$, for any step in a Markov chain is called the *state space* of the process. If $\mathscr{S}=E_1 \cup E_2 \cup \ldots \cup E_n$, where $E_i \cap E_j=\emptyset$, $i \neq j$, then $\mathscr{S}$ is an *n-state system* and whenever the event E_i occurs the system is said to be in *state i*.

Definition 4.2.9
The probability, p_{ij}, that state j succeeds state i, that is, that the event E_j occurs immediately after the event E_i, is called a *transition probability*. The matrix P given by

$$P=\begin{bmatrix} p_{11} & p_{12} & \cdots & p_{1n} \\ p_{21} & p_{22} & \cdots & p_{2n} \\ \vdots & \vdots & & \vdots \\ p_{n1} & p_{n2} & \cdots & p_{nn} \end{bmatrix}$$

is called the *transition matrix* of the process.
Then, for each i

$$\sum_{j=1}^{n} p_{ij}=1$$

since some outcome must occur after E_i.

Example 4.2.9

In an attempt to keep fit, a computer science lecturer occasionally cycles to work, rather than going by car. Not wishing to overdo things, he never cycles to work two days in succession, but if he goes by car then he is just as likely to go by car the next day.

Since the mode of transport for any day depends only on the mode of transport used the previous day, this stochastic process is a Markov chain.

Let E_1 be the event: 'the lecturer cycles to work' and E_2 the event: 'the lecturer drives to work'. Then $\mathscr{S}=E_1 \cup E_2$ and the transition matrix is

$$P=\begin{bmatrix} 0 & 1 \\ 1/2 & 1/2 \end{bmatrix}$$

Definition 4.2.10
Let $q_j^{(0)}$ denote the probability that the system is in state j when observation of the system begins and let $q_j^{(r)}$ denote the probability that

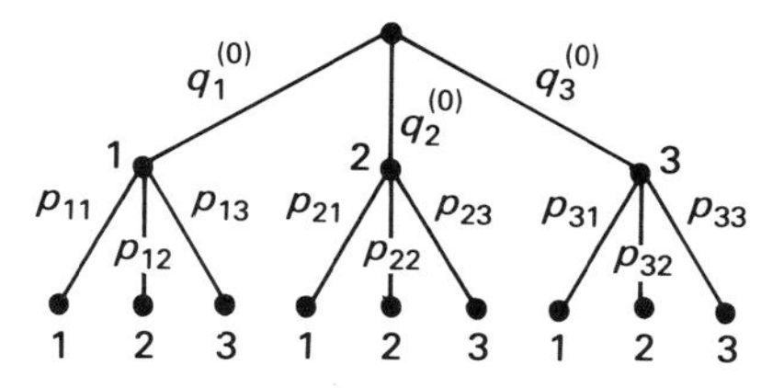

Figure 4.7 Theorem 4.2.3

the system is in state j after the first r steps. Then

$$\boldsymbol{q}^{(0)} = [q_1^{(0)} q_2^{(0)} \ldots q_n^{(0)}]$$

is called the *initial probability vector* and

$$\boldsymbol{q}^{(r)} = [q_1^{(r)} q_2^{(r)} \ldots q_n^{(r)}]$$

is called the rth *step probability vector.*

Since the system must be in some state at each step,

$$\sum_{j=1}^{n} q_j^{(r)} = 1, \; r = 0,1,2,\ldots$$

Theorem 4.2.3
$\boldsymbol{q}^{(r)} = \boldsymbol{q}^{(0)} P^r$, $r = 1,2,3,\ldots$, where P is the transition matrix.

Proof Consider figure 4.7, which illustrates the probability tree for a three-state process after one step. Each path leading to a terminal node represents a distinct event. The probability of being in state j after one step of this process is thus

$$q_j^{(1)} = q_1^{(0)} \times p_{1j} + q_2^{(0)} \times p_{2j} + q_3^{(0)} \times p_{3j}, \; j = 1,2,3$$

More generally, for an n-state process

$$q_j^{(1)} = \sum_{i=1}^{n} q_i^{(0)} \times p_{ij}, \; j = 1,2,\ldots n$$

that is

$$\boldsymbol{q}^{(1)} = \boldsymbol{q}^{(0)} P$$

The result thus holds for $r = 1$ and may easily be extended for a general value of r using mathematical induction. The details are left as an exercise for the reader.

Example 4.2.10

Three boys, A, B and C, are kicking a football to each other. It is known that A is just as likely to pass to B as to C; B is twice as likely to pass to

C as to A; and C always passes to A. If C has the ball at the start, determine the probability that it will be back with him after three passes.

The transition matrix for this Markov chain is

$$P = \begin{array}{c} \\ A \\ B \\ C \end{array} \begin{array}{c} \begin{array}{ccc} A & B & C \end{array} \\ \begin{bmatrix} 0 & 1/2 & 1/2 \\ 1/3 & 0 & 2/3 \\ 1 & 0 & 0 \end{bmatrix} \end{array}$$

Since C has the ball at the start of play,

$$\boldsymbol{q}^{(0)} = [0 \quad 0 \quad 1]$$

Thus

$$\boldsymbol{q}^{(1)} = [0 \ 0 \ 1] \begin{bmatrix} 0 & 1/2 & 1/2 \\ 1/3 & 0 & 2/3 \\ 1 & 0 & 0 \end{bmatrix} = [1 \ 0 \ 0]$$

$$\boldsymbol{q}^{(2)} = [1 \ 0 \ 0] \begin{bmatrix} 0 & 1/2 & 1/2 \\ 1/3 & 0 & 2/3 \\ 1 & 0 & 0 \end{bmatrix} = [0 \quad 1/2 \quad 1/2]$$

$$q^{(3)} = [0 \quad 1/2 \quad 1/2] \begin{bmatrix} 0 & 1/2 & 1/2 \\ 1/3 & 0 & 2/3 \\ 1 & 0 & 0 \end{bmatrix} = [2/3 \quad 0 \quad 1/3]$$

The probability that C has the ball after three passes is thus equal to 1/3.

Definition 4.2.11

If P^r tends to a matrix Q as $r \to \infty$, where P is the transition matrix of a Markov chain and Q is a matrix with identical rows, then the Markov chain is said to be *ergodic*. The vector, $\boldsymbol{q}$, with components equal to the elements in any row of Q is called the *fixed probability vector* of the process.

Clearly

$$PP^r = P^r P = P^{r+1}$$

and, by letting $r \to \infty$, it follows that for an ergodic Markov chain

$$PQ = QP = Q$$

Hence

$$\boldsymbol{q}P = \boldsymbol{q}$$

that is, $\boldsymbol{q}$ is a *fixed vector* of the matrix P.

Example 4.2.11

A travelling salesman's territory consists of Liverpool, Manchester and Sheffield. He never visits the same city two days in succession. If he visits Liverpool, then the next day he visits Manchester. If he visits Manchester, then the next day he is twice as likely to visit Liverpool as he is Sheffield. Similarly, he is twice as likely to visit Liverpool after visiting Sheffield as he is to visit Manchester. In the long run, how often does he visit each city?

The transition matrix for this problem is

$$P = \begin{array}{c} \\ L \\ M \\ S \end{array} \begin{array}{c} \begin{array}{ccc} L & M & S \end{array} \\ \begin{bmatrix} 0 & 1 & 0 \\ 2/3 & 0 & 1/3 \\ 2/3 & 1/3 & 0 \end{bmatrix} \end{array}$$

Let $\boldsymbol{x} = [x_1 \ x_2 \ x_3]$ be any fixed vector of P. Then

$$[x_1 \ x_2 \ x_3] \begin{bmatrix} 0 & 1 & 0 \\ 2/3 & 0 & 1/3 \\ 2/3 & 1/3 & 0 \end{bmatrix} = [x_1 \quad x_2 \quad x_3]$$

that is

$$\frac{2}{3}x_2 + \frac{2}{3}x_3 = x_1$$

$$x_1 + \frac{1}{3}x_3 = x_2$$

$$\frac{1}{3}x_2 = x_3$$

Any set of values x_1, x_2, x_3 satisfying these equations is a fixed vector of P. Let $x_3 = 1$; then $x_2 = 3$ from the third equation and $x_1 = 8/3$ from the second equation. Thus

$$\boldsymbol{x} = [8/3 \ 3 \ 1]$$

is a solution to the equations. The sum of the components of the fixed probability vector, $\boldsymbol{q}$, must equal one. Thus

$$\boldsymbol{q} = \frac{1}{8/3 + 3 + 1}[8/3 \ 3 \ 1]$$

$$= \frac{3}{20}[8/3 \ 3 \ 1]$$

$$= [2/5 \ 9/20 \ 3/20]$$

In the long run, therefore, the salesman visits Liverpool 40 per cent of the time, Manchester 45 per cent of the time and Sheffield 15 per cent of the time.

Exercise 4.2

1. Using mathematical induction, prove that

$$\mathscr{P}(F_1 \cap F_2 \cap \ldots \cap F_r) = \mathscr{P}(F_1)\mathscr{P}(F_2|F_1)\mathscr{P}(F_3|F_1 \cap F_2)\ldots\mathscr{P}(F_r|F_1 \cap F_2 \cap \ldots \cap F_{r-1})$$

2. A 'fair' die is thrown five times. Given that the first throw does not result in a 6, compute the probability that a 6 does not result from any of the throws.

3. A box contains ten parts, of which three are defective. Compute the probability that three parts picked at random from the box will each be defective.

4. Compute the probability that a bridge hand (that is, 13 playing cards) will:

(a) consist entirely of spades;

(b) consist entirely of one suit.

5. Three machines, *A*, *B* and *C*, produce 40, 35 and 25 per cent, respectively, of the total number of items produced by a factory, while 3, 2 and 1 per cent of the items produced by *A*, *B* and *C*, respectively, are defective. Given that an item selected at random is defective, use Bayes' theorem to compute the probability that the defective item was produced by machine *A*.

6. Boxes *a* and *b* contain 20 and 12 parts, respectively, of which 4 and 2, respectively, are defective. If a part is selected at random from each box, what is the probability that at least one of the parts is defective?

7. Two university departments, *A* and *B*, are to acquire computers. A certain computer manufacturer bids for both contracts. The probability that this company will get the contract for department *A* is 0.8. If it gets this contract, then the probability that the company will also get the contract for department *B* is 0.7. Otherwise, the probability that this company will get the latter contract is 0.4.

Draw the probability tree representing the above two-step random experiment. Compute the probability that this company

(a) gets neither contract;

(b) gets precisely one contract;

(c) gets at least one contract;

(d) gets the contract for department B.

8. A two-state Markov chain has transition matrix

$$\begin{bmatrix} 1/2 & 1/2 \\ 1/3 & 2/3 \end{bmatrix}$$

Given the initial probability vector [1/2 1/2], determine the probability that the system is in state 1 after

(a) the first transition;

(b) the second transition.

9. Determine the unique fixed probability vector of the transition matrix

$$P = \begin{bmatrix} 1/2 & 1/4 & 1/4 \\ 1/2 & 0 & 1/2 \\ 0 & 1 & 0 \end{bmatrix}$$

(Hint: seek the vector $\boldsymbol{q} = [u \ \ v \ \ 1-u-v]$ such that $\boldsymbol{q}P = \boldsymbol{q}$.)

10. For the Markov chain described in example 4.2.10, determine how often, in the long run, A, B and C each receives the ball.

4.3 INDEPENDENT TRIALS. DISCRETE PROBABILITY DISTRIBUTIONS

Independent Trials

A stochastic process is called a *repeated random experiment* if each successive step in the process is the repetition of a particular random experiment. A Markov chain is thus an example of such an experiment. If a repeated random experiment consists of M repeated trials of an experiment with sample space $\mathscr{S}$, then the repeated experiment has sample space $\mathscr{S}^M = \mathscr{S} \times \mathscr{S} \times \ldots \times \mathscr{S}$.

Definition 4.3.1
A repeated random experiment consists of *repeated independent trials* if the outcome of each trial is independent of the outcome of any other trial in the experiment.

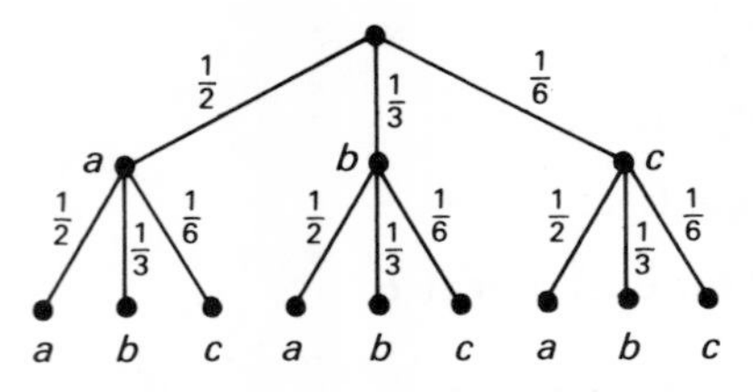

Figure 4.8 Example 4.3.1.2

Let the sample space, $\mathscr{S}$, for an experiment be given by the union of n mutually exclusive events, $E_j=\{s_j\}$, $j=1, 2, \ldots, n$. Suppose that the event E_{j_i} is the outcome of the ith trial in a sequence of M repeated trials of this experiment, where $j_i \in \{1, 2, \ldots, n\}$. If the M trials are independent, then the probability of the event $F=\{(s_{j_1}, s_{j_2}, \ldots, s_{j_M})\}$ in $\mathscr{S}^M$ is given by

$$\mathscr{P}(F)=\mathscr{P}(E_{j_1})\mathscr{P}(E_{j_2})\ldots\mathscr{P}(E_{j_M})$$

Example 4.3.1

(1) The steps in a Markov chain are not independent trials since the outcome of each successive trial in the process generally depends on the outcome of the immediately preceding trial.

(2) Figure 4.8 illustrates the probability tree for a stochastic process consisting of two repeated independent trials of an experiment with sample space $\mathscr{S}=\{a, b, c\}$, in which the events $\{a\}$, $\{b\}$ and $\{c\}$ have probabilities 1/2, 1/3 and 1/6, respectively. Note that any branch leading to the outcome a is labelled with a probability of 1/2, irrespective of the path followed up to the start of the branch. Similarly, any branch leading to b is labelled 1/3 and any branch leading to c is labelled 1/6.

(3) An experiment consists of tossing a coin three times in succession. What is the probability that three 'heads' will show? The repeated trials are clearly independent and the probability of the event $F=\{h, h, h\}$ is thus given by

$$\mathscr{P}(F)=\frac{1}{2}\times\frac{1}{2}\times\frac{1}{2}=\frac{1}{8}$$

Definition 4.3.2
Repeated independent trials in which each trial has only one of two possible outcomes are called *Bernoulli trials.*

It is customary to refer to one of the two possible outcomes of such a trial as a *success* and to the other as a *failure.*

Theorem 4.3.1
If p denotes the probability of a success in a Bernoulli trial, then the probability of exactly k successes in a sequence of n such trials is denoted by $b(k, n, p)$ and is given by

$$b(k, n, p) = {}^nC_k(1-p)^{n-k}p^k$$

Proof The number of ways in which k successes can occur is equal to the number of ways in which k elements can be selected from n elements, that is ${}^nC_k$. Since the probability of a success is equal to p, the probability of a failure is equal to $(1-p)$. The probability of each event in which k successes occur is thus equal to $p^k(1-p)^{n-k}$. Since all of the ${}^nC_k$ possible events are mutually exclusive, the required probability is given by the sum of the probabilities of the individual events, that is

$$b(k, n, p) = {}^nC_k(1-p)^{n-k}p^k$$

Example 4.3.2

(1) If a 'fair' die is thrown four times, what is the probability that a 6 will appear:

(a) exactly twice;

(b) at least twice;

(c) at most twice?

(a) The required probability is

$$b\left(2, 4, \frac{1}{6}\right) = {}^4C_2\left(\frac{5}{6}\right)^2\left(\frac{1}{6}\right)^2$$

$$= \frac{4 \times 3 \times 5 \times 5}{2 \times 6 \times 6 \times 6 \times 6} = \frac{25}{216}$$

(b) Let E_1 denote the event: 'a 6 is thrown exactly twice', E_2 the event: 'a 6 is thrown exactly three times' and E_3 the event: 'a 6 is thrown exactly four times'. Then $E_1 \cup E_2 \cup E_3$ is the event: 'a 6 is thrown at least twice'. The required probability is thus

$$\mathscr{P}(E_1 \cup E_2 \cup E_3) = \mathscr{P}(E_1) + \mathscr{P}(E_2) + \mathscr{P}(E_3)$$

$$= b\left(2, 4, \frac{1}{6}\right) + b\left(3, 4, \frac{1}{6}\right) + b\left(4, 4, \frac{1}{6}\right)$$

$$= \frac{25}{216} + \frac{5}{324} + \frac{1}{1296} = \frac{57}{432}$$

(c) Now, $F = E_2 \cup E_3$ is the event: 'a 6 is thrown more than twice'. Thus, F' is the event: 'a 6 is thrown at most twice'. The required probability is given by

$$\mathscr{P}(F') = 1 - \mathscr{P}(F) = 1 - \frac{5}{324} - \frac{1}{1296} = \frac{425}{432}$$

(2) The *white noise model* is often used when considering the problem of errors in a transmitted binary message. This model is based on the following assumptions

(i) there is an equal probability p of an error in each bit of the message;

(ii) errors in different bits of the message are independent.

One way of encoding a binary message so as to make it error-detecting is to count the number of 1s in the message and then append an extra bit to it so that the entire message (including the additional bit) has *even parity*, that is, has an even number of 1s in it. If a message is received with an odd number of 1s in it, the receiver then knows that the message contains an odd number of errors. With this device, however, the receiver cannot detect an even number of errors.

What is the probability of an undetected error in a binary message consisting of n bits (including the parity bit), assuming the white noise model for errors in the message?

The probability of k errors in the message is clearly given by $b(k, n, p)$. In particular, the probability of no error in the n bits is $(1-p)^n$.

If $m = 2 \times (n/2)$, where '/' here denotes integer division, then the probability of an even number of errors is given by

$$b(0, n, p) + b(2, n, p) + \ldots + b(m, n, p)$$

$$= \frac{1}{2}\left\{ \sum_{k=0}^{n} {}^nC_k p^k (1-p)^{n-k} + \sum_{k=0}^{n} (-1)^k \, {}^nC_k p^k (1-p)^{n-k} \right\}$$

$$= \frac{1}{2}\{[(1-p)+p]^n + [(1-p)-p]^n\}$$

by the binomial theorem (see theorem 3.3.4)

$$= \frac{1}{2}\{1 + (1-2p)^n\}$$

This is the probability of the event: 'either there is no error or there are undetected errors'.

Since the probability of no error in the message is $(1-p)^n$, the probability of an undetected error is

$$\frac{1}{2}\{1+(1-2p)^n\}-(1-p)^n=\frac{1}{2}\{1+b(0, n, 2p)\}-b(0, n, p)$$

The interested reader is referred to Hamming (1980) for an excellent introduction to this important area.

Random Variables

It is often useful to refer to the outcome of an experiment numerically, even though the outcome itself may not be a number.

Definition 4.3.3
A *random variable*, X, for an experiment with sample space $\mathscr{S}$ is a function which assigns a real number to each sample point in $\mathscr{S}$.
The image set of X is called the *range space* of X and is denoted by $\mathscr{R}_X$. If $\mathscr{S}$ is a discrete space then so is $\mathscr{R}_X$ and X is called a *discrete random variable*; if $\mathscr{R}_X$ is continuous then X is called a *continuous random variable* (see section 4.4).

If X is a discrete random variable, then either

$$\mathscr{R}_X=\{x_i|i=1, 2, \ldots, r(r\in N)\} \text{ or } \mathscr{R}_X=\{x_i|i\in N\}$$

that is, $\mathscr{R}_X$ is a countable set (see section 1.3).

Example 4.3.3

(1) When a coin is tossed, the outcome 'heads' may be assigned the value 1 and the outcome 'tails' the value 0. This defines a discrete random variable on the sample space $\mathscr{S}=\{$'heads', 'tails'$\}$.

(2) Two dice are thrown and the score for each die noted. One way of assigning a numerical value to each outome is to add the two scores. Note that for this choice of random variable, a number of outcomes are assigned the same value. For example, (1, 5), (2, 4), (3, 3), (4, 2) and (5, 1) are each assigned the value 6. The range space of the random variable is $\{2, 3, 4, 5, 6, 7, 8, 9, 10, 11, 12\}$.

Many random variables may be defined on the same sample space. For the second of the above examples, an alternative random variable is defined by assigning to each outcome the value of the difference between the two scores. The range space for this random variable is $\{0, 1, 2, 3, 4, 5\}$.

Table 4.2

i	1	2	3	4	5	6	7	8	9	10	11
x_i	2	3	4	5	6	7	8	9	10	11	12
p_i	1/36	1/18	1/12	1/9	5/36	1/6	5/36	1/9	1/12	1/18	1/36

Discrete Probability Distributions

Let X be a random variable defined on a sample space $\mathscr{S}$. If each point in $\mathscr{S}$ is assigned a probability, then each value in the range space of X may also be assigned a probability.

For each $x_i \in \mathscr{R}_X = \{x_1, x_2, \ldots, x_r\}$, let p_i denote the sum of the probabilities of those points in $\mathscr{S}$ to which X assigns the value x_i. The probability, $\mathscr{P}(x_i)$, of the outcome x_i in $\mathscr{R}_X$ is then assigned the value p_i. $\mathscr{P}(x_i)$ is sometimes alternatively written as $\mathscr{P}(X = x_i)$.

Definition 4.3.4
The set $\mathscr{P}_X = \{(x_i, p_i) | x_i \in \mathscr{R}_X\}$ is called the (*discrete*) *probability distribution* of X. The function $x_i \mapsto \mathscr{P}(x_i)$, $x_i \in \mathscr{R}_X$, is sometimes called the *probability density function* of the discrete random variable X.

Clearly, $p_i \geqslant 0$ for each $x_i \in \mathscr{R}_X$. Furthermore, if

$$\sum_{x_i \in \mathscr{R}X} p_i$$

denotes the sum of the probabilities of all the elements in $\mathscr{R}_X$, then

$$\sum_{x_i \in \mathscr{R}X} p_i = 1$$

If $\mathscr{S}$ is an equiprobable space, containing n points, and if X is a random variable on $\mathscr{S}$ assigning the value x_i to n_i distinct sample points, then

$$p_i = \frac{n_i}{n}$$

Example 4.3.4

Two dice are thrown and the score from each die is noted. The sample space for this experiment is equiprobable and has 36 elements. Let X be the random variable assigning to each sample point the sum of the two scores. The probability distribution of X is given by table 4.2.

Only the sample point (1, 1) is assigned the value 2. Hence $\mathscr{P}(2) = p_1$

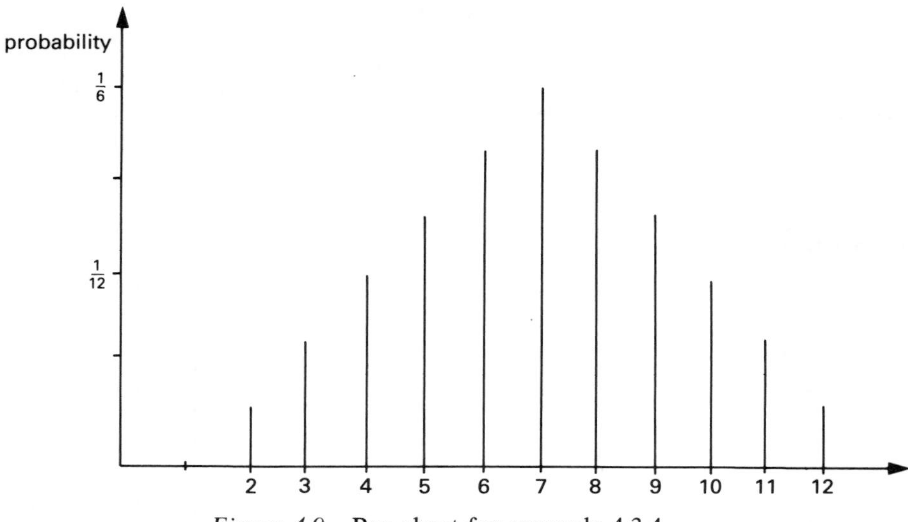

Figure 4.9 Bar chart for example 4.3.4

$=1/36$. Both (1, 2) and (2, 1) are assigned the value 3 and so $\mathscr{P}(3)=p_2$ $=1/18$, and so on.

A discrete probability distribution may be illustrated diagrammatically by means of a *bar chart*. The bar chart corresponding to the distribution given in the last example is shown in figure 4.9.

In computer sampling (see section 4.6), it is more convenient to use the cumulative probability distribution of a random variable, rather than the probability distribution itself. The cumulative probability of $x_i \in \mathscr{R}_X$ is denoted by $\mathscr{U}(x_i)$ and is defined by

$$\mathscr{U}(x_i) \leftarrow \textbf{if } i=1 \textbf{ then } p_1 \textbf{ else } \mathscr{U}(x_{i-1})+p_i \textbf{ endif}$$

Definition 4.3.5
The set $\mathscr{C}_X=\{(x_i, \mathscr{U}(x_i)) | x_i \in \mathscr{R}_X\}$ is called the *cumulative probability distribution* of the random variable X.

Example 4.3.5

The cumulative distribution of the random variable defined in example 4.3.4 is given by table 4.3, and is displayed graphically in figure 4.10.

Table 4.3

i	1	2	3	4	5	6	7	8	9	10	11
x_i	2	3	4	5	6	7	8	9	10	11	12
$\mathscr{U}(x_i)$	1/36	1/12	1/6	5/18	5/12	7/12	13/18	5/6	11/12	35/36	1

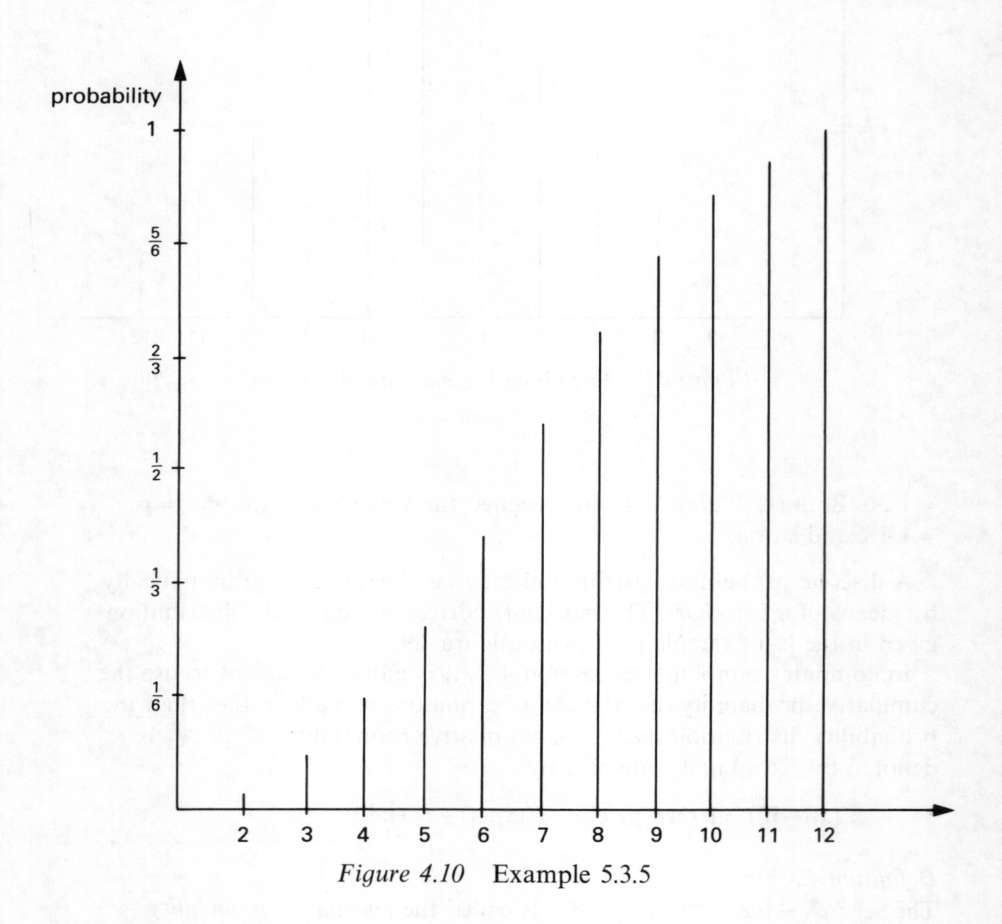

Figure 4.10 Example 5.3.5

Mean and Variance of Probability Distributions

Two measures are commonly used for describing the essential characteristics of the probability distribution of a random variable. These measures are called the mean and the variance of the distribution.

Definition 4.3.6

The *mean*, μ, of the probability distribution $\mathscr{P}_X$ of a discrete random variable X is given by

$$\mu = \sum_{x_i \in \mathscr{R}_X} x_i p_i, \text{ provided this sum exists}$$

This measure is also called the *expectation*, or *expected value*, $\mathscr{E}(X)$, of X.

Example 4.3.6

The mean of the distribution given in example 4.3.4 is equal to 7.

Definition 4.3.7
The *variance*, Var(X), of the probability distribution $\mathscr{P}_X$ is given by

$$\text{Var}(X) = \sum_{x_i \in \mathscr{R}_X} (x_i - \mu)^2 p_i, \text{ when this sum exists}$$

The variance is thus a measure of the 'spread' of the distribution about the expected value.

The square root of Var(X) is usually denoted by σ and is called the *standard deviation* of X.

The calculation of Var(X) using the above formula is described in algorithm 4.3.1, where $\mathscr{R}_X$ is assumed to have cardinality r. There are two **for** loops, and $6r$ arithmetic operations are required.

Algorithm 4.3.1

```
μ←0;
for i from 1 in steps of 1 to r do
    μ←μ+x_i * p_i
endfor;
Var←0;
for i from 1 in steps of 1 to r do
    Var←Var+p_i * (x_i−μ)^2
endfor
```

A more efficient way of computing the variance is described in algorithm 4.3.2. This algorithm uses the following equivalent formula (see exercise 4.3) for Var(X)

$$\text{Var}(X) = \sum_{x_i \in \mathscr{R}_X} x_i^2 p_i - \mu^2$$

Again, $\mathscr{R}_X$ is assumed to have cardinality r. Only one **for** loop is now used and only $4r+2$ arithmetic operations are required.

Algorithm 4.3.2

> $\mu \leftarrow 0$; $x2p \leftarrow 0$;
>
> **for** i **from** 1 **in steps of** 1 **to** r **do**
>
> > $t \leftarrow x_i * p_i$;
> >
> > $\mu \leftarrow \mu + t$;
> >
> > $x2p \leftarrow x2p + t * x_i$;
>
> **endfor**
>
> $\mathrm{Var} \leftarrow x2p - \mu * \mu$

The final arithmetic step in algorithm 4.3.2 will often involve the subtraction of two large and nearly equal numbers. Consequently, loss of significant digits due to cancellation could occur. However, this problem will not arise if the number of significant digits in the given data is substantially less than the maximum number of significant digits to which real numbers may be held in the computer being used. If this is not the case, double-length arithmetic should be used (or else algorithm 4.3.1).

Example 4.3.7

For the distribution in example 4.3.4

$$\sum_{i=1}^{11} x_i^2 p_i = 54.8333$$

and $\mu^2 = 49$. Hence

$$\mathrm{Var}(X) = 5.8333 \text{ and } \sigma = 2.4152$$

Mean and Variance of Frequency Distributions

Suppose that a random variable X is defined on a sample space $\mathscr{S}$. If an experiment consists of M repeated trials, with the outcome of each trial belonging to $\mathscr{S}$, then a sample of size M is said to be taken from X. Let $E_i \subset \mathscr{S}$ consist of the points in $\mathscr{S}$ to which X assigns the value x_i and let f_i denote the relative frequency of E_i with respect to a sample of size M.

Table 4.4

i	1	2	3	4	5	6	7	8	9	10	11
x_i	2	3	4	5	6	7	8	9	10	11	12
f_i	1/50	1/50	1/10	7/50	7/50	3/25	9/50	1/10	2/25	1/10	0

Definition 4.3.8
The set $\mathscr{F}_X = \{(x_i, f_i) | x_i \in R_X\}$ is called the *frequency distribution of* X, for the sample of size M.

Example 4.3.8

Let X be the random variable defined in example 4.3.4. Suppose that the two dice are thrown 50 times, and the following sequence of values obtained when the two scores for each throw are added together: 5, 8, 8, 6, 2, 8, 11, 3, 6, 11, 8, 11, 9, 5, 6, 10, 11, 7, 8, 7, 5, 8, 10, 7, 4, 7, 9, 5, 10, 4, 10, 8, 8, 9, 6, 8, 5, 5, 9, 6, 7, 9, 4, 4, 11, 6, 7, 6, 5, 4.

The frequency distribution of X for this sample of size 50 is given by table 4.4.

This frequency distribution is to be compared with the probability distribution given in example 4.3.4.

The frequency distribution of a random variable, X, for a sample of size M, furnishes an approximation to the underlying probability distribution. In general, the bigger the value of M, the better the approximation to the required distribution. If the probability distribution of a random variable cannot be deduced *a priori* — for example, by symmetry considerations — then it may be obtained as the limit of the frequency distribution for very large sample sizes. This intuitive idea is made precise by the *weak law of large numbers* (theorem 4.5.3).

Definition 4.3.9
The *sample mean*, $\bar{x}$, of the frequency distribution $\mathscr{F}_X = \{(x_i, f_i) | x_i \in \mathscr{R}_X\}$, obtained when a sample of size M is taken from a random variable, X, is given by

$$\bar{x} = \sum_{x_i \in \mathscr{R}_X} x_i f_i$$

Example 4.3.9

For the sample of size 50 given in the last example

$$\bar{x} = 7.12$$

This is to be compared with the value $\mu = 7$ for the underlying probability distribution.

Definition 4.3.10
The *sample variance*, s^2, of the frequency distribution $\mathscr{F}_X$ is given by

$$s^2 = \sum_{x_i \in \mathscr{R}X} (x_i - \bar{x})^2 f_i$$

The square root, s, of the sample variance is called the *standard deviation* of the sample.

In practice, it is more efficient (see algorithm 4.3.2) to use the following equivalent formula for calculating the sample variance

$$s^2 = \sum_{x_i \in \mathscr{R}X} x_i^2 f_i - \bar{x}^2$$

Example 4.3.10

For the sample of size 50 given in example 4.3.8

$$\sum_{i=1}^{11} x_i^2 f_i = 56 \text{ and } \bar{x}^2 = 50.6944$$

Hence

$$s^2 = 5.3056 \text{ and } s = 2.3034$$

These numbers are to be compared with $\sigma^2 = 5.8333$ and $\sigma = 2.4152$, respectively (see example 4.3.7).

Some Important Discrete Distributions

Definition 4.3.11
The probability distribution of a random variable, X, is a *uniform distribution* if, for some non-negative integer, n

(i) $\mathscr{R}_X = \{0, 1, 2, \ldots, n\}$

and

(ii) $\mathscr{P}(k) = 1/(n+1)$, $k \in \mathscr{R}_X$

Theorem 4.3.2
(i) The mean of a uniform distribution is given by $\mu = n/2$.

(ii) The variance of a uniform distribution is given by $\text{Var}(X) = n(n+2)/12$.

Proof (i) From example 3.2.3

$$\mu = \frac{1}{n+1}\sum_{k=0}^{n} k = \frac{1}{(n+1)}\left[\frac{n(n+1)}{2}\right]$$

that is

$$\mu = \frac{1}{2}n$$

(ii)

$$\text{Var}(X) = \frac{1}{n+1}\sum_{k=0}^{n} k^2 - \frac{n^2}{4}$$

$$= \frac{1}{n+1}\left[\frac{n(2n+1)(n+1)}{6}\right] - \frac{n^2}{4}$$

from example 3.2.10

$$= \frac{1}{12}n(n+2)$$

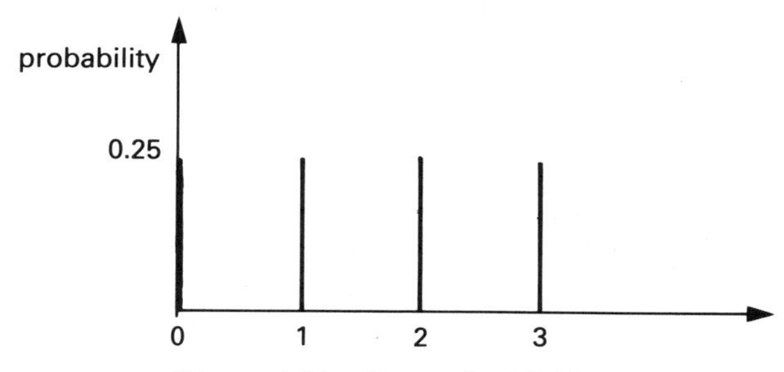

Figure 4.11 Example 4.3.11

Example 4.3.11

In an experiment, two coins are tossed. Let X be the random variable assigning the values 0, 1, 2 and 3, respectively, to the outcomes (h, h), (h, t), (t, h) and (t, t). The probability distribution is given by

$$\mathscr{P}(k) = 0.25,\ k = 0, 1, 2, 3$$

and is thus a uniform distribution, with mean value 1.5 and variance 1.25.
The bar chart corresponding to this distribution is shown in figure 4.11.

Definition 4.3.12
The probability distribution of a random variable, X, is a *geometric distribution* if

(i) $\mathscr{R}_X = Z^+$

and

(ii) $\mathscr{P}(k) = p^k(1-p)$, $k \in \mathscr{R}_X$, where $p \in (0, 1)$.
Clearly, $\mathscr{P}(k) = p^k(1-p) > 0$, for $p \in (0, 1)$. Furthermore

$$\sum_{k=0}^{\infty} p^k(1-p) = (1-p)\sum_{k=0}^{\infty} p^k = (1-p)/(1-p)$$

since

$$\sum_{k=0}^{\infty} p^k$$

is an infinite geometric series (hence the name of the distribution). Thus,

$$\sum_{k=0}^{\infty} \mathscr{P}(k) = 1$$

as required for a valid probability distribution.

Theorem 4.3.3
(i) The mean of a geometric distribution is given by $\mu = p/(1-p)$.
(ii) The variance of a geometric distribution is given by $\mathrm{Var}(X) = p/(1-p)^2$.

Proof (i)

$$\mu = (1-p)\sum_{k=0}^{\infty} kp^k = (1-p)p(1+2p+3p^2+\ldots)$$

$$= (1-p)p/(1-p)^2$$

from example 3.2.6

$$= p/(1-p)$$

(ii) See exercise 4.3.

Example 4.3.12

It is estimated that one in four programs submitted by first-year computer science students compile successfully. A batch of such programs is run by one of the university's computers. Let X be the random variable assigning the value k to the outcome of the batch-run when the first program that fails to compile is the $k+1$st. The probability distribution of this random variable is given by

$$\mathscr{P}(k)=(0.25)^k(0.75),\ k=0,\ 1,\ 2,\ \ldots$$

that is, it is a geometric distribution, with $p=0.25$. The expected value of X is 1/3.

Definition 4.3.13
The probability distribution of a random variable, X, is a *binomial distribution* if for some non-negative integer, n

(i) $\mathscr{R}_X=\{0, 1, 2, \ldots, n\}$;

and

(ii) $\mathscr{P}(k)={}^nC_k(1-p)^{n-k}p^k,\ k\in\mathscr{R}_X$, where $p\in(0, 1)$.

From the binomial theorem (see theorem 3.3.4)

$$\sum_{k=0}^{n}{}^nC_k(1-p)^{n-k}p^k=(1-p+p)^n=1$$

The binomial distribution thus satisfies the conditions necessary for it to be a probability distribution.

From theorem 4.3.1, if $b(k, n, p)$ denotes the probability of exactly k successes in a sequence of n Bernoulli trials, each of which has probability of success p, then

$$\mathscr{P}(k)=b(k, n, p)$$

gives a binomial distribution.

Theorem 4.3.4
(i) The mean of a binomial distribution is given by $\mu=np$;

(ii) The variance of a binomial distribution is given by $\mathrm{Var}(X)=np(1-p)$.

Proof (i) After dropping the $k=0$ term, the expression for μ is

$$\mu=\sum_{k=1}^{n}{}^nC_k(1-p)^{n-k}p^kk$$

$$=\sum_{k=1}^{n}\frac{kn!}{k!(n-k)!}(1-p)^{n-k}p^k$$

$$=np\sum_{k=1}^{n}\frac{(n-1)!}{(k-1)!(n-k)!}(1-p)^{n-k}p^{k-1}$$

$$= np \sum_{k=0}^{n-1} \frac{(n-1)!}{k!(n-k-1)!}(1-p)^{n-1-k}p^k$$

$$= np(1-p+p)^{n-1}$$

by the binomial theorem, and hence

$$\mu = np$$

(ii) See exercise 4.3.

Example 4.3.13

An on-line computer system in a university computer science department has 20 terminals, which operate independently. The probability that any particular terminal is in use is 0.4.

In a random experiment, each terminal is polled and a record kept of those that are not in use. Let X be the random variable assigning to the outcome of this experiment the number of terminals not in use. The probability distribution of X is a binomial distribution given by the expression: for $k = 0, 1, \ldots, 20$

$$\mathscr{P}(k) = {}^{20}C_k(0.4)^{20-k}(0.6)^k$$

Here X has a mean value of 12 ($= 20 \times 0.6$), with standard deviation of $(12 \times 0.4)^{1/2} \approx 2.19$. The probability that m or more terminals are not in use is given by

$$\sum_{k=m}^{20} b(k, 20, 0.4)$$

Note that in the above experiment, a success in the Bernoulli trial corresponds to a terminal being unused when polled.

Definition 4.3.14
The probability distribution of a random variable, X, is a *Poisson distribution* if:

(i) $\mathscr{R}_X = Z^+$

(ii) $\mathscr{P}(k) = \dfrac{\lambda^k e^{-\lambda}}{k!}$, $k \in \mathscr{R}_X$, where $\lambda > 0$.

Since

$$\sum_{k=0}^{\infty} \frac{\lambda^k e^{-\lambda}}{k!} = e^{-\lambda} \sum_{k=0}^{\infty} \frac{\lambda^k}{k!} = e^{-\lambda}e^{\lambda} = 1$$

the Poisson distribution is a valid probability distribution.

Theorem 4.3.5

(i) The expected value of a Poisson distribution is equal to λ;

(ii) The variance of a Poisson distribution is also equal to λ.

Proof (i)

$$\begin{aligned}\mu &= \sum_{k=0}^{\infty} \{k\lambda^k e^{-\lambda}/k!\} = e^{-\lambda} \sum_{k=1}^{\infty} \{\lambda^k/(k-1)!\} \\ &= \lambda e^{-\lambda} \sum_{k=1}^{\infty} \{\lambda^{k-1}/(k-1)!\} \\ &= \lambda e^{-\lambda} e^{\lambda} \\ &= \lambda\end{aligned}$$

(ii) See exercise 4.3.

If an experiment consists of a large number of Bernoulli trials, and if the probability, p, of a successful trial is relatively small, then it may be shown (see exercise 4.3) that

$$b(k, n, p) \approx \frac{\lambda^k e^{-\lambda}}{k!}$$

where $\lambda = np$, that is, for such an experiment, the Poisson distribution approximates the binomial distribution.

The Poisson distribution is one of the two basic distributions arising in *queueing problems* (sometimes called *waiting-line problems*). The other one is the negative exponential distribution (see section 4.4).

Example 4.3.14

1. Once an hour throughout the normal working day, the operator of a 'remote job station' on a university campus inputs to the central computer, via a 'link', each of the jobs that have been submitted during the previous hour. If jobs are submitted at an average rate of 10 per hour, what is the probability that the operator will have: (a) precisely 10 jobs to input? (b) no more than 5 jobs to input?

(a) Since λ is equal to the average rate of submission, which is 10

$$\mathscr{P}(10) = \frac{\lambda^{10} e^{-\lambda}}{10!} = \frac{10^{10} e^{-10}}{10!}$$

Hence, $\mathscr{P}(10) \approx 0.1251$.

(b)

$$\mathscr{U}(5) = \sum_{k=0}^{5} \frac{10^k e^{-10}}{k!} = \frac{4433}{3} e^{-10} \approx 0.0671$$

2. In example 4.3.2(2), the probability of an undetected error in a binary message of length n is shown to be

$$\frac{1}{2}\{1+b(0, n, 2p)\}-b(0, n, p)$$

where p denotes the probability of an error in each bit. If p is relatively small then the above expression is closely approximated by

$$\mathscr{P}(n)=\frac{1}{2}(1+e^{-2np})-e^{-np}$$

Suppose $p=0.01$ and it is required to determine the maximum length n of a message for which the probability of an undetected error is less than 0.005.

Since $\mathscr{P}(10)\approx 0.0045$ and $\mathscr{P}(11)\approx 0.0054$, the required value of n is 10.

Exercise 4.3

1. If a pair of dice are thrown six times, what is the probability that 7 will be obtained exactly twice?

2. A box contains ten parts, of which three are defective. In an experiment, three parts are selected at random from the box. Let X be the random variable assigning to the outcome of the experiment the number of defective parts in the selected sample. By assigning the appropriate probabilities to each of the elements in the range space of X, determine the probability distribution of X and the corresponding cumulative distribution.

3. Verify that the two formulae given in the text for the variance of the distribution of a random variable are equivalent.

4. Calculate the mean and the variance of the probability distribution determined in exercise 4.3.2.

5. Given a discrete random variable, X, for an experiment, let $aX+b$ denote another random variable which assigns the value $ax+b$ to an outcome of the experiment to which X assigns the value x. Show that:

(a) $E(aX+b)=aE(X)+b$;

(b) $\mathrm{Var}(aX+b)=a^2\ \mathrm{Var}(X)$.

6. Prove theorem 4.3.3(ii), that is, show that

$$(1-p)\sum_{k=0}^{\infty} k^2p^k-p^2/(1-p)^2=p/(1-p)^2$$

[Hint: Show that

$$(1-p)s_n = 2\sum_{k=1}^{n} kp^k - \sum_{k=1}^{n} p^k - n^2 p^{n+1}$$

where

$$s_n = \sum_{k=0}^{n} k^2 p^k$$

As $n \to \infty$

$$\sum_{k=1}^{n} p^k \to p/(1-p)$$

and, from example 3.2.6

$$\sum_{k=1}^{n} kp^k \to p/(1-p)^2$$

and $n^2 p^{n+1} \to 0$].

7. Prove theorem 4.3.4(ii), that is, show that

$$\sum_{k=0}^{n} k^2 b(k, n, p) - n^2 p^2 = np(1-p)$$

[Hint: Show that

$$\begin{aligned}\sum_{k=0}^{n} k^2 b(k, n, p) &= np\left\{\sum_{k=0}^{n-1} b(k, n-1, p) + (n-1)p \sum_{k=0}^{n-2} b(k, n-2, p)\right\} \\ &= np\{1+(n-1)p\}]\end{aligned}$$

8. Prove theorem 4.3.5(ii), that is, show that

$$\sum_{k=0}^{\infty} \frac{k^2 \lambda^k e^{-\lambda}}{k!} - \lambda^2 = \lambda$$

9. Show that $b(k, n, p) \approx \lambda^k e^{-\lambda}/k!$, where $\lambda = np$, when n is large and p is small. [Hint: For large n

$$\frac{n!}{(n-k)!n^k} \approx 1 \text{ and } \left(1-\frac{\lambda}{n}\right)^{n-k} \approx e^{-\lambda}]$$

10. On average, a firm's computer fails twice in 100 hours. What is the probability of it failing

(a) exactly once in 48 hours;

(b) more than twice in 72 hours?

11. Show that $E(X)$ is not defined when X has probability distribution $\mathscr{P}_X = \{(2^i, 2^{-i}) | i \in N\}$.

4.4 CONTINUOUS PROBABILITY DISTRIBUTIONS

Continuous Random Variables

The range space of a discrete random variable consists of a discrete set of numbers from the real line. For example, if a computing centre operates a time-sharing system, handling up to n terminals, then an integer in the range 1 to n may be used to indicate the terminal from which a message originates. On the other hand, if an experiment consists of recording the response time when a message is entered via one of the terminals, then any time is theoretically possible. The range space of this random variable is thus the set of non-negative real numbers and the random variable is said to be continuous.

Let a denote a positive real number, chosen at random, and let E denote the event: 'the response time is equal to a'. Intuitively, the probability of E must be infinitesimally small, since a could be *any* positive number. It is clearly more meaningful to consider the probability of the event: 'the response time lies in the interval $[a, b]$', for some $b > a$.

In general, a continuous random variable X may range over the entire real line. Let $\mathscr{P}(\{x | a \leqslant x \leqslant b\})$ or, more briefly, $\mathscr{P}(a \leqslant x \leqslant b)$, denote the probability that the value of X lies within the interval $[a, b]$. Then $\mathscr{P}(a \leqslant x \leqslant b)$ may be defined in terms of a function called the probability density function of X.

Definition 4.4.1
A real function $p: x \mapsto p(x)$ is a valid *probability density function* (p.d.f.) provided it satisfies the following properties

(1) $p(x) \geqslant 0, \forall x \in \mathbb{R}$;

(2) $$\int_{-\infty}^{\infty} p(x) \, dx = 1.$$

Let X be a continuous random variable. Then p is the p.d.f. of X if

$$\mathscr{P}(a \leqslant x \leqslant b) = \int_a^b p(x) \, dx$$

for any interval $[a, b]$ in the range space of X.

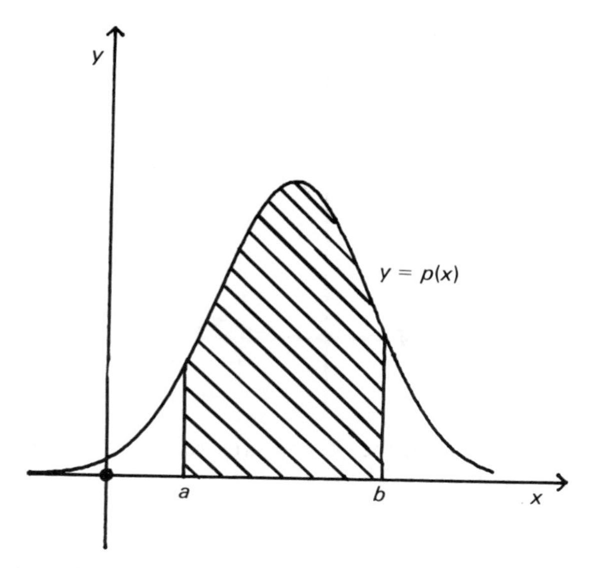

Figure 4.12 A probability density function; $\mathscr{P}(a \leqslant x \leqslant b)$ is represented by the shaded area

The p.d.f. of a continuous random variable X completely defines the probability distribution of X. Such a distribution is called a *continuous probability distribution.*

The graph of a p.d.f. is illustrated in figure 4.12. By definition, the total area beneath the curve $y = p(x)$ is equal to one and the area between $x = a$ and $x = b$ represents the probability that the value of the random variable lies in $[a, b]$.

Example 4.4.1

Let p be defined by

$$p\colon x \mapsto \begin{cases} cx^2, & -3 \leqslant x \leqslant 3 \\ 0, & \text{otherwise} \end{cases}$$

If p is a p.d.f., c must be non-negative. Furthermore

$$\int_{-\infty}^{\infty} p(x)\, \mathrm{d}x = c \int_{-3}^{3} x^2\, \mathrm{d}x = 18c = 1$$

that is, p is a p.d.f. provided $c = 1/18$. The graph of this p.d.f. is illustrated in figure 4.13.

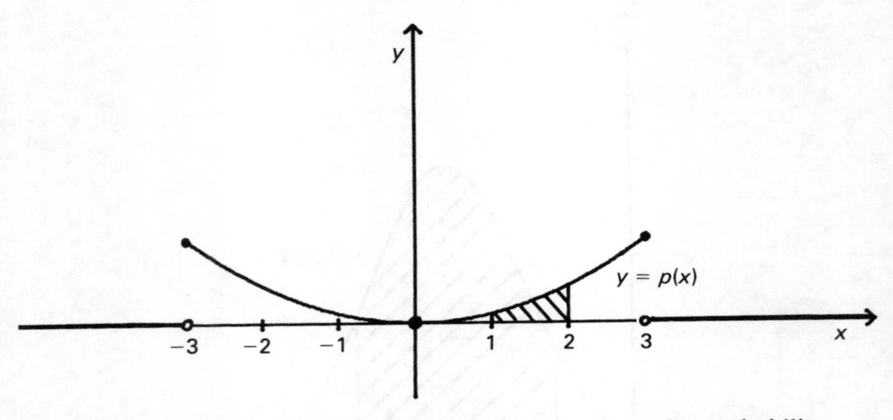

Figure 4.13 Example 4.4.1; the shaded area represents the probability that x lies in the interval [1, 2]

The probability that x lies in the interval [1, 2] is given by

$$\mathscr{P}(1 \leqslant x \leqslant 2) = \frac{1}{18}\int_1^2 x^2 \, dx = \frac{7}{54}$$

This probability is represented in figure 4.13 by the shaded area.

Definition 4.4.2
If p denotes the p.d.f. of a continuous random variable X, then the real function $\mathscr{U}$ defined by

$$\mathscr{U}: x \mapsto \int_{-\infty}^{x} p(t) \, dt$$

is called the *cumulative distribution function* (c.d.f.) of X.

The following two properties of the c.d.f. follow immediately from its definition

(1) $0 \leqslant \mathscr{U}(x) \leqslant 1, \forall x \in \mathbb{R}$;

(2) $\mathscr{P}(a \leqslant x \leqslant b) = \mathscr{U}(b) - \mathscr{U}(a)$.

The graph of a c.d.f. $\mathscr{U}$ is illustrated in figure 4.14, together with the graph of the corresponding p.d.f., p. The area beneath the curve $y = p(x)$ up to the point $x = a$ is equal to the value of $\mathscr{U}(a)$.

From definition 4.4.2, $\mathscr{U}$ is a differentiable function, with

$$\frac{d}{dx} \mathscr{U}(x) = p(x)$$

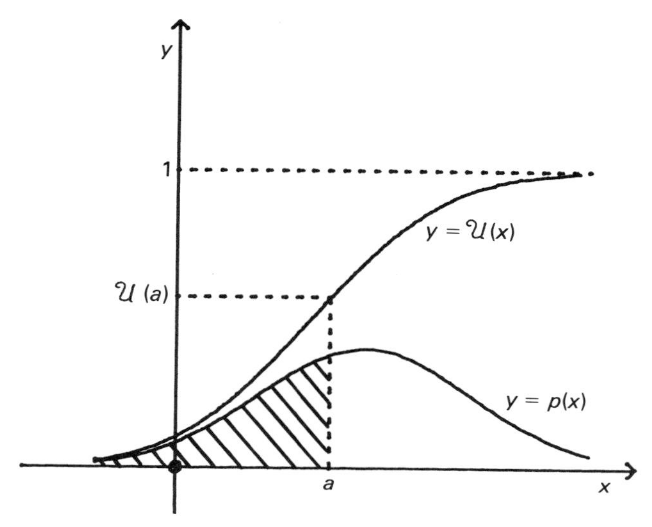

Figure 4.14 A cumulative distribution function $\mathscr{U}(x)$ and its corresponding p.d.f., $p(x)$; the shaded area represents the value of $\mathscr{U}(a)$

The probability distribution of a continuous random variable may thus be completely defined by either its p.d.f. or its c.d.f.

Example 4.4.2

The c.d.f. corresponding to the p.d.f. p given in example 4.4.1 is defined by

$$\mathscr{U}: x \mapsto \begin{cases} 0, & x < -3 \\ \frac{1}{2}(x^3/27 + 1), & -3 \leqslant x \leqslant 3 \\ 1, & x > 3 \end{cases}$$

The graph of this function is illustrated in figure 4.15. *Note* although in this example p is a discontinuous function, the corresponding c.d.f. $\mathscr{U}$ is continuous. The c.d.f. of a continuous random variable is always a continuous function but the p.d.f. may be discontinuous.

Mean and Variance of Continuous Probability Distributions

Definition 4.4.3
The *mean*, μ, of the probability distribution of a continuous random variable, X, is given by

$$\mu = \int_{-\infty}^{\infty} xp(x)\,\mathrm{d}x, \text{ provided this integral exists}$$

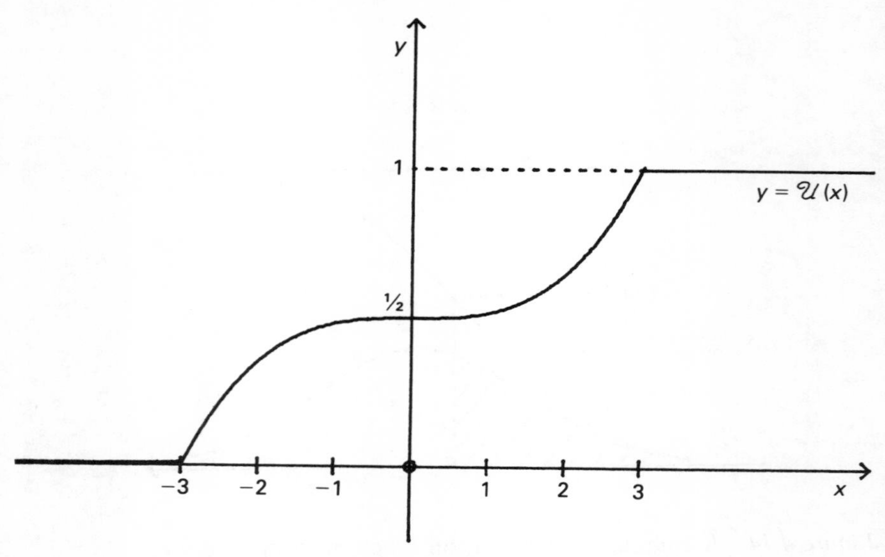

Figure 4.15 Example 4.4.2

where p denotes the p.d.f. of X. The *variance*, Var(X), of the distribution is given by

$$\text{Var}(X) = \int_{-\infty}^{\infty} (x-\mu)^2 p(x)\,\mathrm{d}x, \text{ when this integral exists}$$

The square root of Var(X) is denoted by σ and is called the *standard deviation* of X.

The mean of the distribution is also called the *expected value*, $E(X)$, of X. By analogy with the discrete case, an alternative expression for the variance is given by

$$\text{Var}(X) = \int_{-\infty}^{\infty} x^2 p(x)\,\mathrm{d}x - \mu^2$$

Example 4.4.3

The mean of the probability distribution defined in example 4.4.1 is given by

$$\mu = \int_{-\infty}^{\infty} xp(x)\,\mathrm{d}x = \frac{1}{18}\int_{-3}^{3} x^3\,\mathrm{d}x = 0$$

The variance is given by

$$\mathrm{Var}(X)=\frac{1}{18}\int_{-3}^{3} x^4 \, \mathrm{d}x-\mu^2=\frac{27}{5}$$

Let $aX+b$, where a and b are real constants, denote the random variable assigning the value $ax+b$ to the outcome of a trial of a random experiment to which X assigns the value x. In exercise 4.3(5), the reader is asked to verify the following theorem when X is a discrete random variable.

Theorem 4.4.1

(i) $E(aX+b)=aE(X)+b$;

(ii) $\mathrm{Var}(aX+b)=a^2\mathrm{Var}(X)$.

It is now left as an exercise for the reader to show that the statements in this theorem are also true when X is a continuous random variable.

Moments of a Random Variable and Chebyshev's Inequality

Let X^2 denote the random variable assigning the value x^2 to the outcome of a trial to which X assigns the value x. For a continuous random variable X, the expected value of X^2, if it exists, is given by

$$E(X^2)=\int_{-\infty}^{\infty} x^2 p(x) \, \mathrm{d}x$$

where p is the p.d.f. of X; $E(X^2)$ is often called the mean square value of X. When X is a discrete random variable, the mean square value is given by

$$E(X^2)=\sum_{x_i\in \mathscr{R}_X} x_i^2 p_i$$

which again is the expected value of X^2 and is only defined when the sum of the series exists.

Alternatively, $E(X^2)$ is called the *second moment* of X, and $E(X)$ itself is the *first moment* of X. More generally, the nth *moment* of X is denoted by $E(X^n)$ and defined by

$$E(X^n)=\sum_{x_i\in \mathscr{R}_X} x_i^n p_i$$

when X is discrete, and by

$$E(X^n) = \int_{-\infty}^{\infty} x^n p(x)\, dx$$

when X is continuous. The nth moment of a random variable is only defined when the appropriate series or integral exists and X may not, in fact, have even a first moment [see exercise 4.3(11)].

Theorem 4.4.2

(i) $E(X^2) = \text{Var}(X) + \mu^2$, where $\mu = E(X)$;

(ii) if $(X-\mu)^2$ denotes the random variable assigning the value $(x-\mu)^2$ to the outcome of a trial to which X assigns the value x, then $E((X-\mu)^2) = \text{Var}(X)$;

(iii) for any $\varepsilon > 0$, $\mathscr{P}(|x| \geqslant \varepsilon) \leqslant E(X^2)/\varepsilon^2$.

Proof (i) This statement follows immediately from the alternative definition of $\text{Var}(X)$.

(ii) The proof of this statement is left as an exercise for the reader.

(iii) This is known as *Chebyshev's inequality* and is used to establish the *weak law of large numbers* (theorem 4.5.3). If X is a continuous random variable, then

$$E(X^2) = \int_{-\infty}^{\infty} x^2 p(x)\, dx \geqslant \left[\int_{-\infty}^{-\varepsilon} + \int_{\varepsilon}^{\infty}\right] x^2 p(x)\, dx$$

$$\geqslant \varepsilon^2 \left[\int_{-\infty}^{-\varepsilon} + \int_{\varepsilon}^{\infty}\right] p(x)\, dx$$

$$= \varepsilon^2 \mathscr{P}(|x| \geqslant \varepsilon)$$

The proof when X is discrete is similar and is left as an exercise.

Corollary 4.4.2
For any $\varepsilon > 0$, $\mathscr{P}(|x-\mu| \geqslant \varepsilon) \leqslant \text{Var}(X)/\varepsilon^2$.

Proof Replace x by $x-\mu$ in part (iii), then use part (ii).

Numerical Approximation of Continuous Probability Distributions

If it is not possible to derive a mathematical formula for the p.d.f. p of a continuous distribution, then p may be estimated by performing a sequence of trials. Suppose M such trials are performed, that is, a sample

Table 4.5

Response time (s)	1	2	3	4	5	6	7	8	9	10	11	12	13
Number of messages	0	2	13	34	78	119	176	198	169	129	63	16	3

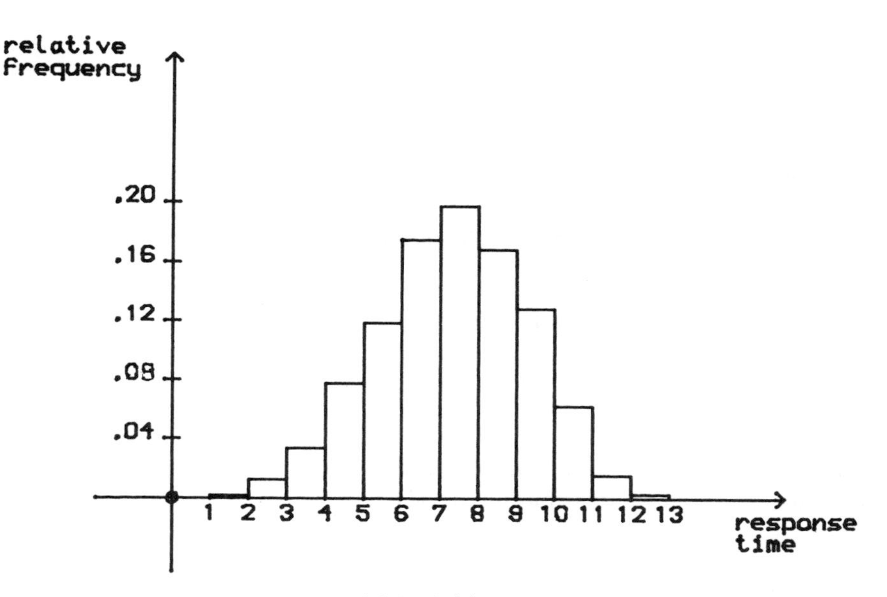

Figure 4.16 A histogram

of size M is taken (see definition 4.1.5). Suppose that r distinct events E_1, $E_2, \ldots, E_r$ occur in the sample, where E_i denotes the event that the value associated with the outcome of a trial lies in the interval $[x_i, x_{i+1}]$. If f_i denotes the relative frequency of E_i with respect to the sample of size M, then f_i is an approximation to $\mathscr{P}(x_i \leqslant x \leqslant x_{i+1})$. The frequency distribution $\{(x_i, f_i) | i = 1, 2, \ldots, r\}$ may be displayed graphically by means of a *histogram* (see, for example, figure 4.16), which may be regarded as an approximation to the graph of the underlying p.d.f., p.

Example 4.4.4

The response time for each of 1,000 messages entered at a terminal in a time-sharing computer system is recorded, rounded up to the nearest second. The resulting frequency distribution is given by table 4.5 and the corresponding histogram is displayed in figure 4.16.

Some Important Continuous Distributions

Definition 4.4.4
A continuous distribution is *rectangular*, or *uniform*, if its p.d.f. p is defined by

$$p: x \mapsto \begin{cases} 1/(b-a), & a \leqslant x \leqslant b \\ 0, & \text{otherwise} \end{cases}$$

for some a, $b \in \mathbb{R}$.

Since each value in the interval $[a, b]$ is equally likely to occur, this type of distribution is just the extension of the uniform discrete distribution (see definition 4.3.11).

From definition 4.4.2, the c.d.f. $\mathscr{U}$ of a uniform distribution is given by

$$\mathscr{U}: x \mapsto \begin{cases} 0, & x < a \\ (x-a)/(b-a), & a \leqslant x \leqslant b \\ 1, & x > b \end{cases}$$

The graphs of the p.d.f. and the c.d.f. of a uniform distribution are illustrated in figures 4.17a and b, respectively.

It is left as an exercise for the reader to verify that the mean and variance of a uniform distribution are given, respectively, by

$$\mu = \frac{1}{2}(a+b)$$

and

$$\operatorname{Var}(X) = (b-a)^2/12$$

Definition 4.4.5
The p.d.f. of the *standardised* (or *normalised*) *normal distribution* is denoted by φ and defined by

$$\varphi: x \mapsto \exp\left(-\frac{1}{2}x^2\right) \Big/ \sqrt{(2\pi)}$$

If the graph of φ is plotted, the resulting curve is bell-shaped and symmetric about $x=0$; see figure 4.18.

The c.d.f. corresponding to φ is denoted by Φ and is given by

$$\Phi: x \mapsto \frac{1}{\sqrt{(2\pi)}} \int_{-\infty}^{x} \exp\left(-\frac{1}{2}t^2\right) \mathrm{d}t$$

The integral in the expression for $\Phi(x)$ can only be evaluated numerically.

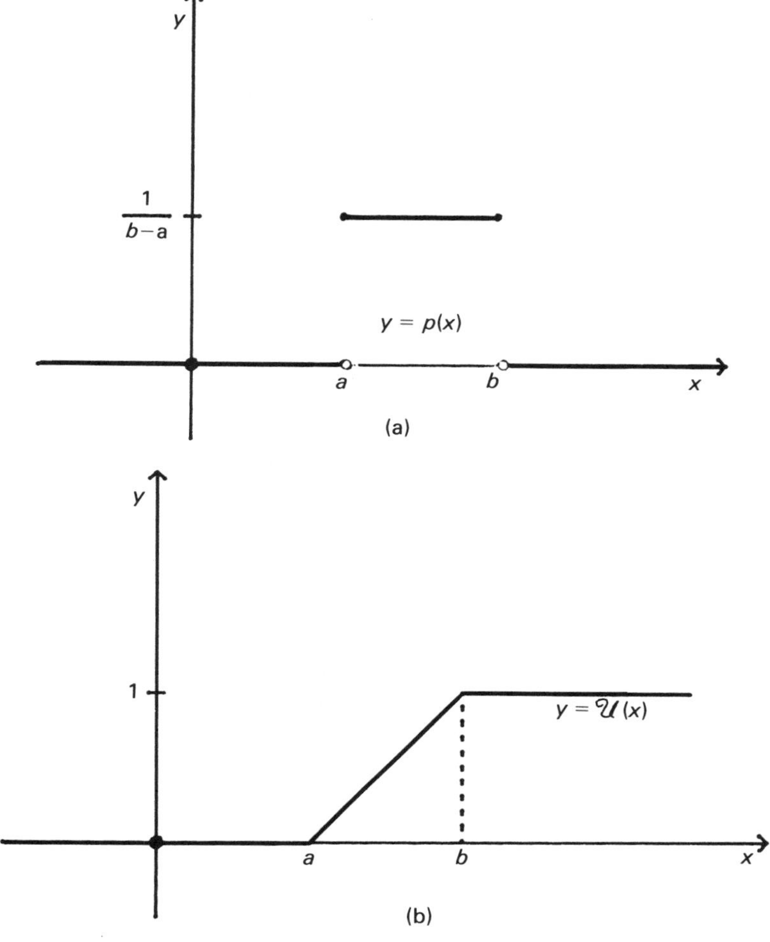

Figure 4.17 The p.d.f. (a) and c.d.f. (b) of a uniform distribution

Tables of values for both φ and Φ are included in most books of statistical tables.

Since

$$\int_{-\infty}^{\infty} x \exp\left(-\frac{1}{2}x^2\right) \mathrm{d}x = \left[-\exp\left(-\frac{1}{2}x^2\right)\right]_{-\infty}^{\infty}$$

the mean of the standardised normal distribution is equal to 0. It may also be shown that the variance of such a distribution is equal to 1 (see, for example, Arthurs, 1965).

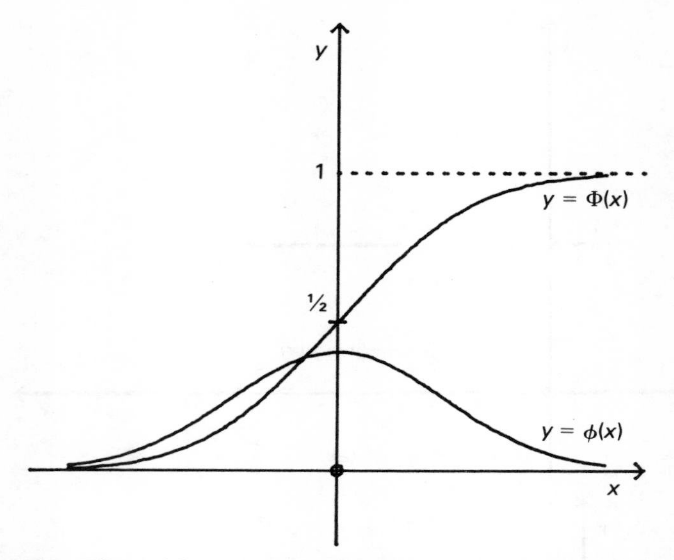

Figure 4.18 The p.d.f., $\varphi(x)$, and c.d.f., $\Phi(x)$, of the standardised normal distribution

A very important and frequently occurring type of continuous distribution is one whose p.d.f. p has the general shape of φ but which has mean μ and variance σ^2. Such a distribution is called, simply, a normal distribution.

Definition 4.4.6
A continuous distribution is a *normal distribution* if its p.d.f. p is defined by

$$p\colon x \mapsto \frac{1}{\sigma}\varphi\left(\frac{x-\mu}{\sigma}\right)$$

that is

$$p(x)=\frac{1}{\sigma\sqrt{(2\pi)}}\exp[-(x-\mu)^2/2\sigma^2]$$

The c.d.f. $\mathscr{U}$ of a normal distribution is related to Φ by

$$\mathscr{U}(x)=\Phi\left(\frac{x-\mu}{\sigma}\right)$$

and it is easy to show that

$$\mathscr{P}\left(a\leqslant\frac{x-\mu}{\sigma}\leqslant b\right)=\Phi(b)-\Phi(a)$$

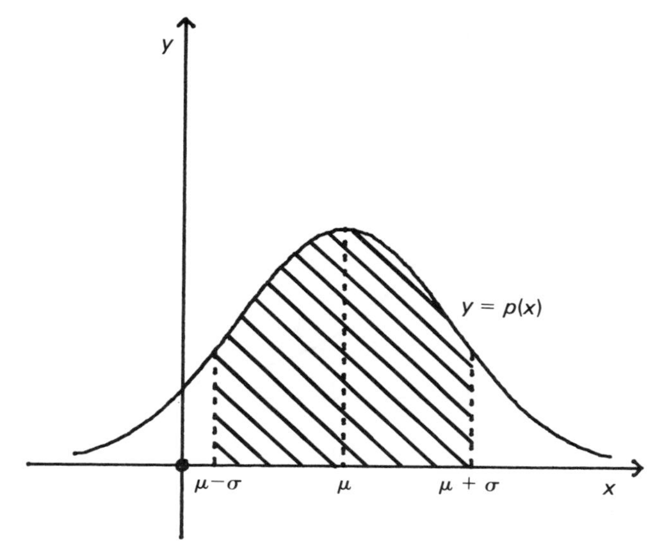

Figure 4.19 Example 4.4.5; the shaded area represents the probability of a value falling within one standard deviation of the mean

Example 4.4.5

A random variable X has a normal distribution, with mean μ and standard deviation σ. Given that $\Phi(1)=0.8413$, compute the probability that the value of X lies within one standard deviation of μ, that is, compute $\mathscr{P}(\mu-\sigma\leqslant x\leqslant\mu+\sigma)$.

The required probability is represented by the shaded area in figure 4.19. From above

$$\mathscr{U}(\mu+\sigma)=\Phi(1)=0.8413$$

Hence

$$\mathscr{P}(\mu+\sigma\leqslant x<\infty)=1-0.8413=0.1587$$

Since the graph of p is symmetric about $x=\mu$

$$\mathscr{P}(-\infty<x\leqslant\mu-\sigma)=\mathscr{P}(\mu+\sigma\leqslant x<\infty)=0.1587$$

Thus

$$\mathscr{P}(\mu-\sigma\leqslant x\leqslant\mu+\sigma)=1-2\times 0.1587=0.6826$$

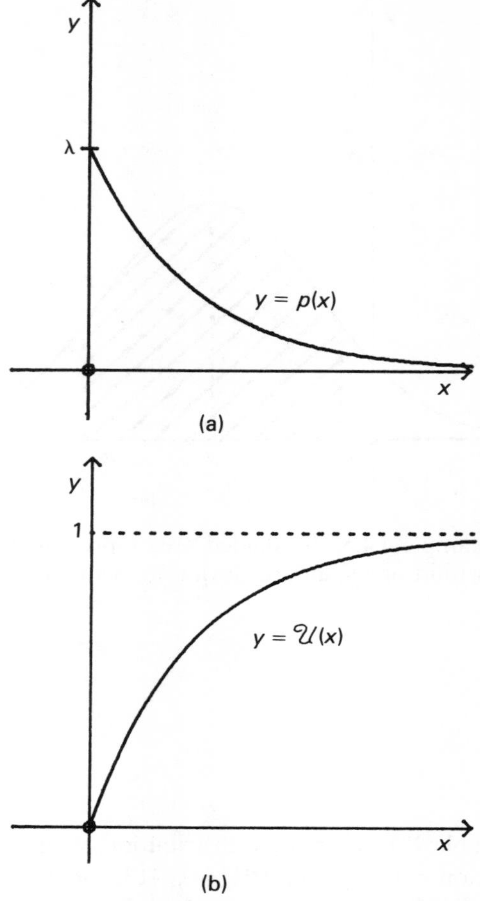

Figure 4.20 The p.d.f. (a) and c.d.f. (b) of a negative exponential distribution

Definition 4.4.7
A continuous distribution is a *negative exponential distribution* if its p.d.f. p is defined by

$$p\colon x \mapsto \lambda \exp(-\lambda x),\ x>0$$

where λ is a real, positive constant.

The graph of such a distribution is illustrated in figure 4.20a. Since

$$\int_0^x \lambda \exp(-\lambda t)\,\mathrm{d}t = 1-\exp(-\lambda x)$$

the c.d.f., $\mathscr{U}$, of a negative exponential distribution is given by

$$\mathscr{U}: x \mapsto 1-\exp(-\lambda x),\ x>0$$

The graph of this function is illustrated in figure 4.20b.

The negative exponential distribution and the discrete Poisson distribution (see definition 4.3.14) are the most frequently encountered distributions in *queueing theory*. Both distributions may be used to describe the same random situation: if the Poisson distribution describes the number of random events occurring per unit time (for example, the number of people joining the end of a queue at a supermarket cashpoint), then the exponential distribution will describe the *inter-arrival time*, that is, the time between two successive arrivals. If the average inter-arrival time is equal to μ, then the constant λ in the negative exponential distribution is equal to $1/\mu$.

A good introduction to queueing theory, with many worked examples, is given by Murdoch (1978). It is easy to show that the mean of the negative exponential distribution is $1/\lambda$ and the variance is $1/\lambda^2$.

Example 4.4.6

Customers arrive randomly at a service point with an average rate of 25 per hour. Compute the probability of an inter-arrival time exceeding 5 minutes.

Since the average inter-arrival time is 2.4 minutes, the required probability is given by

$$\int_5^\infty \frac{1}{2.4}\exp(-t/2.4)\,\mathrm{d}t=\exp(-5/2.4)\approx 0.125$$

Exercise 4.4

1. For what value of k is the following function a p.d.f.?

$$p: x \mapsto \begin{cases} k/x^2, & 10\leqslant x\leqslant 20 \\ 0, & \text{otherwise} \end{cases}$$

2. Obtain the c.d.f. corresponding to the p.d.f. determined in exercise 4.4(1).

3. Determine the mean and variance of the probability distribution determined in exercise 4.4(1).

4. Given a continuous random variable, X, let $aX+b$ denote the random variable which has value $ax+b$ when X has value x. Show that:

(a) $E(aX+b)=aE(X)+b$;

(b) $\text{Var}(aX+b)=a^2\text{Var}(X)$.

5. The c.d.f. of a continuous random variable X is given by

$$\mathscr{U}: x \mapsto \begin{cases} 1-e^{-2x}, & x \geqslant 0 \\ 0, & \text{otherwise} \end{cases}$$

(a) calculate $\mathscr{P}(5 \leqslant x \leqslant 10)$;

(b) determine the corresponding p.d.f.

6. Show that the mean and the variance of a continuous random variable uniformly distributed over the interval $[a, b]$, are given by $\frac{1}{2}(a+b)$ and $(b-a)^2/12$, respectively.

7. Given that $\Phi(1.96)=0.975$, show that 95% of the area under the curve of the p.d.f. for a normal distribution lies within 1.96 standard deviations of the mean.

8. On average, a firm's computer fails twice in 100 hours. What is the probability of it failing within 4 hours?

9. Derive the alternative expression for $\text{Var}(X)$ given after definition 4.4.3.

4.5 INDEPENDENT RANDOM VARIABLES

Discrete Joint Probability Distributions

Suppose a random experiment consists of two parts, with each of which a random variable is associated. Let X and $\mathscr{Y}$ denote these random variables. Assume first that X and $\mathscr{Y}$ are discrete, with range spaces $\mathscr{R}_X$ and $\mathscr{R}_{\mathscr{Y}}$, respectively, and let $\mathscr{P}(x_i, y_j)$ denote the probability that, in a trial of the experiment, x_i occurs in $\mathscr{R}_X$ and y_j occurs in $\mathscr{R}_{\mathscr{Y}}$.

Definition 4.5.1
The function p: $\mathscr{R}_X \times \mathscr{R}_{\mathscr{Y}} \to [0, 1]$, defined by

$$p: (x_i, y_j) \mapsto \mathscr{P}(x_i, y_j)$$

is called the *joint probability density function* (or the *joint probability mass function*) *of X and* $\mathscr{Y}$.

If $\mathscr{P}(x_i)$ denotes the probability that in any trial x_i occurs in $\mathscr{R}_X$ and if $\mathscr{P}(y_j)$ denotes the probability that in any trial y_j occurs in $\mathscr{R}_{\mathscr{Y}}$, then the functions

$$p_X: x_i \mapsto \mathscr{P}(x_i) \quad (x_i \in \mathscr{R}_X)$$

and

$$p_{\mathscr{Y}}: y_j \mapsto \mathscr{P}(y_j) \quad (y_j \in \mathscr{R}_{\mathscr{Y}})$$

are called the *marginal probability density* (or *mass*) *functions* of X and $\mathscr{Y}$, respectively. The expressions relating p_X and $p_{\mathscr{Y}}$ to p are

$$p_X(x_i) = \sum_{y_j \in \mathscr{R}_{\mathscr{Y}}} p(x_i, y_j), \quad \forall x_i \in \mathscr{R}_X$$

$$p_{\mathscr{Y}}(y_j) = \sum_{x_i \in \mathscr{R}_X} p(x_i, y_j), \quad \forall y_j \in \mathscr{R}_{\mathscr{Y}}$$

Definition 4.5.2
If $p(x_i, y_j) = p_X(x_i)p_{\mathscr{Y}}(y_j)$, $\forall x_i \in \mathscr{R}_X$, $y_j \in \mathscr{R}_{\mathscr{Y}}$, then X and $\mathscr{Y}$ are said to be *independent random variables.*

Example 4.5.1

(1) A particular random experiment consists of tossing two fair coins and throwing a pair of fair dice. Let X be the random variable assigning the value 1 to a trial if both coins show heads and the value 0 otherwise. Let $\mathscr{Y}$ be the random variable assigning to a trial the sum of the values on the uppermost faces of the thrown dice. Then

$$\mathscr{R}_X = \{0, 1\}$$

$$\mathscr{R}_{\mathscr{Y}} = \{2, 3, 4, 5, 6, 7, 8, 9, 10, 11, 12\}$$

The two parts of this experiment are clearly independent. Thus

$$p(x_i, y_j) = p_X(x_i)p_{\mathscr{Y}}(y_j)$$

The joint probability density function of X and $\mathscr{Y}$ is given in table 4.6

Table 4.6

x_i \ y_j	2	3	4	5	6	7	8	9	10	11	12	p_X
1	$\frac{1}{144}$	$\frac{1}{72}$	$\frac{1}{48}$	$\frac{1}{36}$	$\frac{5}{144}$	$\frac{1}{24}$	$\frac{5}{144}$	$\frac{1}{36}$	$\frac{1}{48}$	$\frac{1}{72}$	$\frac{1}{144}$	$\frac{1}{4}$
0	$\frac{1}{48}$	$\frac{1}{24}$	$\frac{1}{16}$	$\frac{1}{12}$	$\frac{5}{48}$	$\frac{1}{8}$	$\frac{5}{48}$	$\frac{1}{12}$	$\frac{1}{16}$	$\frac{1}{24}$	$\frac{1}{48}$	$\frac{3}{4}$
$p_{\mathscr{Y}}$	$\frac{1}{36}$	$\frac{1}{18}$	$\frac{1}{12}$	$\frac{1}{9}$	$\frac{5}{36}$	$\frac{1}{6}$	$\frac{5}{36}$	$\frac{1}{9}$	$\frac{1}{12}$	$\frac{1}{18}$	$\frac{1}{36}$	

together with the marginal probability density functions, p_X and $p_{\mathscr{Y}}$.

Note that for any value of y_j, $p_{\mathscr{Y}}(y_j)$ is equal to the sum of the entries in the column corresponding to y_j. Similarly, for either value of x_i, $p_X(x_i)$ is equal to the sum of the entries in the row corresponding to x_i.

Table 4.7

x_i \ y_j	2	3	4	Row sum
1	0.2	0.2	0	0.4
2	0.3	0.1	0.2	0.6
Column sum	0.5	0.3	0.2	

(2) The range spaces of two discrete random variables X and $\mathscr{Y}$ are $\{1,2\}$ and $\{2,3,4\}$, respectively. The joint p.d.f. of X and $\mathscr{Y}$ is given in table 4.7. Determine whether or not X and $\mathscr{Y}$ are independent.

The marginal distributions p_X and $p_{\mathscr{Y}}$ may be obtained by taking the row sums and the column sums, respectively, in table 4.7. X and $\mathscr{Y}$ are not independent since, for example

$$p_X(1)p_{\mathscr{Y}}(3) = 0.12 \neq 0.2 = p(1,3)$$

Continuous Joint Probability Distributions

Definition 4.5.3

A function p: $\mathbb{R}^2 \rightarrow \mathbb{R}$ is a valid *joint probability density function* (*joint p.d.f.*) provided

(a) $p(x, y) \geqslant 0, \forall x, y \in \mathbb{R}$

(b) $$\int_{-\infty}^{\infty}\int_{-\infty}^{\infty} p(x, y)\, \mathrm{d}x\, \mathrm{d}y = 1$$

Let X and $\mathscr{Y}$ be two continuous random variables and let A be any region in the cartesian plane lying in the cross product of the range spaces of X and $\mathscr{Y}$ (A will generally be $\mathbb{R}^2$). Then p is the *joint p.d.f. of* X and $\mathscr{Y}$ provided

$$\mathscr{P}(A) = \iint_A p(x, y)\,\mathrm{d}x\,\mathrm{d}y, \quad \forall A \subset \mathscr{R}_X \times \mathscr{R}_{\mathscr{Y}}$$

Given the joint p.d.f. p of two continuous random variables X and $\mathscr{Y}$, the *marginal probability density functions*, p_X and $p_{\mathscr{Y}}$, are given by

$$p_X(x) = \int_{-\infty}^{\infty} p(x, y) \geqslant 0, \ \forall x, y \in \mathbb{R}$$

and

$$p_{\mathscr{Y}}(y) = \int_{-\infty}^{\infty} p(x, y)\,\mathrm{d}x, \quad \forall y \in \mathbb{R}$$

Definition 4.5.4
The function $\mathscr{U}: \mathbb{R}^2 \to \mathbb{R}$ defined by

$$\mathscr{U}: (x, y) \mapsto \int_{-\infty}^{y} \int_{-\infty}^{x} p(s, t)\,\mathrm{d}s\,\mathrm{d}t$$

is called the *joint cumulative distribution function* (joint c.d.f.) *of* X and $\mathscr{Y}$.
It follows immediately from this definition that

(i) $P(\{(s, t) | s \leqslant x \wedge t \leqslant y\}) = \mathscr{U}(x, y)$

(ii) $P(\{(s,t) | a \leqslant s \leqslant b \wedge c \leqslant t \leqslant d\}) = \mathscr{U}(b, d) - \mathscr{U}(a, c)$.

If $\mathscr{U}_X$ and $\mathscr{U}_{\mathscr{Y}}$ denote the c.d.f.s of X and $\mathscr{Y}$, respectively, then

$\mathscr{U}_X(x) = \lim_{y\to\infty} \mathscr{U}(x, y), \forall x \in \mathbb{R}$ and $\mathscr{U}_{\mathscr{Y}}(y) = \lim_{x\to\infty} \mathscr{U}(x, y), \forall y \in \mathbb{R}$. In the context of joint probability distributions, $\mathscr{U}_X$ and $\mathscr{U}_{\mathscr{Y}}$ are called the *marginal c.d.f.s of X and* $\mathscr{Y}$, respectively.

Definition 4.5.5
If $p(x, y) = p_X(x) p_{\mathscr{Y}}(y), \forall x, y \in \mathbb{R}$, then X and $\mathscr{Y}$ are *independent random variables.*
If follows that if X and $\mathscr{Y}$ are independent random variables, then

$$\mathscr{U}(x, y) = \mathscr{U}_X(x)\ \mathscr{U}_{\mathscr{Y}}(y), \quad \forall x, y \in \mathbb{R}$$

Example 4.5.2

Let p: $\mathbb{R}^2 \to \mathbb{R}$ be given by

$$p: (x, y) \mapsto \begin{cases} \exp(-x-y), & \forall x>0 \wedge y>0 \\ 0, & \text{otherwise} \end{cases}$$

(1) Show that p is a valid joint p.d.f.;

(2) determine the marginal p.d.f.s, $p_{\mathscr{X}}$ and $p_{\mathscr{Y}}$;

(3) determine the joint c.d.f., $\mathscr{U}$;

(4) compute the probability of the event $\{(x, y)|1 \leqslant x \leqslant 4 \wedge 3 \leqslant y \leqslant 5\}$;

(5) show that X and $\mathscr{Y}$ are independent random variables.

(1) To be a valid joint p.d.f., p must satisfy properties (a) and (b) in definition 4.5.3. Since $\exp(x) > 0 \; \forall x \in \mathbb{R}$, property (a) is clearly satisfied

$$\int_{-\infty}^{\infty} \int_{-\infty}^{\infty} p(x, y) \, dx \, dy = \int_{0}^{\infty} \int_{0}^{\infty} \exp(-x-y) \, dx \, dy = \int_{0}^{\infty} \exp(-y) \, dy = \exp(0) = 1$$

Thus p is a valid joint p.d.f.

(2)

$$p_X(x) = \int_{-\infty}^{\infty} p(x, y) \, dy = \begin{cases} \int_0^{\infty} \exp(-x-y) dy, & \forall x>0 \\ 0, & \forall x \leqslant 0 \end{cases}$$

$$= \begin{cases} \exp(-x), & \forall x>0 \\ 0, & \forall x \leqslant 0 \end{cases}$$

Similarly

$$p_{\mathscr{Y}}(y) = \begin{cases} \exp(-y), & \forall y>0 \\ 0, & \forall y \leqslant 0 \end{cases}$$

(3)

$$\mathscr{U}(x, y) = \int_{-\infty}^{y} \int_{-\infty}^{x} p(s, t) \, ds \, dt = \int_{0}^{y} \int_{0}^{x} \exp(-s-t) \, ds \, dt$$

$$= \int_{0}^{y} [-\exp(-x-t) + \exp(-t)] \, dt$$

$$=\exp(-x-y)-\exp(-x)-\exp(-y)+1$$

$$=[1-\exp(-x)][1-\exp(-y)]$$

(4)

$$\mathscr{P}(\{(x, y)|1\leqslant x\leqslant 4 \wedge 3\leqslant y\leqslant 5\})= \mathscr{U}(4, 5)- \mathscr{U}(1, 3)$$

$$=[1-\exp(-4)][1-\exp(-5)]-[1-\exp(-1)][1-\exp(-3)]$$

$$\approx 0.97507-0.60065=0.37442$$

(5) From part (2)

$$p_X(x)p_{\mathscr{Y}}(y)=\begin{cases}\exp(-x)\exp(-y), & \forall x>0 \wedge \forall y>0\\ 0, & \text{otherwise}\end{cases}$$

But

$$\exp(-x)\exp(-y)=\exp(-x-y)$$

Hence

$$p_X(x)p_{\mathscr{Y}}(y)=p(x, y), \quad \forall x, y\in\mathbb{R}$$

Thus, X and $\mathscr{Y}$ are independent random variables.

Some Important Theorems

Definition 4.5.6
Let f be a function from $\mathbb{R}^2$ to $\mathbb{R}$. Given two continuous random variables, X and $\mathscr{Y}$, let $f(X, \mathscr{Y})$ denote the random variable assigning the value $f(x, y)$ to the outcome of a trial to which X assigns the value x and to which $\mathscr{Y}$ assigns the value y. The *expected value* of $f(X, \mathscr{Y})$ is denoted by $E(f(X, \mathscr{Y}))$ and defined by

$$E(f(X, \mathscr{Y}))=\int_{-\infty}^{\infty}\int_{-\infty}^{\infty} f(x, y)p(x, y)\,\mathrm{d}x\,\mathrm{d}y$$

where p is the joint p.d.f. of X and $\mathscr{Y}$.

When X and $\mathscr{Y}$ are discrete random variables, the domain of f need only be $\mathscr{R}_X\times \mathscr{R}_{\mathscr{Y}}$ and the expected value of $f(X, \mathscr{Y})$ is defined by

$$E(f(X, \mathscr{Y}))=\sum_{x_i\in \mathscr{R}_X}\ \sum_{y_j\in \mathscr{R}_{\mathscr{Y}}} f(x_i, y_j)p(x_i, y_j)$$

In both the continuous and the discrete cases, the variance of $f(X, \mathscr{Y})$, $\mathrm{Var}(f(X, \mathscr{Y}))$, is defined by

$$\mathrm{Var}(f(X, \mathscr{Y}))=E((f(X, \mathscr{Y})-\mu)^2)$$

where

$$\mu = E(f(X, \mathscr{Y}))$$

Although joint probability distributions have been presented in terms of just two random variables, X and $\mathscr{Y}$, the idea generalises in a straightforward manner to any finite number of random variables. Definition 4.5.6 may then also be generalised. For example, if $X_1, X_2, \ldots, X_n$ are continuous variables, with joint p.d.f. $p: \mathbb{R}^n \to \mathbb{R}$ then, provided f is defined from $\mathbb{R}^n$ to $\mathbb{R}$

$$E(f(X_1, \ldots, X_n)) = \int_{-\infty}^{\infty} \cdots \int_{-\infty}^{\infty} f(x_1, \ldots, x_n) p(x_1, \ldots, x_n) \, dx_1 \ldots dx_n$$

In all of the above cases, the expected value of f is only defined when the appropriate integral or series exists.

For simplicity of notation, the following theorem is presented for two random variables, X and $\mathscr{Y}$. However, each statement generalises very easily to the case of n random variables.

Theorem 4.5.1

(i) $E(X + \mathscr{Y}) = E(X) + E(\mathscr{Y})$;

(ii) provided X and $\mathscr{Y}$ are independent and $E(f(X))$, $E(g(\mathscr{Y}))$ exist

$$E(f(X)g(\mathscr{Y})) = E(f(X))E(g(\mathscr{Y}))$$

(iii) if X and $\mathscr{Y}$ are independent, then

$$\text{Var}(X + \mathscr{Y}) = \text{Var}(X) + \text{Var}(\mathscr{Y})$$

(iv) let X and $\mathscr{Y}$ be two independent random variables and let $Z \equiv m(X, \mathscr{Y})$ be the random variable defined by

$$z \equiv m(x, y) = \max\{x, y\}$$

Then

$$\mathscr{U}_{\mathscr{Z}}(z) = \mathscr{U}_X(z)\mathscr{U}_{\mathscr{Y}}(z)$$

Proof The proof for each statement is given for the continuous case. The proofs when X and $\mathscr{Y}$ are discrete are very similar and are left as exercises.

(i)

$$E(X + \mathscr{Y}) = \int_{-\infty}^{\infty} \int_{-\infty}^{\infty} (x + y) p(x, y) \, dx \, dy$$

$$= \int_{-\infty}^{\infty} x \left[\int_{-\infty}^{\infty} p(x, y) \, dy \right] dx + \int_{-\infty}^{\infty} y \left[\int_{-\infty}^{\infty} p(x, y) \, dx \right] dy$$

$$= \int_{-\infty}^{\infty} x p_X(x) \mathrm{d}x + \int_{-\infty}^{\infty} y p_{\mathscr{Y}}(y)\, \mathrm{d}y$$

$$= E(X) + E(\mathscr{Y})$$

(ii)

$$E(f(X)g(\mathscr{Y})) = \int_{-\infty}^{\infty}\int_{-\infty}^{\infty} f(x)g(y)p(x, y)\, \mathrm{d}x\, \mathrm{d}y$$

$$= \int_{-\infty}^{\infty}\int_{-\infty}^{\infty} f(x)g(y)p_X(x)p_{\mathscr{Y}}(y)\, \mathrm{d}x\, \mathrm{d}y$$

since X and $\mathscr{Y}$ are independent

$$= \int_{-\infty}^{\infty} f(x)p_X(x)\, \mathrm{d}x \int_{-\infty}^{\infty} g(y)p_{\mathscr{Y}}(y)\, \mathrm{d}y$$

$$= E(f(X))E(g(\mathscr{Y}))$$

An important corollary of this result is that if X and $\mathscr{Y}$ are independent then

$$E(X\mathscr{Y}) = E(X)E(\mathscr{Y})$$

(iii)

$$\mathrm{Var}(X + \mathscr{Y}) = E[(X + \mathscr{Y} - \mu)^2]$$

where $\mu = E(X + \mathscr{Y})$

$$= E[X^2 + \mathscr{Y}^2 + \mu^2 + 2X\mathscr{Y} - 2\mu(X + \mathscr{Y})]$$

$$= E(X^2) + E(\mathscr{Y}^2) + \mu^2 + 2E(X\mathscr{Y}) - 2\mu E(X + \mathscr{Y})$$

by part (i) together with theorem 4.4.1(i)

$$= \mathrm{Var}(X) + [E(X)]^2 + \mathrm{Var}(\mathscr{Y}) + [E(\mathscr{Y})]^2 + \mu^2 + 2E(X)E(\mathscr{Y}) - 2\mu^2$$

by theorem 4.4.2(i) and by part (ii)

$$= \mathrm{Var}(X) + \mathrm{Var}(\mathscr{Y}) + [E(X) + E(\mathscr{Y})]^2 - \mu^2$$

$$= \mathrm{Var}(X) + \mathrm{Var}(\mathscr{Y})$$

since $E(X) + E(\mathscr{Y}) = E(X + \mathscr{Y}) = \mu$.

(iv) By definition

$$\mathscr{U}_{\mathscr{X}}(z) = \mathscr{P}(\{u | u \leqslant z\})$$

Let s and t be the values assigned by X and $\mathscr{Y}$, respectively, to the outcome of a trial and let $u=\max\{s, t\}$. Then

$$u \leqslant z \text{ iff } s \leqslant z \wedge t \leqslant z$$

But $\mathscr{P}(\{(s, t) | s \leqslant z \wedge t \leqslant z\}) = \mathscr{U}(z, z) = \mathscr{U}_X(z)\ \mathscr{U}_{\mathscr{Y}}(z)$, since X and $\mathscr{Y}$ are independent.

Example 4.5.3

Fair Foods Ltd own a chain of supermarkets. Each supermarket can re-order goods from a centrally situated warehouse via the company's on-line computer system. The latter incorporates two identical computers and the probability that either computer will 'go down' is exponentially distributed in time, with a mean failure time of 1700 hours. Stock re-ordering only becomes impossible when both computers are down simultaneously. Making the somewhat unrealistic but simplifying assumption that if either computer goes down during a week then it stays down for the rest of the week, compute the probability that supermarkets will be able to re-order stock at any time during a period of one week (168 hours).

Let X and $\mathscr{Y}$ be the random variables giving the times to failure of the first and second computers, respectively, and left $\mathscr{X}$ be the random variable giving the time to failure of the complete system.

Under the simplifying assumption stated above, $\mathscr{X}$ is defined by $z=\max\{x, y\}$. Since the two computers are identical

$$\mathscr{U}_X(t) = \mathscr{U}_{\mathscr{Y}}(t) = 1 - \exp(-\lambda t)$$

where $\lambda = 1/1700$. By theorem 4.5.1(iv)

$$\mathscr{U}_{\mathscr{X}}(t) = \mathscr{U}_X(t)\ \mathscr{U}_{\mathscr{Y}}(t)$$

Hence

$$\mathscr{U}_{\mathscr{X}}(168) = [1 - \exp(-168/1700)]^2 \approx 0.00885$$

The required probability is thus 0.99115, that is, there is less than a 1 per cent chance of a complete system failure during the week.

Definition 4.5.7
The *covariance* of two random variables, X and $\mathscr{Y}$, is denoted by $\mathrm{Cov}(X, \mathscr{Y})$ and defined by

$$\mathrm{Cov}(X, \mathscr{Y}) = E(X\mathscr{Y}) - E(X)E(\mathscr{Y})$$

If $\mathrm{Cov}(X, \mathscr{Y}) = 0$, then X and $\mathscr{Y}$ are said to be *uncorrelated*.

If follows from theorem 4.5.1(ii) that if X and $\mathscr{Y}$ are independent then they are uncorrelated. However, if two random variables are uncorrelated, they are not necessarily independent — see exercise 4.5(2).

Repeated Independent Trials

Definition 4.5.8
If two or more random variables have identical probability distributions, they are said to be *identically distributed.*

Corollary 4.5.1

(i) If $X_1, \ldots, X_n$ are identically distributed random variables, each having the same distribution as a random variable X, then

$$E(X_1 + \ldots + X_n) = nE(X)$$

(ii) If $X_1, \ldots, X_n$ are identically distributed, independent random variables, each having the same distribution as a random variable X, then

$$\text{Var}(X_1 + \ldots + X_n) = n\text{Var}(X)$$

The most important application of identically distributed, independent random variables occurs in *sampling*, when a single experiment is repeated a number of times and the outcome of each trial is independent of the outcome of any other trial. Suppose a single trial of a repeated experiment consists of a sequence of n independent trials of some other experiment. If X is a random variable associated with the latter experiment, then a random variable X_i, having exactly the same distribution as X, may be associated with the ith trial in the repeated experiment.

Let $\mathscr{Y}_n = (1/n)(X_1 + \ldots + X_n)$ denote the random variable assigning to a trial of the repeated experiment the average of the values assigned to the sequence of independent trials.

Theorem 4.5.2

(i) $E(\mathscr{Y}_n) = E(X)$;

(ii) $\text{Var}(\mathscr{Y}_n) = \frac{1}{n}\text{Var}(X)$.

Proof (i)

$$E(\mathscr{Y}_n) = E\left[\frac{1}{n}(X_1 + \ldots + X_n)\right]$$

$$= \frac{1}{n}E(X_1 + \ldots + X_n)$$

by theorem 4.4.1(i)

$$=\frac{1}{n}[nE(X)]$$

by corollary 4.5.1(i)

$$=E(X)$$

(ii)

$$\mathrm{Var}(\mathscr{Y}_n)=\mathrm{Var}\left[\frac{1}{n}(X_1+\ldots+X_n)\right]$$

$$=\frac{1}{n^2}\mathrm{Var}(X_1+\ldots+X_n)$$

by theorem 4.4.1(ii)

$$=\frac{1}{n}\mathrm{Var}(X)$$

by corollary 4.5.1(ii).

In practice, the underlying probability distribution of a random variable is often approximated by a frequency distribution obtained by taking a sufficiently large sample from the distribution (see examples 4.3.8 and 4.4.4). It seems reasonable to expect that the larger the size of the sample, the better will be the approximation. That this is in fact the case is established by the following theorem, which is known as the *weak law of large numbers.*

Theorem 4.5.3
Suppose a sequence of n independent trials of a random experiment are performed (that is, a sample of size n is taken). Let f_n denote the relative frequency of an event E in the sample. Then, for any $\varepsilon > 0$

$$\lim_{n\to\infty} \mathscr{P}(|f_n - \mathscr{P}(E)| \geqslant \varepsilon) = 0$$

Proof Since the sequence consists of independent trials (that is, each trial has no memory of any previous trial), it may be regarded as a sequence of Bernoulli trials (see definition 4.3.2), with a 'success' corresponding to E occurring and a 'failure' to E not occurring in a trial.

Consider the sequence of n independent trials to be a single trial in a repeated random experiment. Let $\mathscr{Y}_n$ denote the random variable assigning to the outcome of a trial of the repeated experiment the number, y_n, of successes in the sequence and let F_n be the random variable assigning the value f_n to the outcome of a trial of the repeated experiment.

By theorem 4.3.4

$$E(\mathscr{Y}_n)=np \text{ and } \mathrm{Var}(\mathscr{Y}_n)=np(1-p)$$

where $p=\mathscr{P}(E)$. Since $f_n=y_n/n$

$$E(F_n)=E\left(\frac{1}{n}\mathscr{Y}_n\right)=\frac{1}{n}E(\mathscr{Y}_n)=p$$

by theorem 4.4.1(i).

$$\mathrm{Var}(F_n)=\mathrm{Var}\left(\frac{1}{n}\mathscr{Y}_n\right)=\frac{1}{n^2}\mathrm{Var}(\mathscr{Y}_n)$$

by theorem 4.4.1(ii)

$$=p(1-p)/n$$

By corollary 4.4.2

$$\mathscr{P}(|f_n-p|\geqslant\varepsilon)\leqslant\frac{\mathrm{Var}(F_n)}{\varepsilon^2}$$

$$=\frac{p(1-p)}{n\varepsilon^2}$$

Hence, for any $\varepsilon>0$, $\mathscr{P}(|f_n-p|\geqslant\varepsilon)\to 0$ as $n\to\infty$.

Although the weak law of large numbers confirms the intuitive feeling that the frequency distribution obtained by taking a random sample will *probably* be a reasonable approximation to the underlying probability distribution, it gives no indication of the confidence that can be placed in the approximation. This problem is clearly important in the context of computer sampling and is returned to in the next section, where it is shown that *confidence limits* may be derived by applying one of the most important theorems in probability theory, the *central limit theorem.*

Exercise 4.5

1. The joint p.d.f. of two discrete random variables X and $\mathscr{Y}$ is given in the following table.

x_i \ y_j	1	2
1	0.06	0.14
2	0.15	0.35
3	0.09	0.21

Determine:(a) the marginal p.d.f.s of X and $\mathscr{Y}$, (b) whether or not X and $\mathscr{Y}$ are independent.

2. The probability distribution of a random variable X is given in the following table.

x_i	-2	-1	1	2
p_X	0.25	0.25	0.25	0.25

(a) Determine the probability distribution of the random variable X^2;

(b) determine the joint p.d.f. of X and X^2 and show that the latter are not independent;

(c) compute the first, second and third moments of X;

(d) compute the covariance of X and X^2 and hence show that the latter are uncorrelated.

3. Let p: $\mathbb{R}^2 \to \mathbb{R}$ be given by

$$p: (x, y) \mapsto \begin{cases} 2(x+y-2xy), & \forall x, y \in [0, 1] \\ 0, & \text{otherwise} \end{cases}$$

(a) Show that p is a valid joint p.d.f.;

(b) determine the marginal p.d.f.s, p_X, $p_{\mathscr{Y}}$;

(c) determine the joint c.d.f., $\mathscr{U}$;

(d) determine whether or not X and $\mathscr{Y}$ are independent;

(e) compute the probability of the event $\{(x, y) | 0 \leqslant x \leqslant \frac{1}{2} \wedge 0 \leqslant y \leqslant \frac{1}{4}\}$.

4. Prove each of the statements in theorem 4.5.1 when X and $\mathscr{Y}$ are discrete random variables.

5. Let X and $\mathscr{Y}$ be two independent random variables and let $\mathscr{X}$ be another random variable defined by

$$z = \min(x, y)$$

where (x, y) is any element in the cross product of the range space of X and the range space of $\mathscr{Y}$. Show that

$$\mathscr{U}_{\mathscr{X}}(z) = [1 - \mathscr{U}_X(z)][1 - \mathscr{U}_{\mathscr{Y}}(z)]$$

4.6 COMPUTER SAMPLING

Random Numbers

Probability distributions have been introduced as a means of describing the behaviour of random variables. Typical problems have been to determine the probability that a random variable will have a given value (in the discrete case), or will lie in a given interval (in the continuous case). The converse of such problems is to generate a sequence of random numbers having a given probability distribution.

Because of the traditional use of random numbers in gambling, any algorithm employing a sequence of random numbers is called a *Monte Carlo method*. Such methods are applied to many practical problems, of both a deterministic and a probabilistic nature. The numerical approximation of multiple integrals, to which most of the techniques used for one-dimensional integrals are not easily extended, is an important deterministic application of Monte Carlo methods. Applications to intrinsically probabilistic problems arise in the simulation of complicated real-world systems on a computer. The latter are often called *Monte Carlo simulations*.

Generation of Random Numbers

Methods for directly generating a sequence of random numbers with a non-uniform distribution are generally not available. However, such sequences may be derived indirectly providing uniformly distributed random numbers can be generated on demand. There are three possible ways of generating random numbers, with a uniform distribution, on a computer.

Method 1 Store a *random-number table*, generated by a special electronic device, in the memory of the computer.

Storage limitations present an obvious drawback to this method. Furthermore, the table may be too short for a given Monte Carlo simulation.

Method 2 Attach an electronic random-number-generating device to the computer.

A well-known example of such a device in the United Kingdom is a machine called ERNIE, which picks the winning numbers in the weekly draw for Premium Savings Bond prizes. (ERNIE is an acronym for Electronic Random Number Indicator Equipment.)

It would be very expensive to attach a machine such as ERNIE to every computer on which it is required to run programs using Monte

Carlo methods. The main objection to this approach, however, is that it would be impossible to produce exactly the same numbers during two runs of a program, making it extremely difficult to test the program.

Method 3 Utilise the arithmetic operations of the computer.

This is the approach that is generally used. Since computers are deterministic machines, they clearly cannot produce a truly random sequence. However, they can produce sequences of *pseudo-random numbers* that pass certain statistical tests. For a discussion of these tests, the reader is referred to the comprehensive treatment of pseudo-random numbers given by Knuth (1969).

The most commonly implemented pseudo-random number generators are special cases of the *linear congruence method*, in which a sequence of pseudo-random numbers is generated recursively by

$$r_{n+1} \leftarrow (cr_n + d)(\text{mod } m), \; n = 1, 2, \ldots$$

where c, d and m are certain constants.

This just means that r_{n+1} is assigned the value of the remainder when $cr_n + d$ is divided by m.

Definition 4.6.1

The numbers c, d and m in the linear congruence method are called the *multiplier*, the *increment* and the *modulus*, respectively. If $d = 0$, the method is said to be a *multiplicative congruence method*, while, if $c = 1$, it is said to be an *additive congruence method*. Otherwise, it is a *mixed congruence method*.

The starting value r_1 is called the *seed*.

Example 4.6.1

If $m = 8$, $c = 5$, $d = 7$ and $r_1 = 4$, then the first eight terms in the resulting pseudo-random number sequence are

4, 3, 6, 5, 0, 7, 2, 1

Each number in a pseudo-random number sequence is an integer in the range 0 to $m - 1$. This means that ultimately the sequence must begin to repeat itself.

Definition 4.6.2

The *cycle length* of a pseudo-random number sequence is equal to the number of terms in the sequence before it begins repeating itself.

Example 4.6.2

The ninth term in the sequence defined in example 4.6.1 is 4, which is the same as the first term. The cycle length of this sequence is thus equal to 8.

The above sequence has the maximum possible cycle length, that is, m. In general, this desirable property for a pseudo-random number sequence can be ensured by careful choice of the constants c and d.

By taking $m = 2^n$, where n is the word length of the computer, the sequence may be generated extremely efficiently. This is illustrated in algorithm 4.6.1, which gives the assembly-language instructions for implementing the multiplicative congruence method on a particular computer. On this machine, when an integer held in accumulator *acc.i* is multiplied by an integer held in another word of store, the result is held double-length in accumulators *acc.i* and *acc.i + 1*. Since division of a binary integer by 2^n is equivalent to truncating the n least significant bits, these bits represent the required remainder from the division. It is thus unnecessary to perform the division.

Algorithm 4.6.1

```
LDX   1   C     [acc.1←C]
MPY   1   RN    [accs.1,2←acc.1 * RN]
STO   2   RN    [RN←acc.2]
```

The sequence $\{r_n\}$ produced by a pseudo-random number generator is usually normalised to the range [0, 1]. To sample values $\{s_n\}$ uniformly distributed in a given interval $[a, b]$ it is merely necessary to apply a linear scaling

$$s_n \leftarrow (b - a)r_n + a$$

As mentioned above, sequences of random numbers with non-uniform distributions, which are usually called *random deviates* or *random variates*, may be generated indirectly, using sequences of uniformly distributed random numbers.

Sampling from a Non-uniform Discrete Distribution

The technique used is known as the *tabular method* and is best illustrated by an example.

Table 4.8

x_i	1	2	3	4
$\mathscr{U}(x_i)$	0.10	0.60	0.85	1

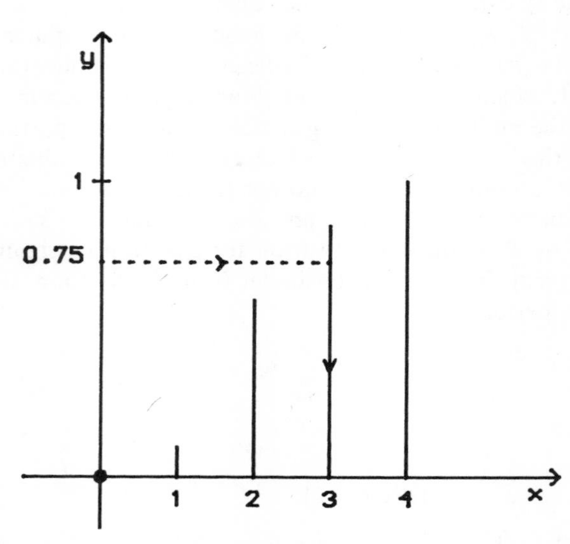

Figure 4.21 Example 4.6.3; $r_n = 0.75$ gives rise to a random deviate $s_n = 3$

Example 4.6.3

The cumulative distribution $\mathscr{U}$ of a discrete random variable X is given by table 4.8 and is shown in figure 4.21. Let $\{r_n\}$ denote a sequence of normalised pseudo-random numbers. A sequence of random deviates $\{s_n\}$, with the required distribution may be defined by

$s_n \leftarrow$ **if** $r_n \leqslant 0.1$ **then** 1 **elseif** $r_n \leqslant 0.60$ **then** 2
elseif $r_n \leqslant 0.85$ **then** 3 **else** 4 **endif**

Sampling from a Negative Exponential Distribution

If $\mathscr{U}$ denotes the c.d.f. of a negative exponential distribution, then from section 4.4

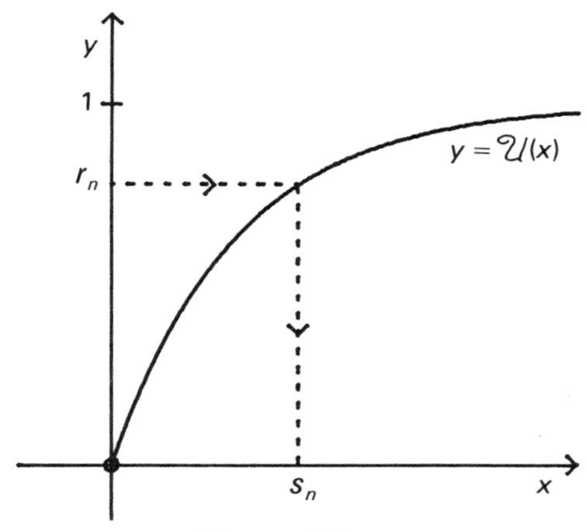

Figure 4.22

$$\mathscr{U}(x) = 1 - \exp(-\lambda x)$$

Hence

$$x = -\frac{1}{\lambda} \log_e (1 - r)$$

where $r = \mathscr{U}(x)$. If $\{r_n\}$ is a sequence of normalised pseudo-random numbers, then the sequence $\{s_n\}$ of random deviates with the required distribution is given by (figure 4.22)

$$s_n \leftarrow -\frac{1}{\lambda} \log_e (1 - r_n), \quad n = 1, 2, \ldots$$

Sampling from a Normal Distribution

This is facilitated by applying an extremely beautiful and remarkable theorem, called the *central limit theorem*, which may be stated as follows.

Theorem 4.6.1
Let $X_1, \ldots, X_n$ be identically distributed, independent random variables, each having the same distribution as a random variable X. Let $\mathscr{Y}_n \equiv (1/n)(X_1 + \ldots + X_n)$ denote the random variable defined by taking the average of the values assigned by $X_1, \ldots, X_n$. Then

$$\lim_{n \to \infty} \mathscr{P}\left(a \leqslant \frac{\mathscr{Y}_n - E(X)}{[\mathrm{Var}(X)/n]^{1/2}} \leqslant b \right) = \Phi(b) - \Phi(a)$$

where Φ is the c.d.f. of the standardised normal distribution.

A proof of this theorem is given by Parzen (1960).

Now, if $\mathscr{X}$ is a normally distributed random variable, with mean μ and standard deviation σ (see definition 4.4.6), then

$$\mathscr{P}\left(a \leqslant \frac{z-\mu}{\sigma} \leqslant b\right) = \Phi(b) - \Phi(a)$$

Thus, the surprising result embodied in the central limit theorem is that no matter what the distribution of X is, the distribution of $\mathscr{Y}_n$ tends to a normal distribution with mean $E(X)$ and standard deviation $\sqrt{[\text{Var}(X)/n]}$.

Now let $\mathscr{R}$ be a continuous random variable on [0, 1], with a uniform distribution. Thus, $E(\mathscr{R}) = 1/2$ and $\text{Var}(\mathscr{R}) = 1/12$. Let $\mathscr{Y}_m$ denote the random variable defined by taking the average of m values sampled from $\mathscr{R}$. From theorem 4.5.2

$$E(\mathscr{Y}_m) = E(\mathscr{R}) = 1/2 \text{ and } \text{Var}(\mathscr{Y}_m) = \text{Var}(\mathscr{R})/m = 1/12m$$

By the central limit theorem, the distribution of $\mathscr{Y}_m$ may be regarded as an approximation to a normal distribution, with mean $\mu = 1/2$ and standard deviation $\sigma = 1/\sqrt{(12m)}$. Thus, if $\mathscr{U}$ denotes the c.d.f. of $\mathscr{Y}_m$, then

$$\mathscr{U}(y) \approx \Phi(s)$$

where

$$s = (y - \mu)/\sigma$$

Since

$$y = \frac{1}{m}\sum_{i=1}^{m} r_i$$

$$s = \left(\frac{1}{m}\sum_{i=1}^{m} r_i - \frac{1}{2}\right)\sqrt{(12m)} = \left(\sum_{i=1}^{m} r_i - m/2\right)\sqrt{(12/m)}$$

If, as is usual, m is taken to be 12, then

$$s = \sum_{i=1}^{12} r_i - 6$$

For each successive random deviate s_n, it is thus necessary to generate 12 pseudo-random numbers. The resulting sequence $\{s_n\}$ consists of a sample of random values from an approximation to the standardised normal distribution. To obtain a sequence $\{t_n\}$ from a normal distribution, with mean *mu* and standard deviation *sigma*, say, a linear scaling is necessary, that is

$$t_n \leftarrow sigma * s_n + mu$$

The whole procedure is summarised in algorithm 4.6.2, in which rand denotes a function that returns the next term in a sequence of pseudo-random numbers, given the previous term.

Algorithm 4.6.2

```
last←seed;
s← −6;
for i from 1 in steps of 1 to 12 do
    last←rand(last);
    s←s+last
endfor;
t←sigma * s+mu
```

Confidence Limits

If a random variable X has a normal distribution, with mean μ and standard deviation σ, then from example 4.4.5

$$\mathscr{P}(\mu-\sigma \leqslant x \leqslant \mu+\sigma) \leqslant -1+2\Phi(1)=0.6826$$

that is, approximately 68 per cent of the area beneath the curve $y=p(x)$ lies within one standard deviation of the mean. Alternatively, if x is a value sampled from the distribution, then

$$\mathscr{P}(x-\sigma \leqslant \mu \leqslant x+\sigma)=0.6826$$

More generally

$$\mathscr{P}(x-\alpha \leqslant \mu \leqslant x+\alpha) = -1+2\Phi(\alpha/\sigma)$$

where α is any non-negative real number. In particular

$$\mathscr{P}(x-1.96\sigma \leqslant \mu \leqslant x+1.96\sigma) \approx 0.95$$

[see exercise 4.4(7)], that is, for any value x sampled from the distribution, there is a 95 per cent chance that the interval $[x-1.96\sigma,\ x+1.96\sigma]$ contains the mean of the distribution.

Suppose now that it is required to determine the mean value μ of a random variable $\mathscr{Y}$ having an unknown probability distribution. If a sample of size n is taken (see definition 4.1.5), then

$$\mu \approx \bar{y} = \frac{1}{n}\sum_{i=1}^{n} y_i$$

If $\mathscr{Y}_n$ denotes the random variable assigning the value $\bar{y}$ to the outcome of a sequence of n trials of the given random experiment, then as above

$$E(\mathscr{Y}_n) = \mu$$

$$\text{Var}(\mathscr{Y}_n) = \text{Var}(\mathscr{Y})/n$$

The distribution of $\mathscr{Y}_n$ is approximately a normal distribution, with mean μ and standard deviation $\sigma = \sqrt{[\text{Var}(\mathscr{Y})/n]}$.

There is thus a 95 per cent chance that the required value μ lies in the interval

$$I_{\mathscr{Y}}[0.95] \equiv [\bar{y} - 1.96\sqrt{\{\text{Var}(\mathscr{Y})/n\}}, \bar{y} + 1.96\sqrt{\{\text{Var}(\mathscr{Y})/n\}}]$$

Definition 4.6.3
The interval $I_{\mathscr{Y}}[0.95]$ is called a 95 per cent *confidence interval* for the mean value of the random variable, $\mathscr{Y}$, and $\bar{y} \pm 1.96\sqrt{[\text{Var}(\mathscr{Y})/n]}$ are called the 95 per cent *confidence limits*.

Similarly, since $\Phi(2.58) = 0.995$, taking $\alpha = 2.58$ instead of 1.96 gives rise to 99 per cent confidence limits.

Of course, in practice, $\text{Var}(\mathscr{Y})$ will not be known. However, it may be estimated by the sample variance

$$\text{Var}(\mathscr{Y}) \approx \frac{1}{n} \sum_{i=1}^{n} (y_i - \bar{y})^2$$

Note for relatively small values of n, a better estimate for the variance is generally given by replacing n by $n-1$ in the denominator of the expression for the sample variance.

Example 4.6.4

The *crude Monte Carlo method* estimates the value of an integral $\int_0^1 f(x)\,dx$

by evaluating

$$\left\{\sum_{i=1}^{n} f(x_i)\right\} \Big/ n$$

where $\{x_i\}$ is a sequence of pseudo-random numbers. For a particular integral, a sample of size 100 is taken and a sample mean $\bar{y} = 0.952$ and a sample variance 0.0017 are obtained.

The 95 per cent confidence limits are given by

$$0.952 \pm 1.96\sqrt{(0.0017/100)}$$

$$\approx 0.952 \pm 0.008$$

that is, there is a 95 per cent chance that the value of the integral lies in the interval [0.944, 0.960].

In order to obtain two-decimal-place accuracy, with 99 per cent confidence, a sample of size n is required, such that

$$2.58\sqrt{[\text{Var}(\mathscr{Y})/n]} \leqslant 0.005$$

Using the above estimate for the variance, Var($\mathscr{Y}$), gives

$$2.58\sqrt{(0.0017/n)} \leqslant 0.005$$

that is

$$n \geqslant 453$$

Exercise 4.6

1. Using the following pseudo-random number generator, with $y_1 = 9$, determine the ten values $r_1, r_2, \ldots, r_{10}$

$$y_i \leftarrow (101 y_{i-1} + 1)(\text{mod } 10^4)$$
$$r_i \leftarrow y_i / 10^4$$

2. Using the numbers obtained in exercise 4.6(1), sample ten values $s_1, s_2, \ldots, s_{10}$ from the discrete distribution given in example 4.6.3.

3. Cars arrive at a petrol station, on average, every 2 minutes. Using the numbers obtained in exercise 4.6(1), sample ten values from the appropriate negative exponential distribution in order to simulate the inter-arrival times of eleven successive customers at the petrol station.

4. Write a computer program to generate twenty numbers having a normal distribution with mean value 50 and standard deviation 5. Obtain a plot of the distribution of the numbers.

5. The outcome of an experiment is an integer in the range 1 to 365, inclusive, representing a day in a normal year. Each outcome is equally likely to occur. A discrete random variable, X, assigns an integer in the range 1 to 12, inclusive, to each outcome of the experiment. The value assigned by X to an outcome of the experiment represents the month in which the given day falls. For example, X assigns a value 3 to the outcome 84, since the 84th day of the year falls in month 3.

Determine the cumulative probability distribution of X. Using the pseudo-random numbers obtained in exercise 4.6(1), sample ten random values from the range space of X.

5 Algebraic Structures

5.1 RELATIONS

Introduction

In section 1.5, real functions of more than one argument were defined using the concept of the *cartesian product*, $A \times B$, of two sets, A and B, where

$$A \times B = \{(a, b) | a \in A, b \in B\}$$

Definition 5.1.1
A *relation in* $A \times B$ is a subset of $A \times B$. A relation in $A \times A$ is sometimes simply called a relation on A.

If (a, b) is an element of a relation R, rather than write $(a, b) \in R$, it is common practice to write aRb (read: a is related to b). Similarly, if $(a, b) \notin R$, then this may be written $a \not R b$ (read: a is not related to b).

Example 5.1.1

(1) $L = \{(a, b) | a < b \text{ and } a, b \in \mathbb{R}\}$ is a relation on $\mathbb{R}$.

(2) $\emptyset$ is the empty relation.

(3) Let $A = \{1, 2, 3, 4\}$ and $B = \{1, 2, 3, 4, 5\}$; if a relation R in $A \times B$ is defined by aRb iff $2a \leqslant b$, then

$$R = \{(1, 2), (1, 3), (1, 4), (1, 5), (2, 4), (2, 5)\}$$

(4) If H denotes the set of all human beings that have ever lived, then $S = \{(a, b) | a \text{ is the offspring of } b\}$ is a relation on H.

Definition 5.1.2
The *domain* of a relation R in $A \times B$ is the subset of A consisting of the first components of the ordered pairs in R. The *range* of R is the subset of B consisting of the second components of the ordered pairs in R. The

domain and range of R are usually denoted by Domain(R) and Range(R), respectively.

Example 5.1.2

Using the relations defined in example 5.1.1

(1) Domain(L) = Range(L) = $\mathbb{R}$;

(2) Domain($\emptyset$) = Range($\emptyset$) = $\emptyset$;

(3) Domain(R) = $\{1, 2\}$; Range(R) = $\{2, 3, 4, 5\}$;

(4) Domain(S) = H, Range(S) = {all parents that have ever lived}.

Definition 5.1.3
The *inverse*, R^{-1}, of a relation R in $A \times B$ is a relation in $B \times A$, where $bR^{-1}a$ iff aRb.

Example 5.1.3

Again using the relations defined in example 5.1.1

(1) $L^{-1} = \{(a, b) | a > b \text{ and } a, b \in \mathbb{R}\}$;

(2) $\emptyset^{-1} = \emptyset$;

(3) $R^{-1} = \{(2, 1), (3, 1), (4, 1), (5, 1), (4, 2), (5, 2)\}$;

(4) $S^{-1} = \{(a, b) | a \text{ is the parent of } b\}$.

Functions as Relations

Definition 5.1.4
If a function $f: X \to Y$ is defined by the rule $x \mapsto f(x)$ then the *graph* of f is the set of ordered pairs $\{(x, f(x)) | x \in X\}$.

The graph of a function is simply a relation but a relation is not necessarily the graph of some function. The definition of a function from X to Y requires that, for any $x \in X$, there is only one pair (x, y) in the graph of the function. Thus, the following theorem may be formulated.

Theorem 5.1.1
A relation f is the graph of a function provided $(x, y_1) \in f$ and $(x, y_2) \in f$ implies $y_1 = y_2$.

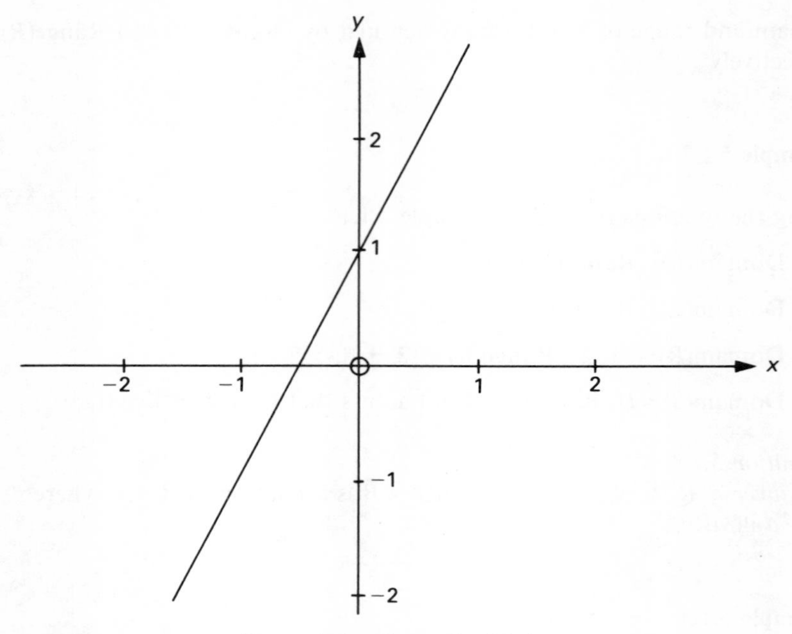

Figure 5.1 Example 5.1.4

Example 5.1.4

If f: $x \mapsto 2x+1$ ($x \in \mathbb{R}$) then the graph of f is $\{(x, y) | y=2x+1, x \in \mathbb{R}\}$, which is a relation on $\mathbb{R}$. Representing the graph diagrammatically results in figure 5.1.

The reader should note that the definitions of domain, range and inverse function given in section 1.5 correspond with the definitions obtained by representing the function in terms of its graph and using the definitions of this section. In fact, many authors choose to introduce functions in terms of sets of ordered pairs and to base the whole theory of functions on that of relations.

Representing Relations

If A and B are finite, then a relation R in $A \times B$ can be represented using a matrix of Boolean values. If $A=\{a_1, \ldots, a_m\}$ and $B=\{b_1, \ldots, b_n\}$ then R can be represented by an $m \times n$ Boolean matrix, M, where

$$M_{ij}=\begin{cases} \text{T} & \text{if } a_i R b_j \\ \text{F} & \text{otherwise} \end{cases}$$

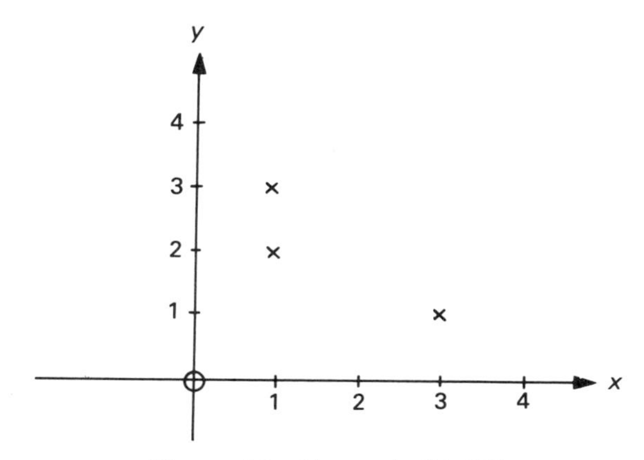

Figure 5.2 Example 5.1.5(1)

The Boolean matrix corresponding to R in example 5.1.1(3) is given by

$$M=\begin{bmatrix} F & T & T & T & T \\ F & F & F & T & T \\ F & F & F & F & F \\ F & F & F & F & F \end{bmatrix}$$

If a relation is defined in $\mathbb{R}\times\mathbb{R}$, then it consists of a set of ordered pairs of real numbers. Each of these ordered pairs can be represented as a point in the cartesian plane. Any relation defined in the reals can thus be represented as a set of points in the cartesian plane. Individual points are marked with a cross and areas are shaded. A solid line delimiting an area means that points on the line are part of the relation, while a dashed line means they are not.

Example 5.1.5

(1) $\{(1, 2), (3, 1), (1, 3)\}$ (figure 5.2).

(2) $\{(x, y)|2x \geqslant y\}$ (figure 5.3).

(3) $\{(x, y)|x^2+y^2<4\}$ (figure 5.4).

(4) Let $R=\{(x, y)|x+2y\geqslant 3\}$, $S=\{(x, y)|2y-x>-6\}$ and $T=\{(x, y)|2x+y\leqslant 6\}$. Then

$$\begin{aligned} S^{-1} &= \{(y, x)|2y-x>-6\} \\ &= \{(x, y)|2x-y>-6\} \end{aligned}$$

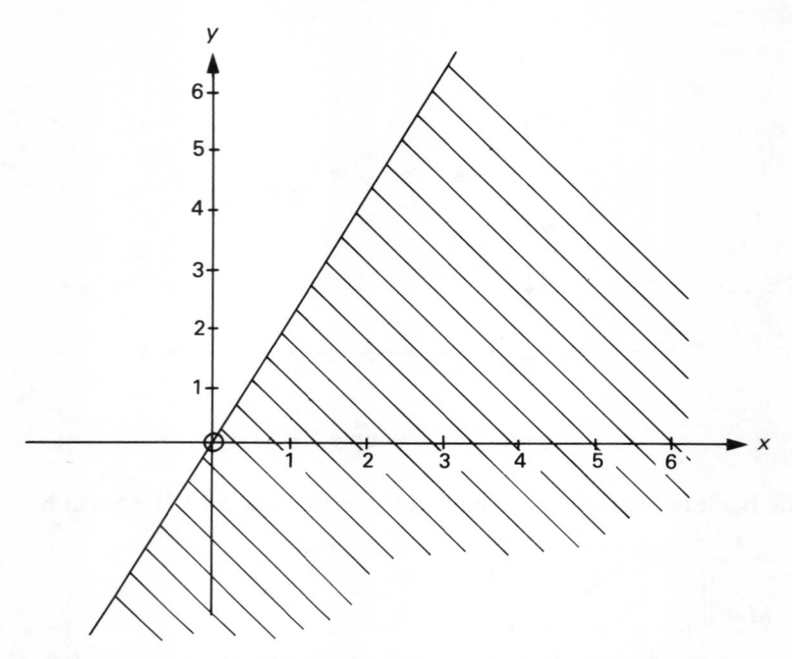

Figure 5.3 Example 5.1.5(2); every point on or below the line $y = 2x$ represents an element of the relation

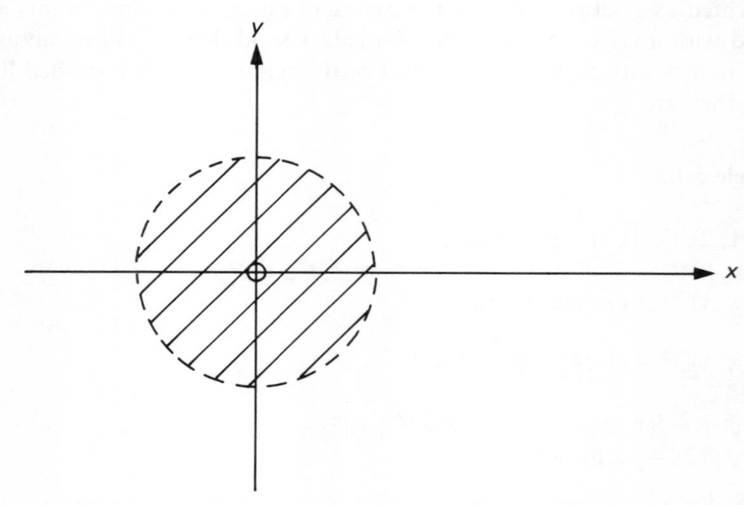

Figure 5.4 Example 5.1.5(3); every point in, but *not* on, the circle $x^2 + y^2 = 4$ represents an element of the relation

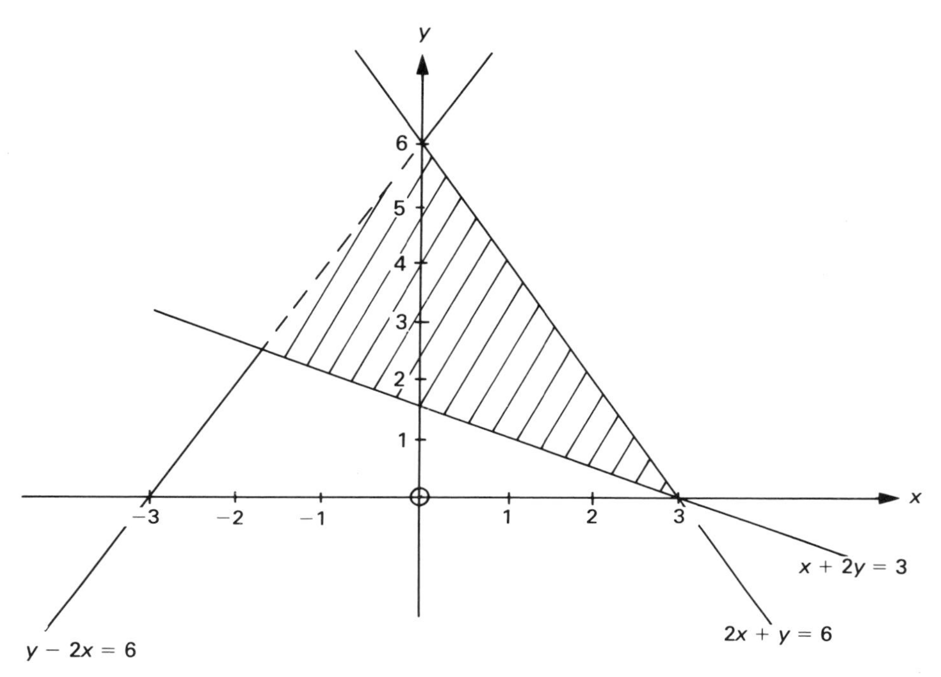

Figure 5.5 Example 5.1.5(4); the shaded area represents the relation $R \cap S^{-1} \cap T$, but points at the intersection of the broken line with the solid lines are not included

The shaded area in figure 5.5. represents the relation $R \cap S^{-1} \cap T$. Since (0, 6) and $(-9/5, 12/5)$ are on the intersection of a broken line and a solid line, they are not included in the relation. Only when two lines are both solid is their intersection included in the relation.

Representing real relations in the cartesian plane can be used in the solution of two-dimensional linear-programming problems. A typical problem of this kind takes the form

$$\begin{aligned}
&\text{maximise (or minimise) } c_1x+c_2y \\
&\text{subject to } a_{11}x+a_{12}y \geqslant b_1 \\
&\qquad\qquad a_{21}x+a_{22}y \geqslant b_2 \\
&\qquad\qquad \vdots \\
&\qquad\qquad a_{n1}x+a_{n2}y \geqslant b_n
\end{aligned}$$

where $c_1, c_2, a_{11}, a_{12}, \ldots, a_{n1}, a_{n2}, b_1, \ldots, b_n$ are all constants.

In the real world, problems involving larger numbers of variables occur, and these are solved with the aid of a computer. The theory behind the algorithms used to solve these harder problems is based on a generalisation of the method used in example 5.1.6 to solve a two-dimensional example.

Example 5.1.6

A businessman has the opportunity to buy cheap sugar and salt. Sugar costs £18 per sack and salt costs £12 per sack. The businessman can only store 30 sacks and cannot afford to spend more than £420. If he expects to make a profit of £11 per sack of sugar and £9 per sack of salt, how many sacks of each product should be purchased to maximise his total profit?

Let x be the number of sacks of sugar and y be the number of sacks of salt.

The cost constraint is

$$18x+12y \leqslant 420$$

that is $3x+2y \leqslant 70$, and the storage constraint is

$$x+y \leqslant 30$$

The linear programming problem is thus

$$\begin{array}{ll} \text{maximise} & 11x+9y \\ \text{subject to} & 3x+2y \leqslant 70 \\ \text{and} & x+y \leqslant 30 \end{array}$$

Clearly, $x \geqslant 0$ and $y \geqslant 0$.

In figure 5.6, the constraints are satisfied by each point in the shaded area. This area is called the *feasible region.*

If £k is the businessman's total profit, then

$$11x+9y=k$$

which can be rewritten as

$$y=\frac{1}{9}(k-11x)$$

The graphs of this equation for different values of k form a family of parallel lines, all with gradient $-11/9$. Some of these are shown in figure 5.6 as broken lines. The optimal solution to the problem must lie within the feasible region and must also lie on the line of the form $y=(1/9)(k-11x)$ with maximum possible k. The intercept of a line of this form with the y axis is $k/9$ and thus the solution should lie on the line

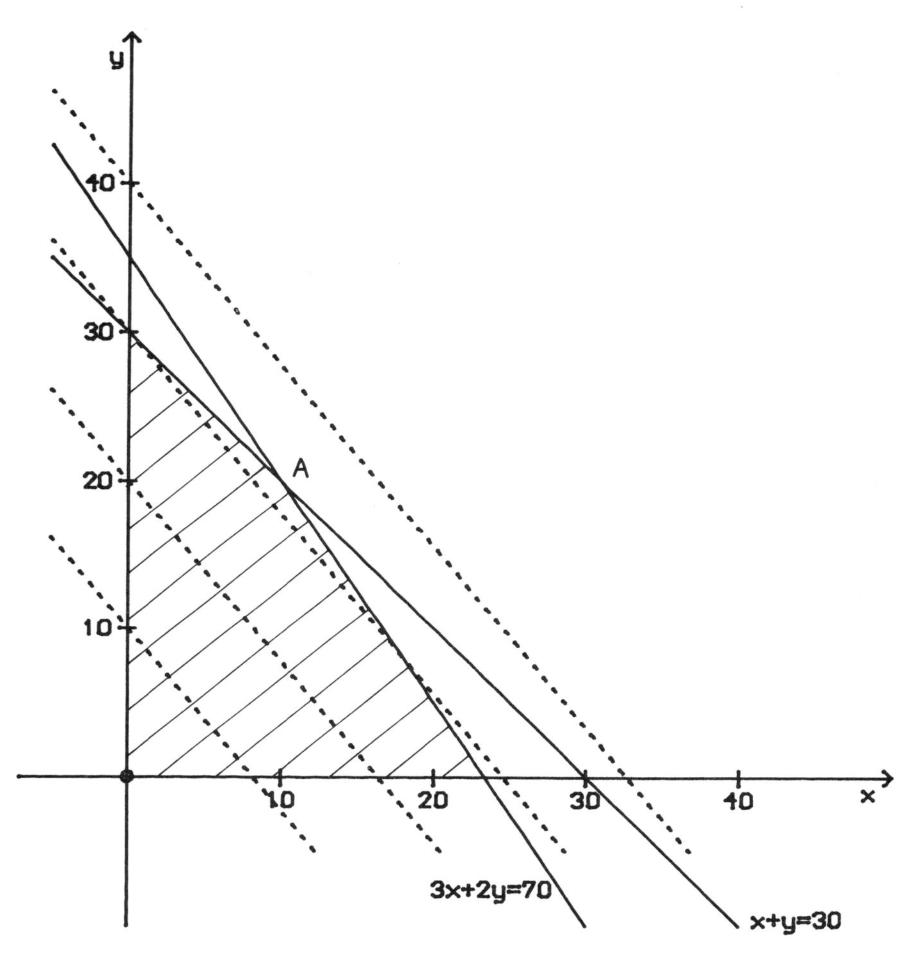

Figure 5.6 Example 5.1.6: the shaded area represents the feasible region

with as great an intercept with the y axis as possible. Thus, if a solution can be found at the point A then that is the best that can be achieved. Here A is the intersection of $3x+2y=70$ and $x+y=30$, that is, the point (10, 20). This is the required solution — the businessman should purchase 10 sacks of sugar and 20 of salt to obtain a profit of £290.

Properties of Relations

Definition 5.1.5
Let R be a relation on A, then

(a) R is *reflexive* iff aRa, $\forall a \in A$;

(b) R is *symmetric* iff $aRb \supset bRa$, $\forall a, b \in A$;

(c) R is *transitive* iff $aRb \wedge bRc \supset aRc$, $\forall a, b, c \in A$;

(d) R is *antisymmetric* iff $aRb \wedge bRa \supset a = b$, $\forall a, b \in A$.

Example 5.1.7

(1) $\emptyset$ is a symmetric and transitive relation on $\mathbb{R}$ but is not reflexive.

(2) $F = \{(a, b) | a \text{ is a factor of } b\}$ is a reflexive and transitive relation on N but is not symmetric.

(3) $C = \{(a, b) | a \text{ owns a car of the same make as } b\}$ is a reflexive and symmetric relation on the set of all human beings but is not transitive. (Remember that is is possible to own two cars.)

(4) $\emptyset$ and F are antisymmetric but C is not.

Definition 5.1.6
If R is any relation on A then the *reflexive closure* of R is the smallest reflexive relation on A containing R. Similarly, the *symmetric closure* of R is the smallest symmetric relation on A containing R and the *transitive closure* is the smallest transitive relation on A containing R.

Example 5.1.8

Let L be the relation on $\mathbb{R}$ where $L = \{(a, b) | a < b\}$. The reflexive closure of L is $\{(a, b) | a \leqslant b\}$, the symmetric closure of L is $\{(a, b) | a \neq b\}$ and the transitive closure of L is L itself.

If A is finite, then efficient algorithms exist to compute the reflexive, transitive and symmetric closures of an arbitrary relation, R, on A. Assuming that the cardinality of A is n, then R can be represented by an $n \times n$ Boolean matrix, M. The algorithms to compute the reflexive, symmetric and transitive closures operate on this matrix, M.

Algorithm 5.1.1

(a) To compute the matrix, M_R, representing the reflexive closure of R

$M_R \leftarrow M$;

for i **from** 1 **in steps of** 1 **to** n **do** $M_R[i, i] \leftarrow T$ **endfor**

(b) To compute the matrix, M_S, representing the symmetric closure of R

$M_S \leftarrow M$;

for i **from** 2 **in steps of** 1 **to** n **do**

for j **from** 1 **in steps of** 1 **to** $i-1$ **do**

$M_S[i, j] \leftarrow M_S[i, j] \vee M_s[j, i]$;

$M_S[j, i] \leftarrow M_S[i, j]$

endfor

endfor

(c) *Warshall's algorithm.* To compute the matrix, M_T, representing the transitive closure of R

$M_T \leftarrow M$;

for k **from** 1 **in steps of** 1 **to** n **do**

for i **from** 1 **in steps of** 1 **to** n **do**

for j **from** 1 **in steps of** 1 **to** n **do**

if $M_T[i, k] \wedge M_T[k, j]$ **then** $M_T[i, j] \leftarrow \mathrm{T}$ **endif**

endfor

endfor

endfor

Definition 5.1.7
A relation R on A is called an *equivalence relation* in A if it is reflexive, symmetric and transitive.

Example 5.1.7 shows that the three conditions required for a relation to be an equivalence relation are independent, that is, that having just two of the conditions does not imply the third.

Definition 5.1.8
For an equivalence relation R on A, the *equivalence class* to which $a \in A$ belongs is defined to be the set

$$\bar{a} = \{x \in A \mid (a, x) \in R\}$$

Theorem 5.1.2
The equivalence classes of an equivalence relation, R on A, partition A into a number of disjoint subsets.

Proof Every element $a \in A$ lies in some equivalence class, in particular, $a \in \bar{a}$ since R is reflexive. It is now necessary to prove that the equivalence classes are disjoint, that is, to show that for any $a, b \in A$, either $\bar{a} = \bar{b}$ or $\bar{a} \cap \bar{b} = \emptyset$. Assuming $\bar{a} \cap \bar{b} \neq \emptyset$, there exists $y \in \bar{a} \cap \bar{b}$, so $(a, y) \in R$ and $(b, y) \in R$. Since R is reflexive, $(y, b) \in R$. Thus, $(a, y) \in R \wedge (y, b) \in R$ and since R is transitive it follows that $(a, b) \in R$, that is, $b \in \bar{a}$. Now, if $x \in \bar{b}$, $(b, x) \in R$ and, since $(a, b) \in R$ and R is transitive, it follows that $(a, x) \in R$, and so $x \in \bar{a}$. Thus $\bar{b} \subset \bar{a}$. Showing that $\bar{a} \subset \bar{b}$ is left as an exercise for the reader.

Example 5.1.9

Let R be a relation on the natural numbers where aRb iff $a \equiv b \pmod 4$, that is, iff a and b have the same remainder when divided by 4. The reader is invited to verify that R is reflexive, symmetric and transitive. The relation R partitions N into four disjoint equivalence classes

$$\bar{1} = \{1, 5, 9, \ldots\}, \bar{2} = \{2, 6, 10, \ldots\}, \bar{3} = \{3, 7, 11, \ldots\}, \bar{4} = \{4, 8, 12, \ldots\}$$

Ordering Relations and Lattices

Definition 5.1.9
A relation on A is *an ordering relation* on A if it is reflexive, antisymmetric and transitive. If such a relation exists for a nonempty set A then A is said to be *partially ordered* with respect to that relation. In a partially ordered set, A, with ordering relation $\sqsubseteq$, there may exist elements $a, b \in A$ such that neither $a \sqsubseteq b$ nor $b \sqsubseteq a$. Such elements are said to be *incomparable*. A partially ordered set is said to be *totally ordered* if it has no incomparable elements.

Example 5.1.10

(1) $\mathbb{R}$ is a totally ordered set with respect to $\leqslant$.

(2) The power set of a set is partially, but not totally, ordered with respect to set inclusion, $\subset$.

Definition 5.1.10
Let X be a subset of a set A that is partially ordered with respect to $\sqsubseteq$. $t \in A$ is an *upper bound* of X iff $\forall x \in X$, $x \sqsubseteq t$. $t \in A$ is a *least upper bound* (lub) of X if t' is an upper bound of X implies $t \sqsubseteq t'$. In this case t is written as $\bigsqcup X$.

$b \in A$ is a *lower bound of* X iff $\forall x \in X$, $b \sqsubseteq x$. b is a *greatest lower bound* (glb) of X if b' is a lower bound of X implies $b' \sqsubseteq b$. In this case, b is written as $\bar{\sqcap} X$.

It is not necessarily the case that a given subset of a partially ordered set will have an lub or a glb. However, if it does, the antisymmetric property of partial orderings guarantees its uniqueness. A partially ordered set where every finite subset has an lub and glb is called a lattice. Lattice theory has become important to computer scientists because of its use in the theory for defining the *semantics* (that is, meaning) of programs.

Definition 5.1.11
A *lattice*, $\mathscr{L}$, is a partially ordered set with ordering $\sqsubseteq_{\mathscr{L}}$ such that each pair of elements x, $y \in \mathscr{L}$ has an lub, $x \sqcup_{\mathscr{L}} y$, in $\mathscr{L}$ and a glb, $x \sqcap_{\mathscr{L}} y$, in $\mathscr{L}$.

By induction every finite subset of a lattice $\mathscr{L}$ must also have an lub and glb. The lattice is said to be *complete* if all subsets of $\mathscr{L}$, both finite and infinite, have an lub and glb.

In particular, if $\mathscr{L}$ is complete, $\mathscr{L}$ itself must have an lub and a glb. The lub of $\mathscr{L}$ is called the *top element* and is denoted by $\top_{\mathscr{L}}$ and the glb of $\mathscr{L}$ is called the *bottom element* and is denoted by $\bot_{\mathscr{L}}$.

The subscript, $\mathscr{L}$, on $\sqsubseteq_{\mathscr{L}}$, $\sqcup_{\mathscr{L}}$, $\sqcap_{\mathscr{L}}$, $\top_{\mathscr{L}}$ and $\bot_{\mathscr{L}}$ is omitted when it is clear to which lattice the ordering, etc., refers.

Example 5.1.11

(1) The power set of a set, A, under the ordering of set inclusion is a complete lattice, where the lub of a set of subsets is their union and glb is their intersection. In this case, $\top$ is the set A and $\bot = \emptyset$.

(2) A finite lattice can be described diagrammatically. Each element of the lattice is represented by a node and $a \sqsubseteq_{\mathscr{L}} b$ is signified by having node a below node b and connecting the nodes by a line. As an example, see figure 5.7, which represents a complete lattice with top element a_0 and bottom element a_8.

(3) If $\mathscr{L}$ is a complete lattice then $\mathscr{L}^n$ denotes the complete lattice whose elements are n-tuples of $\mathscr{L}$ and whose ordering is defined by

$$(x_1, \ldots, x_n) \sqsubseteq_{\mathscr{L}^n} (y_1, \ldots, y_n) \text{ iff } x_1 \sqsubseteq_{\mathscr{L}} y_1, \ldots, x_n \sqsubseteq_{\mathscr{L}} y_n$$

The lattice $\mathscr{L}^n$ is called the nth *power of* $\mathscr{L}$.

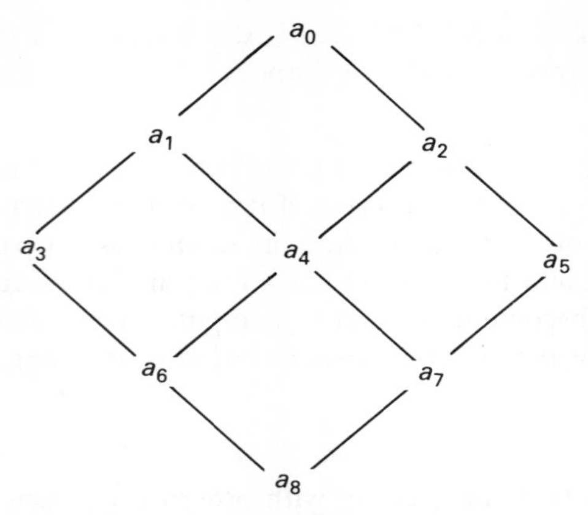

Figure 5.7 Example 5.1.11(2)

Exercise 5.1

1. Let $X = \{2, 3, 4, 6, 8, 9\}$. For each of the following relations, list them as ordered pairs and state the domain and range

(a) 'is equal to';

(b) 'is greater than';

(c) 'is a factor of';

(d) 'is 4 less than';

(e) 'is at least 3 greater than'.

2. Which of the relations in exercise 5.1(1) are: (a) reflexive; (b) symmetric; (c) antisymmetric; (d) transitive? Which then are equivalence relations?

3. If $X = \{x|2x - 1 > 0\}$ and $Y = \{y|2 \leqslant y < 4\}$, display $X \times Y$ as a shaded region in the cartesian plane.

4. Let R be some relation in $\mathbb{R} \times \mathbb{R}$ that is represented by a set of points in the cartesian plane. If the line joining the points (x_1, y_1) and (x_2, y_2) is a subset of R for all $(x_1, y_1) \in R$ and $(x_2, y_2) \in R$, then R is said to be a *convex set*.

Represent each of the following on the cartesian plane and state which are convex sets

(a) $R = \{(x, y)|x + y \geqslant 2 \text{ and } x \geqslant 0, y \geqslant 0\}$;

(b) $S = \{(x, y)|\ |x| < 2 \text{ and } |y| < 2\}$;

(c) $T = \{(x, y)|y \leqslant 2x + 4, y < 4 - 2x, x^2 + y^2 \geqslant 4, y \geqslant 0\}$.

5. Compute R^{-1}, S^{-1} and T^{-1} where R, S and T are defined as in exercise 5.1(4).

6. Let R and R' be relations on a set A. Prove

(a) that, if R and R' are both symmetric, then $R \cup R'$ is symmetric;

(b) that, if R is reflexive and R' is any relation then $R \cup R'$ is reflexive.

7. Solve the linear programming problem

$$\begin{array}{ll} \text{minimise} & 5x_1 + 2x_2 \\ \text{subject to} & 3x_1 - 2x_2 \geqslant 6 \\ & x_1 + 2x_2 \geqslant 8 \\ & x_1 + x_2 \leqslant 7 \\ \text{and} & x_1, x_2 \geqslant 0 \end{array}$$

8. Show that the transitive closure of a symmetric relation is symmetric.

9. Complete the proof of theorem 5.1.2.

10. If $\mathscr{L}$ is a lattice with ordering $\sqsubseteq_{\mathscr{L}}$, show that the partially ordered set, $\hat{\mathscr{L}}$, which has the same elements as $\mathscr{L}$ but is ordered by $\sqsubseteq_{\hat{\mathscr{L}}}$ where $x \sqsubseteq_{\hat{\mathscr{L}}} y$ iff $y \sqsubseteq_{\mathscr{L}} x$, is also a lattice. ($\hat{\mathscr{L}}$ is called the *dual* of $\mathscr{L}$)

11. A *monotonic function*, f, from a complete lattice to itself is a function that is order-preserving, that is, $x \sqsubseteq_{\mathscr{L}} y$ implies $f(x) \sqsubseteq_{\mathscr{L}} f(y)$. Prove that a monotonic function from a complete lattice to itself has at least one fixed point, that is, there exists an $x \in \mathscr{L}$ such that $f(x) = x$. [Hint: Let S be the set of f-enlarged points, $S = \{x \in \mathscr{L} |, x \sqsubseteq f(x)\}$. Show $S \neq \emptyset$ and $f(\bigsqcup S) = \bigsqcup S$.]

5.2 DIGRAPHS

Introduction

Confusingly, the term 'graph' is used not only in the context of the graph of a function but also in the context of a directed graph where it has a completely different meaning. Directed graphs (like figure 5.8) consist of a finite, non-empty set of nodes and a finite set of arcs linking these nodes. The theory of directed graphs has applications in computer science, engineering, architecture, resource allocation, scheduling, etc. For a more comprehensive study than that given in this section, the interested reader can confidently be referred to Bondy and Murty (1977).

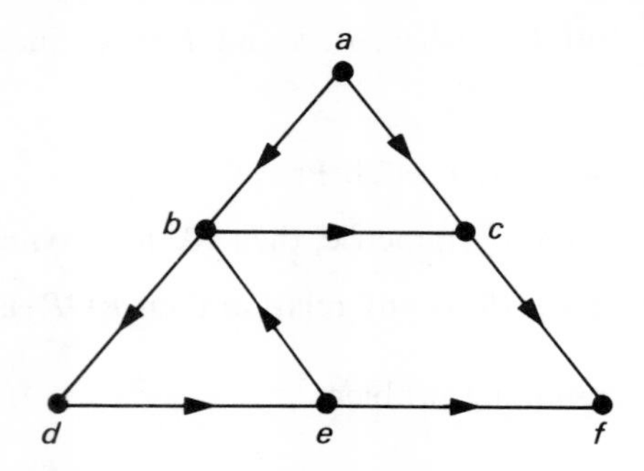

Figure 5.8 Example 5.2.1(1): digraph G_1

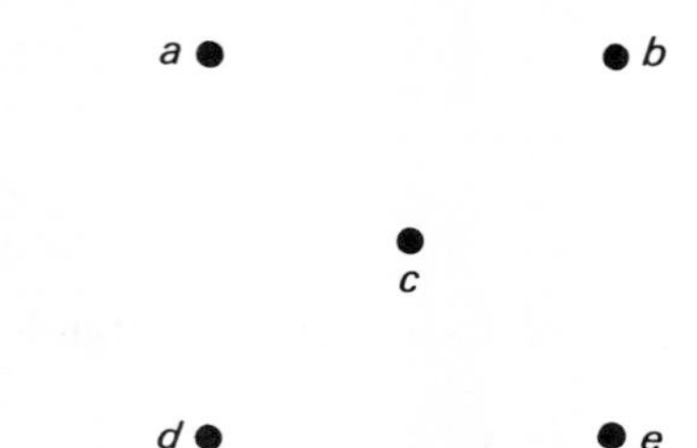

Figure 5.9 Example 5.2.1(2): null graph

Definition 5.2.1
A *directed graph* (or *digraph*), $G=(N_G, A_G)$, is a finite, non-empty set of *nodes*, N_G, together with a relation $A_G \subset N_G \times N_G$. If $(a, b) \in A_G$, there is said to be an *arc* from node a to node b.

Example 5.2.1

(1) The digraph G_1, shown in figure 5.8 has the set of nodes $\{a, b, c, d, e, f\}$ together with arcs defined by the relation $\{(a, b), (a, c), (b, c), (b, d), (c, f), (d, e), (e, b), (e, f)\}$.

(2) If a digraph $G=(N_G, A_G)$ is such that $A_G=\emptyset$, that is, it has no arcs, then it is called a *null graph*. Figure 5.9 illustrates a null graph with nodes $\{a, b, c, d, e\}$.

(3) If a digraph $G=(N_G, A_G)$ is such that $A_G=N_G \times N_G$, that is, if A_G contains (a, b), $\forall a, b \in N_G$, then the graph is said to be *complete*. Figure 5.10 illustrates a complete directed graph with five nodes $\{a, b, c, d, e\}$. This graph has twenty-five arcs since each of the five nodes has to have an

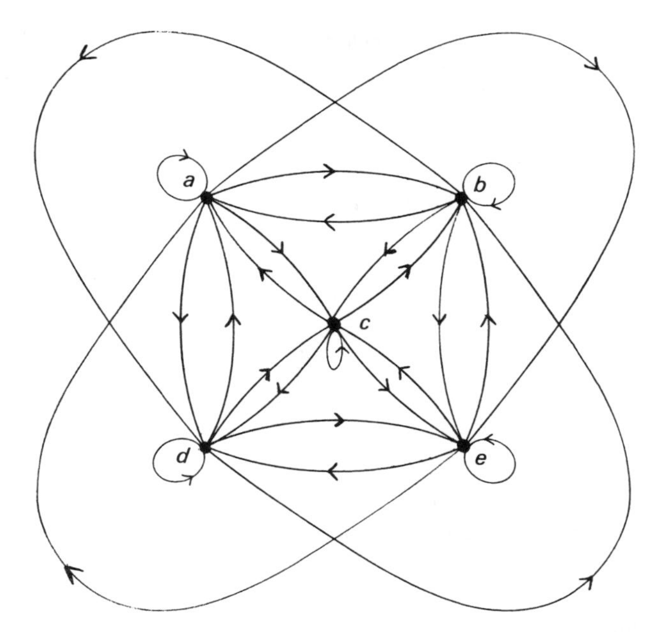

Figure 5.10 Example 5.2.1(3): complete directed graph

arc leading to each of the five other nodes, including itself. In general, a complete directed graph with n nodes must have n^2 arcs.

For a digraph with many arcs, the task of counting them can be quite tricky but this can be simplified by using the concept of indegree and outdegree.

Definition 5.2.2
For any node $n \in N_G$, the *indegree* of n is the number of arcs coming into node n and the *outdegree* of n is the number of arcs leaving node n.

Thus, if $x \in N_G$, indegree$(n) = \#\{(x, n) | (x, n) \in A_G\}$ and outdegree$(n) = \#\{(n, x) | (n, x) \in A_G\}$.

The *degree* of a node n is defined by degree$(n) =$ indegree$(n) +$ outdegree(n).

Theorem 5.2.1
For any directed graph $G = (N_G, A_G)$

$$\#(A_G) = \sum_{n \in N_G} \text{indegree}(n) = \sum_{n \in N_G} \text{outdegree}(n) = \frac{1}{2} \sum_{n \in N_G} \text{degree}(n)$$

Proof Every arc in the digraph contributes 1 to the indegree of the node to which it goes and 1 to the outdegree of the node whence it comes.

Table 5.1

Node	*Indegree*	*Outdegree*	*Degree*
a	0	2	2
b	2	2	4
c	2	1	3
d	1	1	2
e	1	2	3
f	2	0	2
	8	8	16

Example 5.2.2

The degrees of the nodes in the directed graph G_1 (see figure 5.8) can be tabulated as in table 5.1. From theorem 5.2.1, the number of arcs is eight.

Connected Digraphs

There is said to be a *directed route* from a node, *a*, to a distinct node, *b*, in a digraph if either there exists an arc from *a* to *b* or there exist nodes x_1, ..., x_k ($k \geqslant 1$) and arcs *a* to x_1, x_1 to x_2, ..., x_k to *b*. Two nodes *a* and *b* are said to be unilaterally connected if either $a = b$ or there is a directed route from *a* to *b* or from *b* to *a*. More formally, this may be expressed as follows.

Definition 5.2.3
Two nodes *a*, *b* in a digraph $G = (N_G, A_G)$ are *unilaterally connected* if: (i) $a = b$; or (ii) (a, b) *or* (b, a) is an element of the transitive closure of A_G.

The digraph constructed from nodes, N_G, and the transitive closure of the relation A_G is called the *transitive closure* of *G*. It follows from the definition that two distinct nodes are unilaterally connected in a digraph *G* iff there is an arc between them in the transitive closure of *G*.

Example 5.2.3

The relation used to define the digraph in figure 5.8 can be represented by a Boolean matrix as in figure 5.11a. The transitive closure of the relation is then represented by the Boolean matrix in figure 5.11b computed from the original matrix using Warshall's algorithm (algorithm 5.1.1.c). The graph corresponding to this new relation is given in figure 5.11c. Two

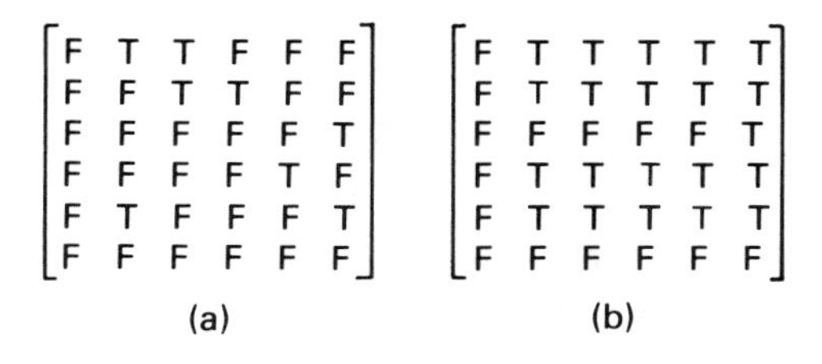

$$\begin{bmatrix} F & T & T & F & F & F \\ F & F & T & T & F & F \\ F & F & F & F & F & T \\ F & F & F & F & T & F \\ F & T & F & F & F & T \\ F & F & F & F & F & F \end{bmatrix} \qquad \begin{bmatrix} F & T & T & T & T & T \\ F & T & T & T & T & T \\ F & F & F & F & F & T \\ F & T & T & T & T & T \\ F & T & T & T & T & T \\ F & F & F & F & F & F \end{bmatrix}$$

(a) (b)

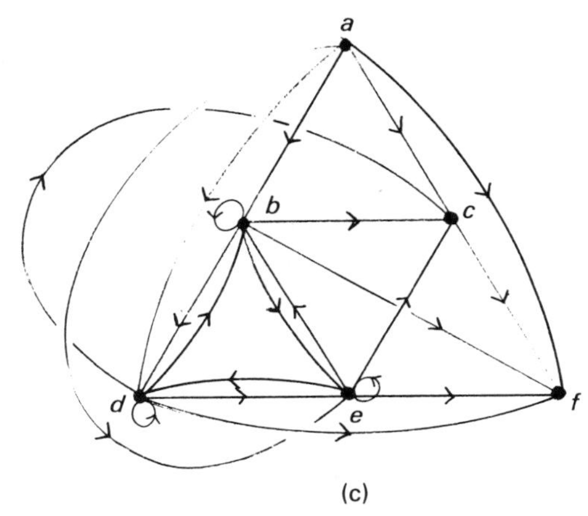

(c)

Figure 5.11 Example 5.2.3: (a) Boolean matrix corresponding to the relation $R=\{(a, b), (a, c), (b, c), (b, d), (c, f), (d, e), (e, b), (e, f)\}$; (b) Boolean matrix corresponding to the transitive closure of R; (c) the transitive closure of the graph G_1

distinct nodes are linked by an arc in this graph iff they are unilaterally connected in the original graph. Thus any two nodes of the original graph are shown to be unilaterally connected.

Definition 5.2.4
Two nodes a, b in a digraph $G=(N_G, A_G)$ are *strongly connected* if: (i) $a=b$; or (ii) both (a, b) and (b, a) are elements of the transitive closure of A_G.

Example 5.2.4

In G_1, the only distinct strongly connected nodes are b and d, b and e and d and e.

Definition 5.2.5
The *symmetric closure* of a directed graph $G=(N_G, A_G)$ has nodes N_G and a relation equal to the symmetric closure of A_G.

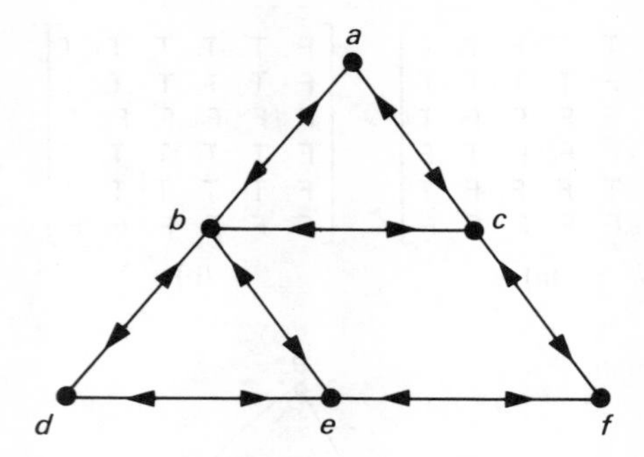

Figure 5.12 The symmetric closure of G_1

The symmetric closure of a digraph G thus differs from G in that for all arcs (a, b) in G there must be two arcs (a, b) and (b, a) in the symmetric closure of G.

Example 5.2.5

Figure 5.12 illustrates the symmetric closure of G_1. In this diagram, the convention has been adopted whereby an arc with two opposing arrows between two nodes a and b indicates the existence of two arcs, one from a to b and one from b to a.

Definition 5.2.6
Two nodes a, b in a digraph G are said to be *weakly connected* if: (i) $a=b$; or (ii) there exists an arc between a and b in the transitive closure of the symmetric closure of G.

Example 5.2.6

If a, b are weakly connected it means that if the directions of the arrows in the digraph are ignored one can still get from a to b. The sequence of arcs used is called the *path between a* and b. Since such a path exists between each pair of nodes in G_1, it follows that every node in G_1 is weakly connected to every other node in G_1.

Definition 5.2.7
A digraph $G=(N_G, A_G)$ is

(a) *weakly connected* if, for every a, $b \in N_G$, a and b are weakly connected;

(b) *unilaterally connected* if, for every a, $b \in N_G$, a and b are unilaterally connected;

(c) *strongly connected* if, for every a, $b \in N_G$, a and b are strongly connected.

An immediate consequence of these definitions is as follows.

Theorem 5.2.2
(i) Every strongly connected digraph is unilaterally connected.

(ii) Every unilaterally connected digraph is weakly connected.

Example 5.2.7

(1) Figure 5.13a illustrates a digraph which is not even weakly connected. Such a graph is called *disconnected.*

(2) Figure 5.13b illustrates a digraph which is weakly connected but not unilaterally connected since node a is not unilaterally connected to node c or to node d.

(3) G_1 is unilaterally connected but not strongly connected.

(4) Every complete digraph is strongly connected but there are also strongly connected digraphs that are not complete, for example, that illustrated in figure 5.13c.

Trees and Subgraphs

Definition 5.2.8
A *subgraph* of a digraph $G=(N_G, A_G)$ is a digraph having all its nodes in N_G and all its arcs in A_G.

If G is a subgraph of G' then G' is a *supergraph* of G.

A *spanning subgraph* of G is a weakly connected subgraph containing all the nodes of G.

Example 5.2.8

In figure 5.14, graphs G_1' and G_1'' are both subgraphs of G_1; G_1'' is a spanning subgraph of G_1.

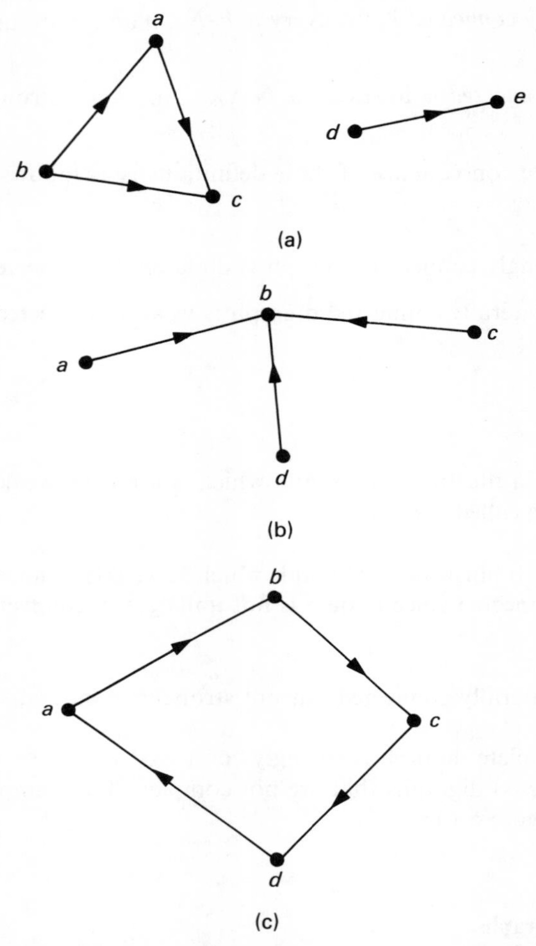

Figure 5.13 Example 5.2.7: (a) disconnected digraph; (b) weakly, but not unilaterally, connected digraph; (c) strongly connected digraph

Definition 5.2.9
A digraph is called a *tree* if any two distinct nodes in the digraph are weakly connected by a unique path.

If a spanning subgraph of a digraph G is a tree it is called a *spanning tree* of G.

Example 5.2.9

(1) Figure 5.14a is not a tree since, for example, nodes a and c are weakly connected by more than one path. If, however, one of the arcs (a, b), (b, c) or (a, c) are removed the resulting digraph is a tree.

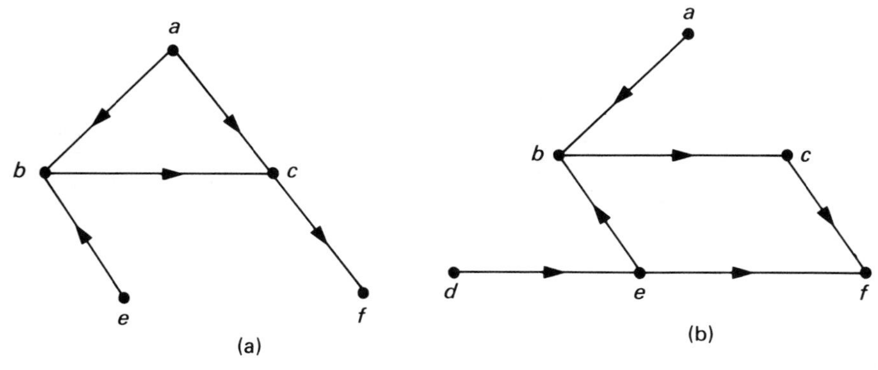

Figure 5.14 Example 5.2.8: (a) graph G_1'; (b) graph G_1''

(2) Similarly, figure 5.14b is not a tree but can be made one by the removal of any one of the arcs (b, c), (c, f), (e, b) or (e, f). The resulting digraph is a spanning tree of G_1 (figure 5.8).

Weighted Digraphs

In the real world, it is a common feature of digraphs that some real value is associated with each of their arcs. For example, if the digraph represents a system of roads, then each arc may be assigned a value representing distance, while if the digraph represents a system of underground pipes, a real value representing capacity may be associated with each arc. Whatever their interpretation, when such real values exist they are called the *weights* of the arcs and the digraph is called a *weighted digraph.*

Definition 5.2.10
A *weighted digraph* (or *network*) is a digraph $G=(N_G, A_G)$ together with a function w: $A_G \rightarrow \mathbb{R}$. This function is called the *weight function* of the digraph.

If the nodes of a weighted digraph are ordered $a_1, \ldots, a_n$, then the weighted digraph can be represented by a *weight matrix*, $[w_{ij}]$, where

$$w_{ij}=w(a_i, a_j)$$

if there is an arc from a_i to a_j, while

$$w_{ij}=0$$

if $a_i=a_j$, and

$$w_{ij}=\infty$$

otherwise.

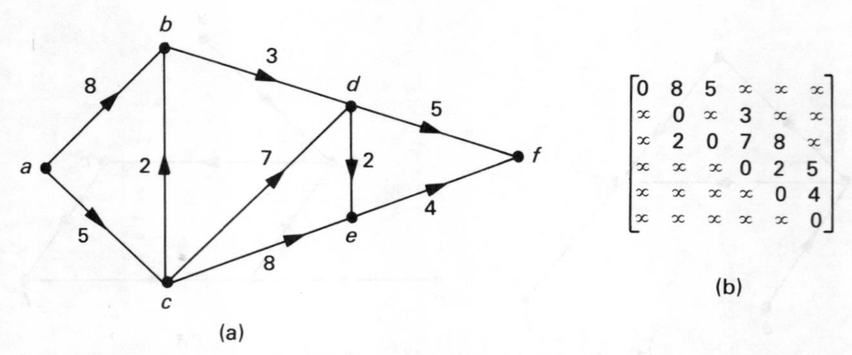

Figure 5.15 Example 5.2.10: (a) weighted digraph; (b) corresponding weight matrix

Example 5.2.10

The weighted digraph illustrated in figure 5.15a has the weight matrix given in figure 5.15b, where the nodes are ordered a, b, c, d, e, f.

Definition 5.2.11
If a directed path from node x_0 to node x_k in a weighted digraph consists of the arcs (x_0, x_1), (x_1, x_2), ..., (x_{k-1}, x_k), then the *weight of the path* is defined to be

$$\sum_{r=0}^{k-1} w(x_r, x_{r+1})$$

Thus, in the above example, the directed path from node a to node f comprising the arcs (a, b), (b, d), (d, e), (e, f) has a weight of $8+3+2+4=17$. However, this is not the directed path of least weight, usually known as the *shortest route*, from a to f. This comprises the arcs (a, c), (c, b), (b, d) and (d, f), with a total weight of $5+2+3+5=15$. Providing none of the weights of the arcs are negative, there is an efficient algorithm, known as *Dijkstra's algorithm*, for finding the shortest route between any two given nodes.

Algorithm 5.2.1 Dijkstra's Shortest-route Algorithm

The algorithm finds the shortest route from node x_0 to node x_k in a weighted digraph. Throughout the algorithm each node, y, carries two labels. The first, $l(y)$, is an upper bound on the weight of the shortest path from node x_0 to node y. Initially, $l(x_0)=0$ and $l(y)=\infty$ for all $y \neq x_0$. The second label attached to y, $P(y)$, is used to indicate the node previous to y on the shortest route found from x_0 to y. Initially, $P(y)$ is undefined (denoted by $-$) for all nodes y in the digraph. In addition, S is used to denote the set of nodes for which the shortest route from x_0 has been found by the algorithm. Initially S is just $\{x_0\}$ but, as soon as x_k is inserted into S, the algorithm terminates and the weight of the shortest route from x_0 to x_k is given by $l(x_k)$. The arcs which comprise this route can be discovered by tracing back from x_k using the P labels. The steps of the algorithm are as follows.

1. (initialisation)

    $l(x_0) \leftarrow 0;$

    $l(y) \leftarrow \infty$, for $y \neq x_0$;

    $S \leftarrow \{x_0\}$; $i \leftarrow 0$;

2.

    while $x_i \neq x_k$ **do**

    (calculate, $\bar{S}$, the set of nodes which can be reached from a node in S by traversing a single arc)

    $\bar{S} \leftarrow \{y | y \notin S$ and there is an arc (x, y) in the weighted digraph with $x \in S\}$;

    (recompute labels)

    for each $y \in \bar{S}$ **do**

        let x' denote the element in S such that

        $l(x') + w(x', y)$ is a minimum;

        if $l(x') + w(x', y) < l(y)$ **then** $P(y) \leftarrow x'$;

        $l(y) \leftarrow l(x') + w(x', y)$

        endif

    endfor;

    (recalculate S)

    compute $\min_{y \in \bar{S}} \{l(y)\}$ and let x_{i+1} denote a node for which that minimum is obtained;

    $S \leftarrow S \cup \{x_{i+1}\}$;

    $i \leftarrow i + 1$

    endwhile

Table 5.2

			Labels $(l(y), P(y))$					
i	*Elements of S*	*Elements of* $\bar{S}$	a	b	c	d	e	f
			(0, −)	(∞, −)	(∞, −)	(∞, −)	(∞, −)	(∞, −)
0	a	b, c	(0, −)	(8, a)	(5, a)	(∞, −)	(∞, −)	(∞, −)
1	a, c	b, d, e	(0, −)	(7, c)	(5, a)	(12, c)	(13, c)	(∞, −)
2	a, c, b	d, e	(0, −)	(7, c)	(5, a)	(10, b)	(13, c)	(∞, −)
3	a, c, b, d	e, f	(0, −)	(7, c)	(5, a)	(10, b).	(12, d)	(15, d)
4	a, c, b, d, e	f	(0, −)	(7, c)	(5, a)	(10, b)	(12, d)	(15, d)

Example 5.2.11

If Dijkstra's algorithm is applied to the weighted digraph of figure 5.15a to find the shortest route from a to f, the steps can be tabulated as in table 5.2.

The shortest path from a to f thus has a cost of 15. Using the P values, the previous node to f is d, which has previous node b, which in turn has previous node c, which has previous node a. Thus, the shortest path comprises (a, c), (c, b), (b, d) and (d, f).

Dijkstra's algorithm depends for its validity on the condition that all the weights be greater than or equal to 0. If this is not the case, other algorithms have to be used. A good survey of such algorithms can be found in Minieka (1978). The discussion of efficient algorithms in graph theory merits a book in itself and it is difficult to illustrate the wealth of results in this short section. The two algorithms chosen, Dijkstra's shortest route and Kruskal's minimal spanning tree (see below) are two of the most well-known and widely applied algorithms in the area.

Definition 5.2.12
In a weighted digraph, the *weight of a subgraph* is the sum of the weights of the arcs in that subgraph. A spanning tree of a digraph of minimum weight is called a *minimal spanning tree.*

Kruskal's algorithm can be applied to any weighted digraph to find a minimal spanning tree.

Algorithm 5.2.2 Kruskal's Algorithm

Let $G=(N_G, A_G)$ be any digraph. The minimal spanning tree of G can be constructed by choosing the arcs (and hence the nodes) as follows.

(a) Choose an arc, $(x_0, y_0) \in A_G$ such that $w(x_0, y_0)$ is as small as possible.

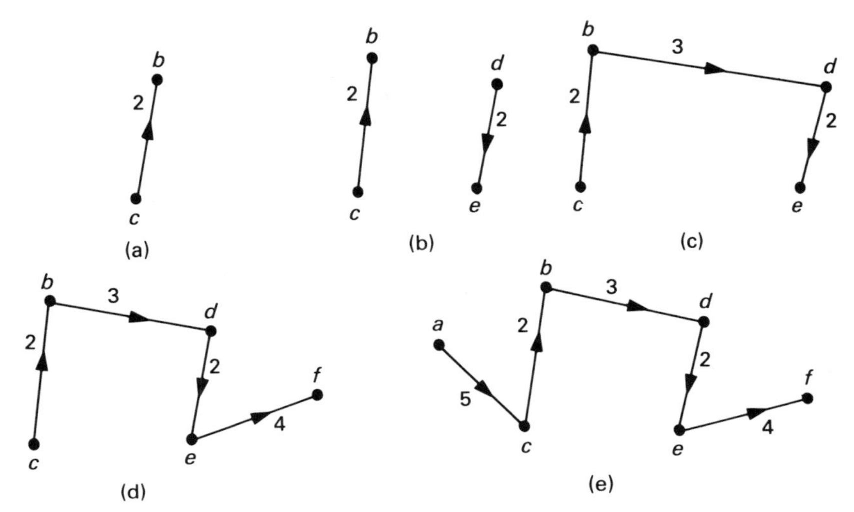

Figure 5.16 Example 5.2.12: (a) first iteration; (b) second iteration; (c) third iteration; (d) fourth iteration; (e) final iteration

(b) If arcs $(x_0, y_0), (x_1, y_1), \ldots, (x_i, y_i)$ have been chosen, choose an arc (x_{i+1}, y_{i+1}) from $A_G \backslash \{(x_0, y_0), (x_1, y_1), \ldots, (x_i, y_i)\}$ such that

(i) no two distinct nodes so far chosen are weakly connected by more than one path;

(ii) $w(x_{i+1}, y_{i+1})$ is as small as possible, subject to (i).

(c) Stop when step (b) cannot be implemented further.

Example 5.2.12

Applying Kruskal's algorithm to the weighted digraph of figure 5.15a results in the sequence of digraphs given in figure 5.16. The algorithm terminates with the construction of the minimal spanning tree as given in figure 5.16e.

Exercise 5.2

1. Use theorem 5.2.1 to count the number of arcs in figure 5.10.

2. Define $a \sim b$ iff nodes a and b are weakly connected nodes in a digraph G. Show that $\sim$ is an equivalence relation. What are the equivalence classes defined by $\sim$?

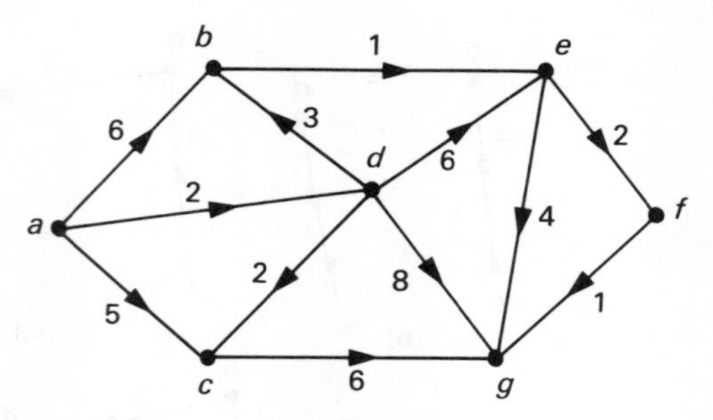

Figure 5.17 Exercise 5.2.6

3. Is 'strongly connected' an equivalence relation? Is 'unilaterally connected' an equivalence relation?

4. If G is a complete digraph of n nodes, how many distinct subgraphs of G are there?

5. Prove that Dijkstra's algorithm really does compute the shortest path from x_0 to x_k. [Hint: Use induction to show that $l(y)$ is the length of the shortest path from x_0 to y for all $y \in S$.]

6. Represent the weighted graph in figure 5.17 by its weight matrix. Use Dijkstra's algorithm to find the shortest route from a to g.

7. Use Kruskal's algorithm to find a minimal spanning tree for the weighted digraph in figure 5.17.

8. Prove that Kruskal's algorithm is correct. [Hint: Suppose that Kruskal's algorithm does not yield a minimal spanning tree and seek a contradiction.]

5.3 GROUPS AND SEMIGROUPS

Binary Operations

The five operations of addition, subtraction, multiplication, division and exponentiation, together with the real numbers, form what is called an *algebraic structure*. This is a term used to describe any set together with a number of operations defined over it. In this book, other algebraic structures have also been discussed, for example, the algebra of sets and the algebra of matrices. In this and the next two sections, a unifying theory of algebraic structures is developed, on the basis of an axiomatic approach.

Definition 5.3.1
A *binary operation on a set A* is a rule whereby each element of $A \times A$ is mapped to an element of A. The adjective *binary* is used because the operations act on *pairs* of elements of A. Let $*$ denote a binary operation in A; then the result of applying this operation to a pair of elements $(a, b) \in A \times A$ will be denoted in the usual way by $a*b$.

Binary operations are the most commonly occurring operations but occasionally one meets *unary operations*, which map A to A, *tertiary operations*, which map $A \times A \times A$ to A and, more generally, *n-ary operations*, which map A^n to A, where A^n denotes the set of all n-tuples of elements of A, as in definition 1.5.8.

Example 5.3.1

(1) In the arithmetic expression $-5+6-7$, the first minus is used to denote a unary operation while the plus and the second minus denote binary operations.

(2) Union and intersection are binary operations on sets but complement is a unary operation.

Definition 5.3.2
Let A be an arbitrary set and let $*$ denote a binary operation on A. Then

(a) A is *closed under* $*$ iff $a*b \in A$, $\forall a, b \in A$;

(b) $*$ is *commutative* (in A) iff $a*b = b*a$, $\forall a, b \in A$;

(c) $*$ is *associative* (in A) iff $(a*b)*c = a*(b*c)$, $\forall a, b, c \in A$.

Example 5.3.2

(1) The integers, Z, are closed under $+$, $\times$ and $-$ but not under $\div$. The operations $+$ and $\times$ are commutative and associative but $-$ and $\div$ are neither commutative nor associative.

(2) The universe of all sets is closed under the set operations of union and intersection, which are both associative and commutative.

(3) In $\mathbb{R}$, let $a*b$ be defined to be the minimum of a and b. Then $\mathbb{R}$ is closed under $*$, which is commutative and associative.

(4) In Q, let $a \oplus b$ be defined to be $(a+b)/2$. Then Q is closed under $\oplus$, which is commutative but not associative.

Definition 5.3.3
Let A be a set with a binary operation denoted by $*$. An element $e \in A$ is an *identity* or *unit element* for $*$ iff $e*a=a*e=a$, $\forall a \in A$.

Theorem 5.3.1
If $e \in A$ is an identity element for $*$ then e is unique, that is, it is the only identity element for $*$ in A.

Proof Assume that e and e' are both identity elements. Then $e=e*e'$, since e' is an identity element. But $e*e'=e'$, since e is an identity element. Therefore, $e=e'$.

Example 5.3.3

(1) In $\mathbb{R}$, 1 is the identity for $*$ but 0 is the identity for $+$. No identities exist for $-$ and $\div$.

(2) In a universe of sets, $\emptyset$ is the identity for union and the universe is the identity for intersection.

Definition 5.3.4
Let A be a set with binary operation denoted by $*$ and let $e \in A$ be the identity for $*$. If $a \in A$ is such that there exists an element $a^{-1} \in A$ where $a^{-1}*a=a*a^{-1}=e$, then a^{-1} is called the *inverse* of a with respect to $*$.

Theorem 5.3.2
If $*$ is associative with respect to A, then the inverse, a^{-1}, of $a \in A$, if it exists, is unique.

Proof Assume that b and c are both inverses of $a \in A$. Then $b*a=a*b=e$ and $c*a=a*c=e$. Now

$$\begin{aligned} b &= b*e && \text{(since } e \text{ is the identity)} \\ &= b*(a*c) && \text{(since } c \text{ is an inverse of } a\text{)} \\ &= (b*a)*c && \text{(since } * \text{ is associative)} \\ &= e*c && \text{(since } b \text{ is an inverse of } a\text{)} \\ &= c && \text{(since } e \text{ is the identity)} \end{aligned}$$

that is

$$b=c$$

Example 5.3.4

(1) Provided $a \neq 0$, the inverse of $a \in \mathbb{R}$ with respect to multiplication exists and is $1/a$. With respect to addition, the inverse of $a \in \mathbb{R}$ always exists and is $-a$.

(2) If S is a set such that $S \neq \emptyset$ and $S \neq \mathscr{U}$, then an inverse of S does not exist with respect to union nor with respect to intersection.

Definition 5.3.5
Let A be a set with binary operation denoted by $*$. $z \in A$ is called a *zero* of A with respect to $*$ iff $a*z = z*a = z$, $\forall a \in A$.

If, given $a \in A$, there exists $a' \in A$ such that $a'*a = a*a' = z$, where z is the zero of A, then a' is called a *complement* of a with respect to $*$.

It is left as an exercise for the reader to show that if a zero of a set A with respect to $*$ exists, then it must be unique. It follows immediately from the definitions that such a zero is a complement of every element $a \in A$. It is thus the existence of non-zero complements that is of primary interest.

Example 5.3.5

(1) The zero of $\mathbb{R}$ with respect to multiplication is 0 but there is no zero of $\mathbb{R}$ with respect to addition. There is no non-zero complement with respect to multiplication for an arbitrary $a \in \mathbb{R}$.

(2) In set theory, the zero with respect to intersection is $\emptyset$ and the complement of an arbitrary set S is as defined in definition 1.2.6.

Groups

A group is a commonly occurring type of simple algebraic structure that has some elegant properties. A group is merely a set with a binary operation that must satisfy some simple axioms. Once these axioms are accepted, a whole theory of groups can be established.

Definition 5.3.6
A *semigroup* consists of a non-empty set, G, closed under some binary operation usually denoted by $*$, which is associative in G.

A group is a semigroup which has a unit element with respect to $*$ and where every element in the group has an inverse in the group. Thus, a *group* is a nonempty set, G, over which a binary operation, denoted by $*$, is defined, where the binary operation satisfies the following rules or *axioms*

(G1) G is closed under $*$ (*closure axiom*);

(G2) $*$ is associative in G (*associative axiom*);

(G3) G contains an identity element for $*$, usually denoted by e (*identity axiom*);

(G4) every element $a \in G$ has a unique inverse for $*$, denoted by a^{-1}, where $a^{-1} \in G$ (*inverse axiom*).

Example 5.3.6

(1) $\mathbb{R}$, with binary operation, $+$, forms a group. The identity element is 0 and the inverse of $a \in \mathbb{R}$ is $-a$.

(2) Z, with binary operation, $+$, forms a group. As with $\mathbb{R}$, the identity element is 0 and the inverse of $a \in Z$ is $-a$.

(3) $\mathbb{R}\backslash\{0\}$ with binary operation, $\times$, forms a group. The identity element is 1 and the inverse of $a \in \mathbb{R}\backslash\{0\}$ is $1/a$.

(4) Let $G = \{p, q, r\}$ and let $*$ be defined by the following table.

$*$	p	q	r
p	p	q	r
q	q	r	p
r	r	p	q

With this operation, G forms a group with identity element p, while $p^{-1} = p$, $q^{-1} = r$ and $r^{-1} = q$.

(5) The integers $\{0, 1, \ldots, m-1\}$ form a group under $\oplus_m$, where $a \oplus_m b = (a+b)(\text{mod } m)$. This group is denoted by Z_m. The identity element is 0 and the inverse of $i \in Z_m$ is $(m-i)(\text{mod } m)$.

(6) Let S be an arbitrary set of cardinality n and let $B(S)$ denote the set of all bijections of S to itself. Then $B(S)$, with function composition as its binary operation, forms a group. The identity element is the identity function on S and the inverse of a bijection in S is the inverse as defined in section 1.5.

Definition 5.3.7
The *order* of a group G is the cardinality of G. A *finite group* has finite order, an *infinite group* has infinite order.

Example 5.3.7

The order of the first three groups described in example 5.3.6 is infinite; for the fourth, fifth and sixth it is 3, m and $n!$, respectively.

Definition 5.3.8
A group G is an *Abelian group* if its binary operation is commutative in G.

Example 5.3.8

Every group in example 5.3.6 is Abelian except the sixth.

Definition 5.3.9
If a subset, H, of G, together with the binary operation of G, is itself a group, then H is called a *subgroup* of G.

It is easy to show that if H is a subgroup of G, then H must contain the identity element of G and that if $a \in H$ then necessarily $a^{-1} \in H$.

Example 5.3.9

(1) Example 5.3.6(2) describes a subgroup of the group described in example 5.3.6(1).

(2) For any group G, $\{e\}$ and G are both subgroups of G. Subgroups of G other than $\{e\}$ and G are called *proper subgroups* of G.

(3) Z_6 has three subgroups which can be tabulated as follows.

$\oplus_6$	0	1	2	3	4	5
0	0	1	2	3	4	5
1	1	2	3	4	5	0
2	2	3	4	5	0	1
3	3	4	5	0	1	2
4	4	5	0	1	2	3
5	5	0	1	2	3	4

$\oplus_6$	0	2	4
0	0	2	4
2	2	4	0
4	4	0	2

$\oplus_6$	0
0	0

There is only one proper subgroup of Z_6, which is identical to the subgroup described in example 5.3.6(4) except that the elements are called 0, 2 and 4 instead of p, q and r and the binary operation is called $\oplus_6$ instead of $*$. Such identical groups are said to be isomorphic (from the Greek: *isos*, meaning equal and *morphe*, meaning form).

Definition 5.3.10
Let G with binary operation, $*$, and H with binary operation, $\circ$, be groups. A (*group*) *homomorphism* of G into H is a function θ: $G \rightarrow H$ such that $\theta(a*b) = \theta(a) \circ \theta(b)$, $\forall a, b \in G$. If the function θ is injective, the homomorphism is called a (*group*) *monomorphism*; if θ is surjective it is a (*group*) *epimorphism* and if θ is bijective it is a (*group*) *isomorphism*.

Theorem 5.3.3
Let θ: $G \rightarrow H$ be an homomorphism from a group G to a group H. Then:

(i) if e_G is the identity element in G, $\theta(e_G)$ is the identity element in H;

(ii) if a^{-1} denotes the inverse of $a \in G$, then $\theta(a^{-1})$ is the inverse of $\theta(a)$ in H.

Proof (i) For any element $a \in G$, $a = e_G * a = a * e_G$ from definition 5.3.3. From definition 5.3.10, $\theta(e_G * a) = \theta(e_G) \circ \theta(a)$ and $\theta(a * e_G) = \theta(a) \circ \theta(e_G)$. Thus $\theta(a) = \theta(e_G) \circ \theta(a) = \theta(a) \circ \theta(e_G)$, that is, $e_H = \theta(e_G)$.
(ii) From the definition, $\theta(a) \circ \theta(a^{-1}) = \theta(a * a^{-1}) = \theta(e_G) = e_H$.

If there exists an isomorphism between G and H, the two groups are said to be *isomorphic*, which is written $G \cong H$. The relation $\cong$ has all the properties required of an equivalence relation, that is: (1) $G \cong G$, for any group G; (2) if $G \cong H$ then $H \cong G$; and (3) if $G \cong H$ and $H \cong K$ then $G \cong K$. A property of groups is said to be *invariant* if, whenever $G \cong H$ and G has the property, then it follows that H also has the property — the order of a finite group is an example of an invariant property. Group theory has been described as a study of invariant properties of groups and the reader is referred to the standard text by Hall (1959). The wealth of results in group theory demonstrates the power of taking an axiomatic approach to algebraic systems. In the rest of this section and in the next section, axiomatic approaches are developed for other algebraic systems.

Algebra of Strings

Let V denote some finite set called the *alphabet*. This section develops some theory concerning strings over V, where a *string over V* consists of a number of elements of V written sequentially.

Example 5.3.10

If $V=\{a, b, c\}$, then a, abb and $bbbca$ are strings over V.

Definition 5.3.11
The length of a string, α, over V, which is denoted by $l(\alpha)$, is the number of elements of V in α counted according to multiplicity. The *empty string* has length 0 and is commonly denoted by ε.

Example 5.3.11

$l(\varepsilon)=0$, $l(a)=1$, $l(abb)=3$.

Definition 5.3.12
Let α, β denote strings over V. The *concatenation*, $\alpha\beta$, of α and β is defined to be the string composed of the elements in α immediately followed by the elements in β.

Example 5.3.12

If $\alpha=ab$ and $\beta=bcc$, then $\alpha\beta=abbcc$ and $\beta\alpha=bccab$.

In general $\alpha\beta\neq\beta\alpha$, so concatenation is not commutative but it is associative since $(\alpha\beta)\gamma=\alpha(\beta\gamma)$, for all strings α, β and γ over V.

Definition 5.3.13
Let A, B be two sets of strings over V. The (*set*) *concatenation*, of A and B is defined by $AB=\{\alpha\beta|\alpha\in A, \beta\in B\}$.

Example 5.3.13

If $A=\{\varepsilon, b, ab\}$ and $B=\{bb, a\}$, then $AB=\{bb, a, bbb, ba, abbb, aba\}$ and $BA=\{bb, a, bbb, ab, bbab, aab\}$.

In general, $AB\neq BA$, so set concatenation is not commutative. However, using the associativity of string concatenation, it is easy to show that set concatenation is associative.

Definition 5.3.14
The set concatenation of the alphabet V with itself, VV, is often written V^2 and denotes the set of strings over V of length 2. Similarly, $V^3=VV^2=V^2V$ denotes the strings over V of length 3. In general, $V^i=VV^{i-1}=V^{i-1}V$ for $i=2, 3, \ldots$ denotes the strings of V of length i.

V^+ is defined to equal $V \cup V^2 \cup V^3 \cup \ldots$, that is, $\bigcup_{i=1}^{\infty} V^i$. Thus V^+ denotes the set of all strings over V of length $\geqslant 1$. V^*, the set of all strings over V is then defined by $V^* = V^+ \cup \{\varepsilon\}$.

There is only one binary operation that is defined in V^* and that is concatenation. V^* is closed under concatenation and has an identity, ε. Concatenation is associative and thus V^* with concatenation is a semigroup with an identity — such a structure is sometimes called a *monoid*. However, V^* is not a group since the inverse axiom is not satisfied.

Exercise 5.3

1. A set $A = \{-1, 0, 1\}$ has a binary operation $\pm$ defined by the following table.

$\pm$	-1	0	1
-1	1	0	-1
0	0	1	1
0	-1	-1	1

(a) Is A closed under $\pm$?

(b) Is $\pm$ associative, is it commutative?

(c) Is there an identity element for $\pm$ in A?

(d) Is there a zero element for $\pm$ in A?

2. Consider the two binary operations $\uparrow$ and $\downarrow$ defined on the set Z^+ of non-negative integers:

(i) $a \uparrow b$ is the highest common factor of $a, b \in Z^+$;

(ii) $a \downarrow b$ is the least common multiple of $a, b \in Z^+$.

For each of the operations, $\uparrow$ and $\downarrow$, defined on Z^+, answer the following questions

(a) is it commutative;

(b) is it associative;

(c) is there a zero;

(d) is there an identity?

3. Prove that if z is a zero of a set A with binary operation $*$ then z must be unique.

4. Prove the following theorems concerning the group G with binary operation $*$:

(a) if $a, b \in G$ then there is one and only one element x of G such that $x * a = b$;

(b) if $x \in G$ then $(x^{-1})^{-1} = x$;

(c) if G is an Abelian group, then for all $a, b \in G$ and $n \in N$, $(a*b)^n = a^n * b^n$;

(d) if $(a*b)^2 = a^2 * b^2$ for all $a, b \in G$ then G must be Abelian.

5. Show that the integers $\{0, 1, 2, 3, 4, 5, 6, 7\}$ under addition modulo 8 form a group, and that this group has a subgroup of order 4 which in turn has a subgroup of order 2.

6. Show that the property of being an Abelian group is invariant.

7. The notation

$$\begin{pmatrix} 1 & 2 & 3 \\ 2 & 3 & 1 \end{pmatrix}$$

means change 1 to 2, 2 to 3 and 3 to 1. Denote this function by P_1. Denote the functions

$$\begin{pmatrix} 1 & 2 & 3 \\ 1 & 2 & 3 \end{pmatrix}, \begin{pmatrix} 1 & 2 & 3 \\ 3 & 1 & 2 \end{pmatrix}, \begin{pmatrix} 1 & 2 & 3 \\ 1 & 3 & 2 \end{pmatrix}, \begin{pmatrix} 1 & 2 & 3 \\ 3 & 2 & 1 \end{pmatrix}, \begin{pmatrix} 1 & 2 & 3 \\ 2 & 1 & 3 \end{pmatrix}$$

by P_0, P_2, P_3, P_4 and P_5, respectively. Show that $\{P_0, P_1, P_2, P_3, P_4, P_5\}$ forms a group under function composition as defined in 1.5.4. Is the group Abelian?

8. Let G be a group with binary operation $*$. If $a*b = a$ or $b*a = a$ for some element $a \in G$, show that b must be the identity element e of G. Show also that if $a*b = c*b$ then $a = c$.

5.4 RINGS, FIELDS AND VECTOR SPACES

Rings

The axioms defining the algebraic structure known as a ring are derived from important properties of the integers. Like the integers, a ring has two binary operations, denoted by $+$ and $\times$, where these two operations are connected by *distributive laws.*

Definition 5.4.1
Let R be any set. An operation $\times$ in R is *left distributive over* an operation $+$ in R if

$$a\times(b+c)=(a\times b)+(a\times c),\ \forall a,\ b,\ c\in R$$

The operation $\times$ in R is *right distributive over* the operation $+$ in R if

$$(a+b)\times c=(a\times c)+(b\times c),\ \forall a,\ b,\ c\in R$$

Example 5.4.1

(1) In Z, the operation of multiplication is both left and right distributive over the operation of addition. However, addition is neither right nor left distributive over multiplication.

(2) In set theory, $\cup$ is left and right distributive over $\cap$ and, similarly, $\cap$ is left and right distributive over $\cup$.

(3) If the binary operation $\otimes$ in $\mathbb{R}$ is defined by $a\otimes b=b$, $\forall a,\ b\in\mathbb{R}$, then $\otimes$ is left distributive but not right distributive over addition.

Definition 5.4.2
A *ring* is a set R over which two binary operations, $+$ and $\times$, are defined such that the following axioms are satisfied

(R1) R is an Abelian group with respect to $+$;

(R2) R is a semigroup with respect to $\times$;

(R3) $\times$ is left and right distributive over $+$.

It is customary in ring theory to follow the conventions of normal arithmetic and write ab for $a\times b$, 0 for the identity with respect to addition, $-a$ for the inverse of $a\in R$ with respect to addition and $a-b$ for $a+(-b)$.

Example 5.4.2

(1) Z is a ring where $+$ denotes addition and $\times$ denotes multiplication. Similarly, the rationals, Q, the reals, $\mathbb{R}$, and the complex numbers, $\mathscr{C}$, are all rings under the operations of addition and multiplication.

(2) $J = \{a+ib | a, b \in Z\}$ is a subset of $\mathscr{C}$ and forms the *ring of Gaussian integers* under the operations of addition and multiplication.

(3) The set of $n \times n$ real matrices forms a ring under matrix addition and matrix multiplication.

Theorem 5.4.1
If R is a ring, then

(i) $a0 = 0a = 0, \forall a \in R$;

(ii) $(-a)b = a(-b) = -(ab), \forall a, b \in R$.

Proof (i) For any $b \in R$, $b+0=b$. Thus, $(b+0)a = ba$. Hence, $ba + 0a = ba$. Adding $-ba$ to both sides gives $0a = 0$. Similarly, $a(b+0) = ab$. Hence $ab + a0 = ab$, and adding $-ab$ to both sides gives $a0 = 0$.
(ii) By definition, $a + (-a) = 0$. Hence $(a+(-a))b = 0b = 0$ for any $a, b \in R$. Thus $ab + (-a)b = 0$. But, by the uniqueness of inverses (theorem 5.3.2), $(-a)b = -(ab)$. Similarly, $b + (-b) = 0$ and thus $ab + a(-b) = 0$. Hence, $a(-b) = -(ab)$.

A ring homomorphism is a mapping of one ring into another which is 'structure preserving'. More formally, it may be defined as follows.

Definition 5.4.3
A *ring homomorphism* of a ring R_1 into a ring R_2 is a map $\theta: R_1 \to R_2$ such that $\forall a, b \in R_1$ the following hold

(RH1) $\theta(a+b) = \theta(a) + \theta(b)$;

(RH2) $\theta(ab) = \theta(a)\theta(b)$.

If the function θ is injective, the ring homomorphism is called a *ring monomorphism*; if θ is surjective it is a *ring epimorphism*; and if θ is bijective, it is a *ring isomorphism.*
Since R_1 and R_2 are Abelian groups under their respective operations $+$, such a ring homomorphism is a homomorphism of R_1 into R_2 (*qua* groups) as defined in definition 5.3.10. Hence a ring homomorphism of R_1 into R_2 maps the identity of R_1 (with respect to $+$), 0_{R_1}, to the identity of R_2 (with respect to $+$), 0_{R_2}, and the inverse of a in R_1, $-a$, to the inverse of $\theta(a)$, $-\theta(a)$, in R_2 (see theorem 5.3.3).

Definition 5.4.4
Let S be a subset of a ring R. Then S is called a *subring* of R if it is itself a ring under the operations of R.

The reader is asked to verify that this is always the case provided *both* of the following statements are true

(i) S contains the unit element with respect to addition, 0_R, of R;

(ii) for any two elements a, $b \in S$, the elements $a+b$, $a-b$ and ab are also in S.

Example 5.4.3

Let θ be a homomorphism from a ring R into a ring R'. Define $S = \{a \in R | \theta(a) = 0\}$. It is easy to show that S contains 0_R and that, $\forall a, b \in S$, $a+b$, $a-b$ and ab are all in S. Thus, S is a subring of R known as the *kernel* of θ.

From the examples of rings already considered, it can be seen that a ring is a useful abstraction of many commonly arising structures. Many of the rings that exist satisfy additional axioms, and there are several special classes of rings that have been defined as extensions of the ring definition.

Definition 5.4.5
A *commutative ring* is a ring where the operation $\times$ is commutative. In such a ring, $\times$ is left-distributive over $+$ if and only if $\times$ is right-distributive over $+$.

A *ring with a multiplicative identity* is a ring, R, which contains an element denoted by e for which $re = er = r$, $\forall r \in R$. By theorem 5.3.1, such an identity must be unique.

An *integral domain* is a commutative ring with a multiplicative identity ($\neq 0$) which has no *zero divisors*. A zero divisor of a commutative ring, R, is an element $r \in R$ such that $r \neq 0$ and $rs = 0$ for some $s \neq 0$ in R.

A *field* is a commutative ring in which the set of nonzero elements forms a group under multiplication. Thus a field, K, is an integral domain with the additional axiom that, for each $a \neq 0$ in K, there exists an inverse element, $a^{-1} \in K$, such that $a^{-1}a = aa^{-1} = e$.

Figure 5.18 illustrates the relationship between these various classes of rings.

Theorem 5.4.2 Cancellation Law
Let R be an integral domain and $a(\neq 0)$, x, $y \in R$. Then $ax = ay \supset x = y$.

Proof If $ax = ay$ then $ax - ay = 0$. But, by the distributivity axiom, $a(x-y) = ax - ay = 0$ and, since R has no zero divisors, it follows that $x - y = 0$. Hence, $x = y$.

Example 5.4.4

(1) Z is an integral domain but is not a field since the only element in Z to have an inverse with respect to multiplication is the multiplicative identity itself, 1.

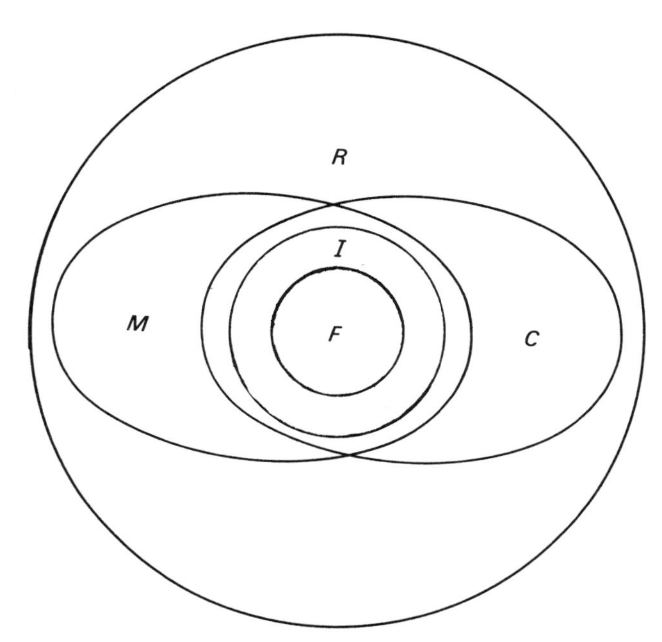

Figure 5.18 Venn diagram showing the relationship between the various classes of rings: R, rings; M, rings with multiplicative identity; C, commutative rings; I, integral domains; F, fields

(2) $\mathbb{R}$ is a field.

(3) The ring of Gaussian integers, J, is an integral domain with identity $1+i0$. The reader should check that J can have no zero divisors.

(4) The set of $n \times n$ real matrices is a field but the set of $n \times n$ integer matrices is only an integral domain.

Vector Spaces

Just as integers motivate the definition of a ring, so vectors motivate the definition of a vector space. Vector spaces are particularly important algebraic structures, because they arise in so many practical branches of mathematics. In particular, many of the results in chapter 2 on linear algebra are specific examples of fundamental results for a general vector space. In this section, the hope is that the reader will see enough of these results to appreciate this statement.

Definition 5.4.6
A non-empty set V is said to be a *vector space over a field* K if

(V1) V is an Abelian group under an operation denoted by $+$;

(V2) for every $\alpha \in K$, $\mathbf{v} \in V$, there is defined an element, denoted by $\alpha\mathbf{v}$, in V such that:

(i) $\alpha(\mathbf{v}+\mathbf{w})=\alpha\mathbf{v}+\alpha\mathbf{w}$, $\forall \alpha \in K$, $\mathbf{v}$, $\mathbf{w} \in V$;

(ii) $(\alpha+\beta)\mathbf{v}=\alpha\mathbf{v}+\beta\mathbf{v}$, $\forall \alpha, \beta \in K$, $\mathbf{v} \in V$;

(iii) $\alpha(\beta\mathbf{v})=(\alpha\beta)\mathbf{v}$, $\forall \alpha, \beta \in K$, $\mathbf{v} \in V$;

(iv) $e\mathbf{v}=\mathbf{v}$, $\forall \mathbf{v} \in V$, where e is the identity element in K.

The reader should note that in axiom V2(ii), the $+$ in $(\alpha+\beta)$ denotes a binary operation in K while the $+$ in $\alpha\mathbf{v}+\beta\mathbf{v}$ denotes a (different) binary operation in V.

Sometimes, the elements of K are called *scalars* and the elements of V are called *vectors*. The binary operation $K \times V \to V$ defined by $(\alpha, \mathbf{v}) \mapsto \alpha\mathbf{v}$ is called *scalar multiplication.*

If V is a vector space over a field K then both V and K contain zero elements, that is, identity elements for their respective operations of addition. The zero element of V is denoted by $\mathbf{0}$ and that of K by 0. Similarly, the inverse of $\mathbf{v}$ with respect to addition in V is denoted by $-\mathbf{v}$ and the inverse of α with respect to addition in K is denoted by $-\alpha$.

Example 5.4.5

(1) A *subfield* of a field F is a subset of F that is itself a field under the binary operations of F. Let K denote such a subfield of F; then F can be regarded as a vector space over K where the $+$ of the vector space denotes the addition of elements of F and where $\alpha\mathbf{v}$ is defined to be $\alpha \times \mathbf{v}$, $\alpha \in K$, $\mathbf{v} \in F$.

(2) Let K be any field and let K^n denote the set of all ordered rows of n elements of K. Two such elements $\boldsymbol{\alpha}=[\alpha_1\ \alpha_2 \ldots \alpha_n]$ and $\boldsymbol{\beta}=[\beta_1\ \beta_2 \ldots \beta_n]$ are equal if $\alpha_i=\beta_i$, $i=1, \ldots, n$. The binary operation $+$ in K^n is defined by $[\alpha_1\ \alpha_2 \ldots \alpha_n]+[\beta_1\ \beta_2 \ldots \beta_n]=[\alpha_1+\beta_1\ \alpha_2+\beta_2 \ldots \alpha_n+\beta_n]$, $\forall \boldsymbol{\alpha}, \boldsymbol{\beta} \in K^n$, and scalar multiplication by $\gamma\boldsymbol{\alpha}=[\gamma\alpha_1\ \gamma\alpha_2 \ldots \gamma\alpha_n]$, $\gamma \in K$, $\boldsymbol{\alpha} \in K^n$. Under these operations, K^n is a vector space over K.

(3) Let K be any field and let $K[x]$ denote the set of polynomials in x over K, that is, all expressions of the form $\alpha_0+\alpha_1x+\alpha_2x^2+\ldots+\alpha_nx^n$, $n \geqslant 0$, where $\alpha_i \in K$, $i=1, \ldots, n$. Any two polynomials in $K[x]$ can be added

together to give another polynomial in $K[x]$. Also, any polynomial can be multiplied, term by term, by a scalar $\gamma \in K$ to get another polynomial. Under these operations, $K[x]$ is a vector space over K.

(4) Let K_{mn} denote the set of $m \times n$ matrices $[\alpha_{ij}]$ where each $\alpha_{ij} \in K$, $i=1, \ldots, m$, $j=1, \ldots, n$. Then under matrix addition and scalar multiplication, K_{mn} forms a vector space over the field K.

The reader should check that each of these examples does indeed satisfy all the axioms required of a vector space.

Theorem 5.4.3
If V is a vector space over a field K, then

(i) $\alpha \mathbf{0} = \mathbf{0}$, $\forall \alpha \in K$;

(ii) $0\boldsymbol{v} = \mathbf{0}$, $\forall \boldsymbol{v} \in V$;

(iii) $(-\alpha)\boldsymbol{v} = -(\alpha \boldsymbol{v})$, $\forall \alpha \in K$, $\boldsymbol{v} \in V$;

(iv) if $\boldsymbol{v} \neq \mathbf{0}$ then $\alpha \boldsymbol{v} = \mathbf{0} \supset \alpha = 0$.

Proof (i) This result follows immediately from $\alpha \mathbf{0} = \alpha(\mathbf{0}+\mathbf{0}) = \alpha \mathbf{0} + \alpha \mathbf{0}$.
(ii) This result follows similarly, since $0\boldsymbol{v} = (0+0)\boldsymbol{v} = 0\boldsymbol{v} + 0\boldsymbol{v}$.
(iii) Since $\mathbf{0} = (\alpha + (-\alpha))\boldsymbol{v} = \alpha \boldsymbol{v} + (-\alpha)\boldsymbol{v}$, $(-\alpha)\boldsymbol{v} = -(\alpha \boldsymbol{v})$.
(iv) If $\alpha \boldsymbol{v} = \mathbf{0}$ and $\alpha \neq 0$ then $\mathbf{0} = \alpha^{-1}\mathbf{0} = \alpha^{-1}(\alpha \boldsymbol{v}) = e\boldsymbol{v} = \boldsymbol{v}$.

Definition 5.4.7
If V is a vector space over K and if $W \subset V$ then W is a *subspace* of V if, under the operations of V, W itself forms a vector space over K.
Equivalently, W is a subspace of V if $\alpha \mathbf{w}_1 + \beta \mathbf{w}_2 \in W$, $\forall \alpha, \beta \in K$, $\mathbf{w}_1, \mathbf{w}_2 \in W$.

Example 5.4.6

(1) The subset of K^n consisting of all the ordered rows of n elements of the form $[\alpha_1 \ \alpha_2 \ \ldots \ \alpha_{n-1} \ 0]$ is a subspace of K^n.

(2) Let $K_n[x]$ denote the set of all polynomials in x over K of degree less than n. Under the operations of polynomial addition and scalar multiplication, this is a vector space over K and is a subspace of $K[x]$.

Definition 5.4.8
Let $\boldsymbol{v}_1, \ldots, \boldsymbol{v}_r$ be elements of a vector space V. A *linear combination of elements* $\boldsymbol{v}_1, \ldots, \boldsymbol{v}_r$ *in* V *with coefficients* in K is an expression of the form $\alpha_1 \boldsymbol{v}_1 + \ldots + \alpha_r \boldsymbol{v}_r$, $\alpha_i \in K$, $i=1, \ldots, r$.
The elements $\boldsymbol{v}_1, \ldots, \boldsymbol{v}_r$ are said to be *linearly dependent* if there exists a

linear combination of $\boldsymbol{v}_1, \ldots, \boldsymbol{v}_r$ with coefficients not all zero which is equal to $\mathbf{0}$, that is, there exist scalars $\lambda_1, \ldots, \lambda_r$ (not all zero) such that $\lambda_1\boldsymbol{v}_1 + \ldots + \lambda_r\boldsymbol{v}_r = \mathbf{0}$. If no such scalars exist, the vectors are said to be *linearly independent.*

If $\boldsymbol{v}_1, \ldots, \boldsymbol{v}_r$ are linearly dependent then, from the definition, there exist scalars $\lambda_1, \ldots, \lambda_r$ (not all zero) such that $\lambda_1\boldsymbol{v}_1 + \ldots + \lambda_r\boldsymbol{v}_r = \mathbf{0}$. Suppose $\lambda_1 \neq 0$. Then, since K is a field, there exists a multiplicative inverse of λ_1, λ_1^{-1}, in K. Multiplying the equation by λ_1^{-1} results in $\boldsymbol{v}_1 + \lambda_1^{-1}\lambda_2\boldsymbol{v}_2 + \ldots + \lambda_1^{-1}\lambda_r\boldsymbol{v}_r = \mathbf{0}$. Hence, $\boldsymbol{v}_1 = -(\lambda_1^{-1}\lambda_2\boldsymbol{v}_2 + \ldots + \lambda_1^{-1}\lambda_r\boldsymbol{v}_r) = -\lambda_1^{-1}\lambda_2\boldsymbol{v}_2 - \ldots - \lambda_1^{-1}\lambda_r\boldsymbol{v}_r$. Thus $\boldsymbol{v}_1$ is a linear combination of $\boldsymbol{v}_2, \ldots, \boldsymbol{v}_r$. The converse of this result is similarly proved. Thus the following theorem may be formulated.

Theorem 5.4.4
The elements $\boldsymbol{v}_1, \ldots, \boldsymbol{v}_r$ are linearly dependent if and only if at least one of them can be expressed as a linear combination of the remaining $r-1$.

Example 5.4.7

(1) In $\mathbb{R}^3$, [1 0 0], [0 1 0] and [0 0 1] are linearly independent but [1 0 0], [0 1 0], [0 0 1] and [1 1 1] are linearly dependent, since the last vector is the sum of the first three. In fact any vector in $\mathbb{R}^3$, $[a\ b\ c]$, can be expressed as a linear combination of [1 0 0], [0 1 0] and [0 0 1], since $[a\ b\ c] = a[1\ 0\ 0] + b[0\ 1\ 0] + c[0\ 0\ 1]$.

(2) One way of checking whether three arbitrary vectors in $\mathbb{R}^3$, $[a_1\ b_1\ c_1]$, $[a_2\ b_2\ c_2]$ and $[a_3\ b_3\ c_3]$ are linearly independent is to use the determinant

$$\begin{vmatrix} a_1 & b_1 & c_1 \\ a_2 & b_2 & c_2 \\ a_3 & b_3 & c_3 \end{vmatrix}$$

This determinant will be zero if and only if any one vector is a linear combination of the other two, that is, if and only if the three vectors are linearly dependent. For example, since

$$\begin{vmatrix} 0 & 3 & 2 \\ 2 & 1 & 4 \\ -2 & 5 & 0 \end{vmatrix} = 0$$

it follows that [0 3 2], [2 1 4] and [−2 5 0] are linearly dependent.

Definition 5.4.9
If S is a non-empty subset of the vector space V then $L(S)$, the *linear span* of S, is the set of all linear combinations of finite sets of elements of S.

Theorem 5.4.5
The linear span $L(S)$ is a subspace of V.

Proof If $\boldsymbol{v}, \boldsymbol{w} \in L(S)$ and $S = \{\boldsymbol{u}_1, \boldsymbol{u}_2, \ldots, \boldsymbol{u}_m\}$, then $\boldsymbol{v} = \lambda_1 \boldsymbol{u}_1 + \lambda_2 \boldsymbol{u}_2 + \ldots + \lambda_m \boldsymbol{u}_m$, for some $\lambda_1, \lambda_2, \ldots, \lambda_m \in K$, and $\boldsymbol{w} = \mu_1 \boldsymbol{u}_1 + \mu_2 \boldsymbol{u}_2 + \ldots + \mu_m \boldsymbol{u}_m$, for some $\mu_1, \mu_2, \ldots, \mu_m \in K$. Hence, if $\alpha, \beta \in K$

$$\alpha \boldsymbol{v} + \beta \boldsymbol{w} = \alpha\lambda_1 \boldsymbol{u}_1 + \ldots + \alpha\lambda_m \boldsymbol{u}_m + \ldots + \beta\mu_1 \boldsymbol{u}_1 + \beta\mu_m \boldsymbol{u}_m$$
$$= (\alpha\lambda_1 + \beta\mu_1)\boldsymbol{u}_1 + \ldots + (\alpha\lambda_m + \beta\mu_m)\boldsymbol{u}_m$$

which is an element of $L(S)$.

Definition 5.4.10
A vector space V over a field K is said to be *finite-dimensional* (*over* K) if there is a finite subset $S \subset V$ such that $L(S) = V$. Such a subset is said to be a *basis of* V (*over* K) if its elements are linearly independent. In the exercises at the end of this section, the reader is asked to prove a series of theorems that yield the important result that any basis of a finite-dimensional vector space V has the same number of elements, this number being called the *dimension* of V.

Example 5.4.8

The vector space $\mathbb{R}^n$ is finite-dimensional and has a basis consisting of the n vectors $\{[1\ 0\ 0\ \ldots\ 0], [0\ 1\ 0\ \ldots\ 0], \ldots, [0\ 0\ 0\ \ldots\ 1]\}$. The dimension of $\mathbb{R}^n$ is thus n.

Theorem 5.4.6
If $\boldsymbol{v}_1, \boldsymbol{v}_2, \ldots, \boldsymbol{v}_r \in V$ are linearly independent, then every element in their linear span can be expressed *uniquely* in the form

$$\lambda_1 \boldsymbol{v}_1 + \lambda_2 \boldsymbol{v}_2 + \ldots + \lambda_r \boldsymbol{v}_r \ (\lambda_i \in K).$$

Proof By definition, every element in the linear span can be expressed as $\lambda_1 \boldsymbol{v}_1 + \lambda_2 \boldsymbol{v}_2 + \ldots + \lambda_r \boldsymbol{v}_r$. The task is thus to show that this expression is unique. Assume there are two such expressions $\lambda_1 \boldsymbol{v}_1 + \lambda_2 \boldsymbol{v}_2 + \ldots + \lambda_r \boldsymbol{v}_r$ and $\mu_1 \boldsymbol{v}_1 + \mu_2 \boldsymbol{v}_1 + \ldots + \mu_r \boldsymbol{v}_r$. Since $\lambda_1 \boldsymbol{v}_1 + \lambda_2 \boldsymbol{v}_2 + \ldots + \lambda_r \boldsymbol{v}_r = \mu_1 \boldsymbol{v}_1 + \mu_2 \boldsymbol{v}_2 + \ldots + \mu_r \boldsymbol{v}_r$, it follows that $(\lambda_1 - \mu_1)\boldsymbol{v}_1 + (\lambda_2 - \mu_2)\boldsymbol{v}_2 + \ldots + (\lambda_r - \mu_r)\boldsymbol{v}_r = \mathbf{0}$. But, since $\boldsymbol{v}_1, \boldsymbol{v}_2, \ldots, \boldsymbol{v}_r$ are linearly independent, this implies $\lambda_1 - \mu_1 = 0$, $\lambda_2 - \mu_2 = 0$, $\ldots$, $\lambda_r - \mu_r = 0$, that is, $\lambda_1 = \mu_1$, $\lambda_2 = \mu_2$, $\ldots$, $\lambda_r = \mu_r$.

The number of interesting theorems concerned with vector spaces is an indication of the richness of this abstract structure. Some of the most important results arise from the consideration of vector-space homomorphisms, that is, functions from one vector space to another that preserve structure.

Definition 5.4.11
If U, V are vector spaces over a field K then the mapping θ of U into V is said to be a (*vector-space*) *homomorphism* if

(i) $\theta(\boldsymbol{u}_1+\boldsymbol{u}_2)=\theta\boldsymbol{u}_1+\theta\boldsymbol{u}_2, \forall \boldsymbol{u}_1, \boldsymbol{u}_2 \in U$;

(ii) $\theta(\alpha\boldsymbol{u})=\alpha(\theta\boldsymbol{u}), \forall \boldsymbol{u}\in U, \alpha\in K$.

If a vector-space homomorphism is injective it is called a (*vector-space*) *monomorphism*; if it is surjective it is a (*vector-space*) *epimorphism*; and if it is bijective it is a (*vector-space*) *isomorphism*. If there exists an isomorphism between two vector spaces, they are said to be *isomorphic*.

Example 5.4.9

(1) K^n is isomorphic to $K_n[x]$ under the mapping $[\alpha_0\ \alpha_1\ \ldots\ \alpha_{n-1}] \mapsto \alpha_0+\alpha_1 x+\ldots+\alpha_{n-1}x^{n-1}$.

(2) There is a homomorphism from K^{n-1} to K^n defined by $[\alpha_1\ \alpha_2\ \ldots\ \alpha_{n-1}] \mapsto [\alpha_1\ \alpha_2\ \ldots\ \alpha_{n-1}\ 0]$.

(3) Let M be a real, $n\times n$ matrix; then M defines a homomorphism $\mathbb{R}^n \to \mathbb{R}^n$ where $\boldsymbol{x}=[x_1\ \ldots\ x_n]$ is mapped to the n-tuple obtained from the matrix multiplication of $\boldsymbol{x}$ and M. For example, if

$$M=\begin{bmatrix} 2 & 1 & 3 \\ -1 & 3 & 6 \\ 2 & -1 & 3 \end{bmatrix}$$

then this defines a homomorphism where $[x_1\ x_2\ x_3]$ is mapped to $[2x_1-x_2+2x_3\quad x_1+3x_2-x_3\quad 3x_1+6x_2+3x_3]$.

The set of all homomorphisms from a vector space U over a field K to a vector space V over K is denoted by Hom(U, V). If $\theta_1, \theta_2 \in \text{Hom}(U, V)$, a binary operation $+$ on Hom(U, V) can be defined, where $\theta_1+\theta_2$: $\boldsymbol{u} \mapsto \theta_1(\boldsymbol{u})+\theta_2(\boldsymbol{u}), \forall \boldsymbol{u}\in U$. Here $\theta_1+\theta_2$ is a homomorphism from U to V, since

$$\begin{aligned}(\theta_1+\theta_2)(\boldsymbol{u}_1+\boldsymbol{u}_2) &= \theta_1(\boldsymbol{u}_1+\boldsymbol{u}_2)+\theta_2(\boldsymbol{u}_1+\boldsymbol{u}_2)=\theta_1\boldsymbol{u}_1+\theta_1\boldsymbol{u}_2+\theta_2\boldsymbol{u}_1+\theta_2\boldsymbol{u}_2 \\ &= (\theta_1+\theta_2)\boldsymbol{u}_1+(\theta_1+\theta_2)\boldsymbol{u}_2\end{aligned}$$

and

$$(\theta_1+\theta_2)(\alpha\boldsymbol{u})=\theta_1(\alpha\boldsymbol{u})+\theta_2(\alpha\boldsymbol{u})=\alpha\theta_1\boldsymbol{u}+\alpha\theta_2\boldsymbol{u}$$
$$=\alpha(\theta_1+\theta_2)\boldsymbol{u}$$

Similarly, scalar multiplication can be defined in Hom(U, V) by $(\alpha\theta)$: $\boldsymbol{u}\mapsto\alpha(\theta\boldsymbol{u})$, $\forall\,\boldsymbol{u}\in U$. It is left as an exercise for the reader to show that $\alpha\theta$ is an element of Hom(U, V).

Theorem 5.4.7
Under the operations described above, Hom(U, V) is a vector space over K. Moreover, if U is of dimension m and V is of dimension n then Hom(U, V) is of dimension mn.

Proof The first part of the theorem is left as an exercise for the reader. The zero homomorphism is the mapping 0: $U\to V$ where $0(\boldsymbol{u})=\boldsymbol{0}$, $\forall\,\boldsymbol{u}\in U$.

For the second part of the theorem, let $\{\boldsymbol{u}_1, \boldsymbol{u}_2, \ldots, \boldsymbol{u}_m\}$ be a basis of U and $\{\boldsymbol{v}_1, \boldsymbol{v}_2, \ldots, \boldsymbol{v}_n\}$ be a basis of V. If $\boldsymbol{u}\in U$ then it can be expressed uniquely in the form

$$\boldsymbol{u}=\sum_{i=1}^{m}\lambda_i\boldsymbol{u}_i$$

where $\lambda_1, \lambda_2, \ldots, \lambda_m\in K$.

Define θ_{ij}: $U\to V$ by

$$\theta_{ij}\left(\sum_{k=1}^{m}\lambda_k\boldsymbol{u}_k\right)=\lambda_i\boldsymbol{v}_j\ (i=1,\ldots,m,\ j=1,\ldots,n)$$

It is claimed that these mn elements form a basis of Hom(U, V).

They are linearly independent since if $\sum_{i=1}^{m}\sum_{j=1}^{n}\alpha_{ij}\theta_{ij}=0$, then

$$\left(\sum_{i=1}^{m}\sum_{j=1}^{n}\alpha_{ij}\theta_{ij}\right)\boldsymbol{u}_k=0(\boldsymbol{u}_k)=\boldsymbol{0}$$

But

$$\left(\sum_{i=1}^{m}\sum_{j=1}^{n}\alpha_{ij}\theta_{ij}\right)\boldsymbol{u}_k=\sum_{i=1}^{m}\sum_{j=1}^{n}(\alpha_{ij}\theta_{ij}\boldsymbol{u}_k)=\sum_{j=1}^{n}\alpha_{kj}\boldsymbol{v}_j$$

since $\theta_{ij}\,\boldsymbol{u}_k=\boldsymbol{0}$ if $i\neq k$ and $\theta_{kj}\boldsymbol{u}_k=\boldsymbol{v}_j$. Thus

$$\sum_{j=1}^{n}\alpha_{kj}\boldsymbol{v}_j=\boldsymbol{0}$$

which implies $\alpha_{k1}=\alpha_{k2}=\ldots=\alpha_{kn}=0$, since $\{\boldsymbol{v}_1, \boldsymbol{v}_2, \ldots, \boldsymbol{v}_n\}$ is a basis of V. This argument holds for $k=1, 2, \ldots, m$.

The elements span Hom(U, V) since if θ is an arbitrary homomorphism, then fo each $\boldsymbol{u}_i$, $\theta\boldsymbol{u}_i=\alpha_{i1}\boldsymbol{v}_1+\alpha_{i2}\boldsymbol{v}_2+\ldots+\alpha_{in}\boldsymbol{v}_n$ since $\boldsymbol{v}_1, \boldsymbol{v}_2, \ldots, \boldsymbol{v}_n$ span V. Thus, for any

$$\boldsymbol{u}=\sum_{i=1}^{m}\lambda_i\boldsymbol{u}_i$$

$$\begin{aligned}\theta\boldsymbol{u}&=\sum_{i=1}^{m}\lambda_i\theta\boldsymbol{u}_i=\sum_{i=1}^{m}\sum_{j=1}^{n}\lambda_i\alpha_{ij}\boldsymbol{v}_j\\&=\sum_{i=1}^{m}\sum_{j=1}^{n}\alpha_{ij}\lambda_i\boldsymbol{v}_j\\&=\sum_{i=1}^{m}\sum_{j=1}^{n}\alpha_{ij}\theta_{ij}(\boldsymbol{u})\end{aligned}$$

that is

$$\theta=\sum_{i=1}^{m}\sum_{j=1}^{n}\alpha_{ij}\theta_{ij}$$

Corollary 5.4.7
If V is of dimension n over K then Hom(V, V) is of dimension n^2 over K.

If U is a vector space over K of dimension m with basis $\{\boldsymbol{u}_1, \ldots, \boldsymbol{u}_m\}$ and V is a vector space of dimension n with basis $\{\boldsymbol{v}_1, \ldots, \boldsymbol{v}_n\}$ then any homomorphism θ: $U\rightarrow V$ is uniquely determined by its operation on the vectors $\boldsymbol{u}_1, \ldots, \boldsymbol{u}_m$. Say $\theta\boldsymbol{u}_i=\alpha_{i1}\boldsymbol{v}_1+\ldots+\alpha_{in}\boldsymbol{v}_n$, $i=1, \ldots, m$. A convenient way of representing θ is by an $m\times n$ matrix with coefficients in K

$$\begin{bmatrix}\alpha_{11} & \alpha_{12} & \cdots & \alpha_{1n}\\ \alpha_{21} & \alpha_{22} & \cdots & \alpha_{2n}\\ \vdots & \vdots & & \vdots\\ \alpha_{m1} & \alpha_{m2} & \cdots & \alpha_{mn}\end{bmatrix}$$

This matrix is called the *matrix of* θ *relative to the bases* $\{\boldsymbol{u}_1, \boldsymbol{u}_2, \ldots, \boldsymbol{u}_m\}$ *and* $\{\boldsymbol{v}_1, \ldots, \boldsymbol{v}_n\}$ *in* U *and* V. It is used to calculate $\theta\boldsymbol{u}$ for any $\boldsymbol{u}\in U$ as follows. If $\boldsymbol{u}=\lambda_1\boldsymbol{u}_1+\ldots+\lambda_m\boldsymbol{u}_m$, then the coefficients of $\boldsymbol{v}_1, \ldots, \boldsymbol{v}_n$ in $\theta\boldsymbol{u}$ are obtained from the matrix multiplication

$$[\lambda_1 \ \ldots \ \lambda_m]\begin{bmatrix}\alpha_{11} & \alpha_{12} & \cdots & \alpha_{1n}\\ \alpha_{21} & \alpha_{22} & \cdots & \alpha_{2n}\\ \vdots & \vdots & & \vdots\\ \alpha_{m1} & \alpha_{m2} & \cdots & \alpha_{mn}\end{bmatrix}$$

Example 5.4.10

Let $U=\mathbb{R}^3$, $V=\mathbb{R}^2$. Let U have basis $\{[1\ 0\ 0], [0\ 1\ 0], [0\ 0\ 1]\}$ and V have basis $\{[1\ 2], [2\ 1]\}$.

Let θ: $U \rightarrow V$ be defined by

$$\theta[1\ 0\ 0]=[3\ 3]=[1\ 2]+[2\ 1]$$

$$\theta[0\ 1\ 0]=[0\ 3]=2[1\ 2]-[2\ 1]$$

$$\theta[0\ 0\ 1]=[1\ \ -4]=-3[1\ 2]+2[2\ 1]$$

Then θ is represented by the matrix

$$\begin{bmatrix} 1 & 1 \\ 2 & -1 \\ -3 & 2 \end{bmatrix}$$

and

$$\begin{aligned} \theta[3\ 2\ 1] &= \theta(3[1\ 0\ 0]+2[0\ 1\ 0]+1[0\ 0\ 1]) \\ &= a[1\ 2]+b[2\ 1] \end{aligned}$$

where

$$[a\ b]=[3\ 2\ 1]\begin{bmatrix} 1 & 1 \\ 2 & -1 \\ -3 & 2 \end{bmatrix}=[4\ 3]$$

An important observation is the following result.

Theorem 5.4.8
Let U, V be vector spaces over a field K of dimensions m, n, respectively. Let $\{\boldsymbol{u}_1, \boldsymbol{u}_2, \ldots, \boldsymbol{u}_m\}$ be a basis for U and $\{\boldsymbol{v}_1, \boldsymbol{v}_2, \ldots, \boldsymbol{v}_n\}$ be a basis for V. Then the mapping that associates each homomorphism θ: $U \rightarrow V$ with its matrix relative to these bases is an isomorphism from the vector space Hom(U, V) to the vector space of $m \times n$ matrices with coefficients in K.

This theorem shows that much of the work done in chapter 2 could have been presented in terms of vector-space homomorphisms.

Exercise 5.4

1. Let A be a finite set of cardinality n. Define the operations $+$ and $\times$ in 2^A by $X+Y=(X\cup Y)\backslash(X\cap Y)$ and $X\times Y=X\cap Y$. Show that, under these operations, 2^A is a ring.

2. Let G be an Abelian group with binary operation $+$. An *endomorphism* of G is group homomorphism of G into itself. Let End(G) denote the set of all such endomorphisms and define the operations $+$ and $\times$ in End(G) by $\theta_1+\theta_2$: $x \mapsto \theta_1(x)+\theta_2(x)$ and $\theta_1 \times \theta_2$: $x \mapsto \theta_1(\theta_2(x))$, $\forall x \in G$. Show that, under these operations, End(G) is a ring.

3. Show that in any ring, R, $(-a)(-b)=ab$, $\forall a, b \in R$.

4. Let θ be a ring homomorphism of a ring, R, with multiplicative identity, e, to an integral domain, R', with multiplicative identity e'. Assuming there exists at least one $a \in R$ such that $\theta(a) \neq 0$, prove that $\theta(e) = e'$.

5. Prove that the statement following definition 5.4.7 is correct.

6. If S, T are subsets of a vector space V, show that

(a) $S \subset T$ implies $L(S)$ is a subspace of $L(T)$;

(b) $L(L(S)) = L(S)$.

7. Show that if $\boldsymbol{v}_1, \boldsymbol{v}_2, \ldots, \boldsymbol{v}_n$ are elements of a vector space V then either they are linearly independent or some $\boldsymbol{v}_k$ is a linear combination of the preceding ones, $\boldsymbol{v}_1, \boldsymbol{v}_2, \ldots, \boldsymbol{v}_{k-1}$. Hence, show that, if V is a finite-dimensional vector space, then it contains a finite basis.

8. Show that if $\{\boldsymbol{v}_1, \boldsymbol{v}_2, \ldots, \boldsymbol{v}_n\}$ is a basis of V over K and if $\boldsymbol{u}_1, \boldsymbol{u}_2, \ldots, \boldsymbol{u}_m$ in V are linearly independent over K then $m \leqslant n$. Hence, show that if V is finite-dimensional over K then any two bases have the same number of elements.

9. Show that K^n is isomorphic to a subspace of K^{n+1}. [Hint: Look at example 5.4.6.]

10. Show that Hom(U, V) is a vector space. Prove theorem 5.4.8.

11. If V is a vector space of dimension n over K, what is the dimension of the vector space Hom(V, K)? This vector space is called the *dual space* of V and denoted by $\hat{V}$. Show that $\hat{\hat{V}}$ is isomorphic to V.

5.5 BOOLEAN ALGEBRAS

Introduction

A Boolean algebra, like rings and vector spaces, is an example of an algebraic structure with more than one binary operation. Boolean algebra merits a whole section of this chapter because of its particular importance to computer science, especially with regard to circuit design and analysis.

Just as the integers motivated the definition of an integral domain, so the algebra of sets motivates the definition of a Boolean algebra. The properties of sets in a given universe, $\mathscr{U}$, are summarised as follows. If A, B, C denote any sets in the universe, $\mathscr{U}$, then

Closure

$A \cup B$ is a set in $\mathscr{U}$ and $A \cap B$ is a set in $\mathscr{U}$

Associativity

$A \cup (B \cup C) = (A \cup B) \cup C$ and $A \cap (B \cap C) = (A \cap B) \cap C$

Commutativity

$A \cup B = B \cup A$ and $A \cap B = B \cap A$

Identity laws

$\emptyset$ satisfies $A \cup \emptyset = \emptyset \cup A = A$ and $\mathscr{U}$ satisfies $A \cap \mathscr{U} = \mathscr{U} \cap A = A$

Distributivity

$A \cup (B \cap C) = (A \cup B) \cap (A \cup C)$ and $A \cap (B \cup C) = (A \cap B) \cup (A \cap C)$

Idempotency

$A \cup A = A$ and $A \cap A = A$

Complement

$A \cup A' = \mathscr{U}$ and $A \cap A' = \emptyset$

Involution law

$(A')' = A$

De Morgan's laws

$(A \cup B)' = A' \cap B'$ and $(A \cap B)' = A' \cup B'$

If $+$ is written for $\cup$, $\times$ for $\cap$, 0 for $\emptyset$ and 1 for $\mathscr{U}$, the first four of these rules have parallels in the algebra of reals, as follows. If a, b, c denote any real numbers, then

Closure

$a + b \in \mathbb{R}$ and $a \times b \in \mathbb{R}$

Associativity

$a + (b + c) = (a + b) + c$ and $a \times (b \times c) = (a \times b) \times c$

Commutativity

$$a+b=b+a \text{ and } a\times b=b\times a$$

Identity laws

0 satisfies $a+0=0+a=a$ and 1 satisfies $a\times 1=1\times a=a$

However, the distributive rules do not have a complete parallel in the algebra of reals since although

$$a\times(b+c)=(a\times b)+(a\times c),\ \forall a, b, c\in\mathbb{R}$$

It is *not* true that

$$a+(b\times c)=(a+b)\times(a+c),\ \forall a, b, c\in\mathbb{R}$$

The remaining rules of Boolean algebra similarly have no parallels in the algebra of reals.

A Boolean algebra is an abstract algebraic structure which encapsulates each of the above rules satisfied by the algebra of sets. It is worthwhile studying Boolean algebra because the theory developed is not only applicable to set theory but also to logic, switching theory and circuit design.

Definition 5.5.1

A Boolean algebra is a non-empty set of elements, B, over which two binary operations, $+$ and $\times$, and one unary operation, $'$, are defined satisfying the following axioms.

Closure

$$a+b\in B \text{ and } a\times b\in B,\ \forall a, b\in B$$

Associativity

$$a+(b+c)=(a+b)+c \text{ and } a\times(b\times c)=(a\times b)\times c,\ \forall a, b, c\in B$$

Commutativity

$$a+b=b+a \text{ and } a\times b=b\times a,\ \forall a, b\in B$$

Identity laws There exists an identity with respect to $+$, $0\in B$, such that

$$a+0=0+a=a,\ \forall a\in B$$

and there exists an identity with respect to $\times$, $1\in B$, such that

$$a\times 1=1\times a=a,\ \forall a\in B$$

Distributivity

$$a+(b\times c)=(a+b)\times(a+c) \text{ and } a\times(b+c)=(a\times b)+(a\times c),\ \forall a, b, c\in B$$

Idempotency

$$a+a=a \text{ and } a\times a=a, \ \forall a\in B$$

Complement

$$a+a'=1 \text{ and } a\times a'=0, \ \forall a\in B$$

Involution law

$$(a')'=a, \ \forall a\in B$$

De Morgan's laws

$$(a+b)'=a'\times b' \text{ and } (a\times b)'=a'+b', \ \forall a, b\in B$$

As in normal algebra, $a\times b$ is often written in Boolean algebra as ab and parentheses are omitted from expressions in Boolean algebra according to the rule that $'$ takes priority over $\times$ which in turn takes priority over $+$.

Example 5.5.1

(1) For any set A, 2^A is a Boolean algebra where $+$ denotes $\cup$, $\times$ denotes $\cap$, 0 denotes $\emptyset$ and 1 denotes A.

(2) It is possible to construct a Boolean algebra with just two elements $\{0, 1\}$, where $0'=1$ and $1'=0$, and $+$ and $\times$ are defined by

$+$	0	1
0	0	1
1	1	1

$\times$	0	1
0	0	0
1	0	1

This Boolean algebra is known as the *binary Boolean algebra.*

(3) Consider the set of all statement forms as defined in definition 1.1.1. Let $\mathscr{T}$ denote a tautology (that is, a statement which is always true) and $\mathscr{F}$ a contradiction (i.e. a statement which is always false). If $P=Q$ is used to denote $P\equiv Q$ and if $+$ denotes $\vee$ and $\times$ denotes $\wedge$, then the set of statement forms is a Boolean algebra. The identity 0 corresponds to $\mathscr{F}$ and 1 corresponds to $\mathscr{T}$. The complement of a statement P is $\sim P$. The reader should check that each of the axioms of definition 5.5.1 is indeed satisfied.

(4) Simple electric circuits can be constructed using on–off switches p, q, r, ... Let $p+q$ denote the switches p and q connected in parallel, as in figure 5.19a, and $p\times q$ denote the switches p and q connected in series, as

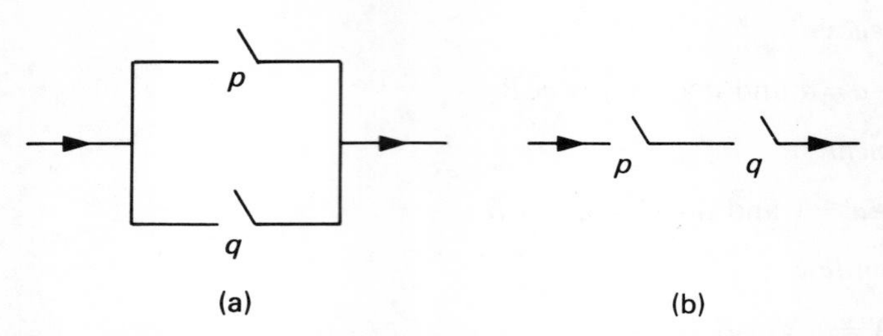

Figure 5.19 Example 5.5.1(4); (a) $p+q$, switches in parallel; (b) $p \times q$, switches in series

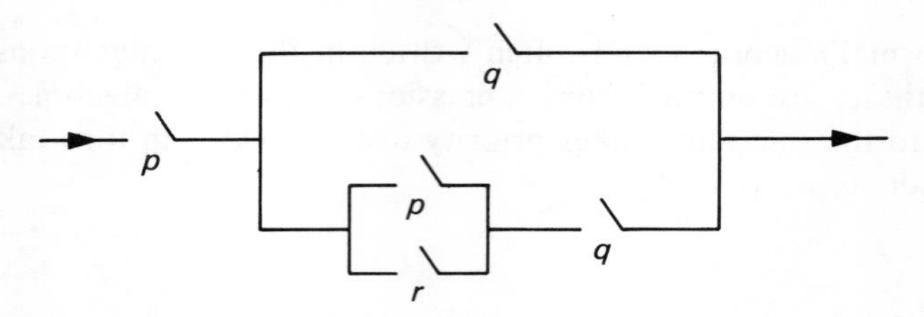

Figure 5.20 Example 5.5.1(4): circuit representing $p \times (q+(p+r) \times q)$

in figure 5.19b. Using the operators $+$ and $\times$, complex switching circuits constructed from the switches p, q, r, ... can be represented by simple expressions. For example, the switching circuit in figure 5.20 can be represented by $p \times (q+(p+r) \times q)$.

Let p' denote a circuit consisting of a switch which is on when p is off and is off when p is on. Let 1 denote the circuit with no switches, which is thus always on, and 0 denote the circuit that is always off. Two circuits are said to be equal if, whatever on–off values are given to the switches p, q, r, ..., the circuits both behave identically, that is, either both allow electricity to flow or both prevent it. With these definitions, the circuits form a Boolean algebra.

Boolean algebra is not only a means of handling sets, propositional calculus and switching circuits but, as shown later, is also a valuable tool in the design of computer circuitry. The manipulation and simplification of expressions in Boolean algebra is thus a useful skill to acquire. Each step of such a manipulation is merely the application of one or more of the axioms given in definition 5.5.1. It is also useful to develop a few general results which can make the task a little easier.

Theorem 5.5.1
For any Boolean algebra, B

(i) $a \times 0 = 0 \times a = 0, \ \forall a \in B,$

(ii) $a+1=1+a=1, \forall a \in B,$

(iii) $a+b\times a=a+a\times b=a, \forall a, b \in B$

(iv) $a\times(b+a)=a\times(a+b)=a, \forall a, b \in B,$

(v) $(a+b)\times c=a\times c+b\times c, \forall a, b, c \in B,$

(vi) $(a\times b)+c=(a+c)\times(b+c), \forall a, b, c \in B,$

(vii) $0'=1,$

(viii) $1'=0.$

Proof (i) This is left as an exercise for the reader.

(ii)

$$
\begin{aligned}
a+1 &= a+(a+a') && \text{(complement)}\\
&= (a+a)+a' && \text{(associativity of } +)\\
&= a+a' && \text{(idempotency of } +)\\
&= 1 && \text{(complement)}\\
1+a &= a+1 && \text{(commutativity of } +)
\end{aligned}
$$

(iii)

$$
\begin{aligned}
a+a\times b &= a\times 1+a\times b && \text{(identity law)}\\
&= a\times(1+b) && \text{(distributivity of } \times \text{ over } +)\\
&= a\times 1 && \text{(theorem 5.5.1(ii))}\\
&= a && \text{(identity law)}\\
a+a\times b &= a+b\times a && \text{(commutativity of } \times)
\end{aligned}
$$

(iv) This is left as an exercise for the reader.

(v)

$$
\begin{aligned}
(a+b)\times c &= c\times(a+b) && \text{(commutativity of } \times)\\
&= c\times a+c\times b && \text{(distributivity of } \times \text{ over } +)\\
&= a\times c+b\times c && \text{(commutativity of } \times)
\end{aligned}
$$

(vi) This is left as an exercise for the reader.

(vii)

$$
\begin{aligned}
0' &= 0+0' && \text{(identity law)}\\
&= 1 && \text{(complement)}
\end{aligned}
$$

(viii) This is left as an exercise for the reader.

Example 5.5.2

Simplify: (1) $(a'bc+a'bc+a'b')'$ and (2) $ab+ab'cd+ad'+a'c+a'bc'+a'b'd'$. In example (1), justification of each step is given. In example (2), this is left as an exercise for the reader who should note that some steps involve the application of more than one axiom or theorem.

(1)

$$
\begin{aligned}
(a'bc+a'bc+a'b')' &= (a'bc+a'b')' && \text{(idempotency of } +) \\
&= (a'(bc+b'))' && \text{(distributivity of } \times \text{ over } +) \\
&= a''+(bc+b')' && \text{(De Morgan's law)} \\
&= a+(bc+b')' && \text{(involution)} \\
&= a+(bc)'b'' && \text{(De Morgan's law)} \\
&= a+(bc)'b && \text{(involution)} \\
&= a+(b'+c')b && \text{(De Morgan's law)} \\
&= a+b(b'+c') && \text{(commutativity of } \times) \\
&= a+bb'+bc' && \text{(distributivity of } \times \text{ over } +) \\
&= a+0+bc' && \text{(complement)} \\
&= a+bc' && \text{(identity law)}
\end{aligned}
$$

(2)

$$
\begin{aligned}
ab+ab'cd+ad'+a'c+a'bc'+a'b'd' &= ab+acd(b+b')+ad'+a'c+ \\
&\quad a'b(c+c')+b'd'(a+a')
\end{aligned}
$$

(Hint: use theorem 5.5.1(iii))

$$
\begin{aligned}
&= ab+acd+ad'+a'c+a'b+b'd' \\
&= b(a+a')+ac(d+d')+ad'+a'c+b'd' \\
&= b+ac+ad'+a'c+b'd' \\
&= b+c+ad'+b'd' \\
&= b+c+ad'+d'(b+b') \\
&= b+c+d'+ad' \\
&= b+c+d'
\end{aligned}
$$

This second result shows that the complex switching circuit of figure 5.21a can be simplified to that of 5.21b.

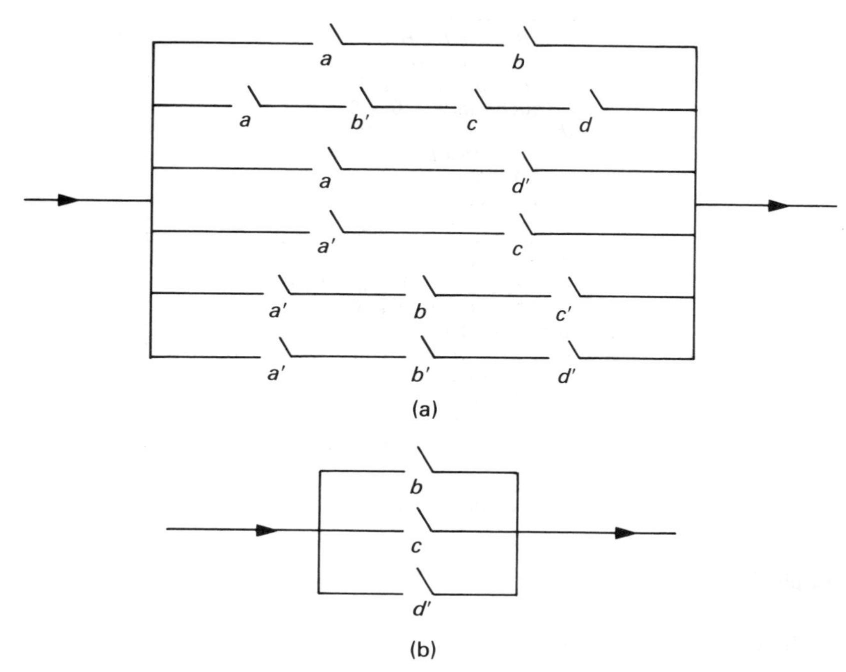

Figure 5.21 (a) Circuit corresponding to $ab+ab'cd+ad'+a'c+a'bc'+a'b'd'$; (b) circuit corresponding to $b+c+d'$

Duality

As illustrated in figure 5.22, the axioms defining a Boolean algebra can be divided into two subsets, the $+$ axioms and the $\times$ axioms, which have an interesting interrelation. Given any $+$ axiom, a $\times$ axiom can be constructed by replacing every $+$ by $\times$, every $\times$ by $+$, every 1 by 0 and every 0 by 1. Similarly, the same technique can be used to construct a $+$ axiom from any $\times$ axiom. An axiom constructed from another axiom, $\mathscr{A}$, in this way is called the *dual* of $\mathscr{A}$ and is denoted by $\hat{\mathscr{A}}$.

Definition 5.5.2
For any fully bracketed Boolean expression, $\mathscr{B}$, the *dual* of $\mathscr{B}$, denoted by $\hat{\mathscr{B}}$, is constructed from $\mathscr{B}$ by replacing each $+$ by $\times$, each $\times$ by $+$, each 1 by 0 and each 0 by 1.

To construct the dual of a Boolean expression that is not fully bracketed, parentheses must be inserted according to the usual precedence of the operators. Similarly, this convention can be used to eliminate redundant parentheses in the constructed dual.

+ AXIOMS	$a+b \in B, \quad \forall a, b \in B$ $a+(b+c)=(a+b)+c, \quad \forall a, b, c \in B$ $a+b=b+a, \quad \forall a, b \in B$ $\exists\, 0 \in B$ such that $a+0=0+a=a, \forall a \in B$ $a+(b \times c)=(a+b) \times (a+c), \quad \forall a, b, c \in B$ $a+a=a, \quad \forall a \in B$ $a+a'=1, \quad \forall a \in B$ $(a')'=a, \quad \forall a \in B$ $(a+b)'=a' \times b', \quad \forall a, b \in B$
× AXIOMS	$a \times b \in B, \quad \forall a, b \in B$ $a \times (b \times c)=(a \times b) \times c, \quad \forall a, b, c \in B$ $a \times b=b \times a, \quad \forall a, b \in B$ $\exists 1 \in B$ such that $a \times 1=1 \times a=a, \quad \forall a \in B$ $a \times (b+c)=(a \times b)+(a \times c), \quad \forall a, b, c \in B$ $a \times a=a, \quad \forall a \in B$ $a \times a'=0, \quad \forall a \in B$ $(a')'=a, \quad \forall a \in B$ $(a \times b)'=a'+b', \quad \forall a, b \in B$

Figure 5.22

Example 5.5.3

(1) The dual of $a+bc$ is $a \times (b+c)$.

(2) The dual of $1+(a0)'$ is $0(a+1)'$.

(3) The dual of $(a(b'+c)+a'b')'$ is $((a+b'c)(a'+b'))'$.

Theorem 5.5.2
(i) For any Boolean expression, $\mathscr{B}, \hat{\hat{\mathscr{B}}}=\mathscr{B}$.

(ii) For any Boolean expressions, $\mathscr{B}$ and $\mathscr{C}, \mathscr{B}=\mathscr{C} \supset \hat{\mathscr{B}}=\hat{\mathscr{C}}$.

Proof (i) This follows immediately from the definition of a dual.
(ii) If an axiom of Boolean algebra is applied to a Boolean expression, $\mathscr{B}_1$, to get another Boolean expression, $\mathscr{B}_2$, then the dual of that axiom can be applied to $\hat{\mathscr{B}}_1$ to get $\hat{\mathscr{B}}_2$. A simple induction argument on the number of steps in the proof that $\mathscr{B}=\mathscr{C}$ can then be used to show that there is a proof of the *dual result*, that is, $\hat{\mathscr{B}}=\hat{\mathscr{C}}$.

Example 5.5.4

(1) In theorem 5.5.1, results (i) and (ii) are duals, (iii) and (iv) are duals, (v) and (vi) are duals and (vii) and (viii) are duals.

(2) The result, given in example 5.5.2a, that $(a'bc+a'bc+a'b')'=a+bc'$

gives the dual result that $((a'+b+c)(a'+b+c)(a'+b'))'=a(b+c')$. This dual result can be proved by taking for each step in the proof the application of an axiom (or theorem) which is a dual axiom (or theorem) to that used in the corresponding step of the original proof. This results in the following proof

$$
\begin{aligned}
((a'+b+c)(a'+b+c)(a'+b'))' &= ((a'+b+c)(a'+b'))' && \text{(idempotency of } \times) \\
&= (a'+(b+c)b')' && \text{(distributivity of } + \text{ over } \times) \\
&= a''((b+c)b')' && \text{(De Morgan's law)} \\
&= a((b+c)b')' && \text{(involution)} \\
&= a((b+c)'+b'') && \text{(De Morgan's law)} \\
&= a((b+c)'+b) && \text{(involution)} \\
&= a(b'c'+b) && \text{(De Morgan's law)} \\
&= a(b+b'c') && \text{(commutativity of } +) \\
&= a(b+b')(b+c') && \text{(distributivity of } + \text{ over } \times) \\
&= a1(b+c') && \text{(complement)} \\
&= a(b+c') && \text{(identity law)}
\end{aligned}
$$

Computer Circuitry

The calculations performed by a computer can be viewed as a series of binary operations on electric signals in two input wires producing an output signal in one or more output wires. An electric signal may be 'on', denoted by 1 or 'off', denoted by 0. Binary operations in a computer are performed by electronic devices known as *logic elements*.

Definition 5.5.3.
The simplest of these logic elements is the *negator*, which has one input and one output. If the input is 1, the output is 0; if the input is 0, the output is 1. Thus, in the binary Boolean algebra the negator computes the complement, x', of the signal, x (figure 5.23).

The action of the negator can be summarised using an *input/output table*, which defines the output for each possible input

x	x'
0	1
1	0

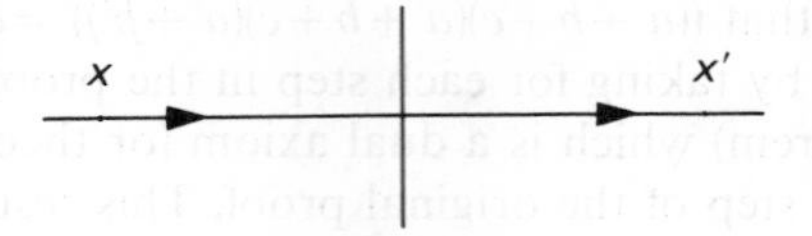

Figure 5.23 The negator

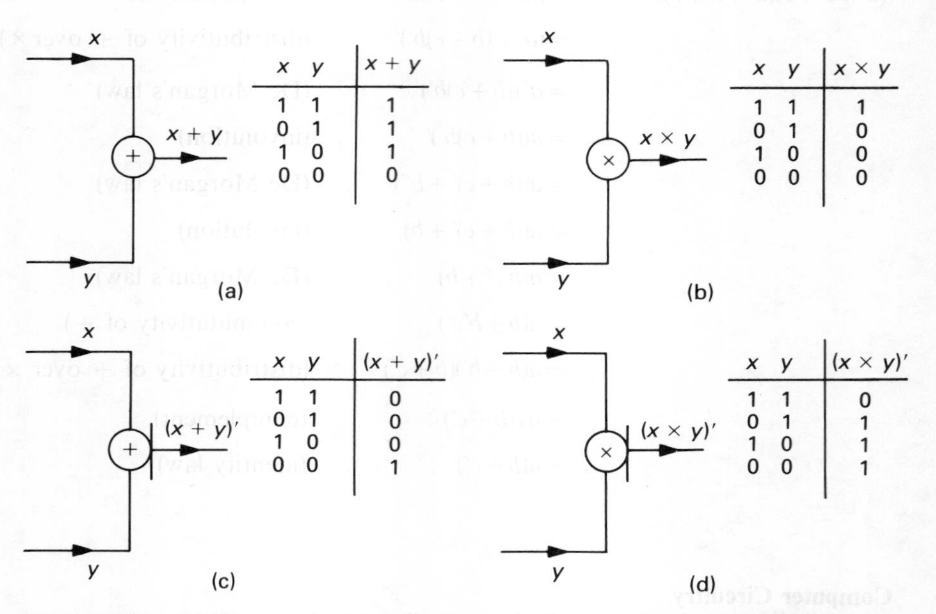

Figure 5.24 (a) The or-element; (b) the and-element; (c) the nor-element; (d) the nand-element

Definition 5.5.4
The *or-element*, *and-element*, *nor-element* and *nand-element* are all logic elements having two inputs and one output (figure 5.24). If x, y denote elements in the binary Boolean algebra, the or-element computes $x+y$, the and-element computes $x \times y$, the nor-element computes $(x+y)'$ and the nand-element computes $(x \times y)'$.

By passing the output from one of the logic elements to the input of another, logic circuits of varying complexity can be constructed. Figure 5.25 shows such a circuit constructed from two negators, one or-element and two and-elements. Figure 5.26 is constructed from one nand-element, one or-element and one and-element.

Definition 5.5.5
Two logic circuits are said to be *equivalent* if, given identical input signals,

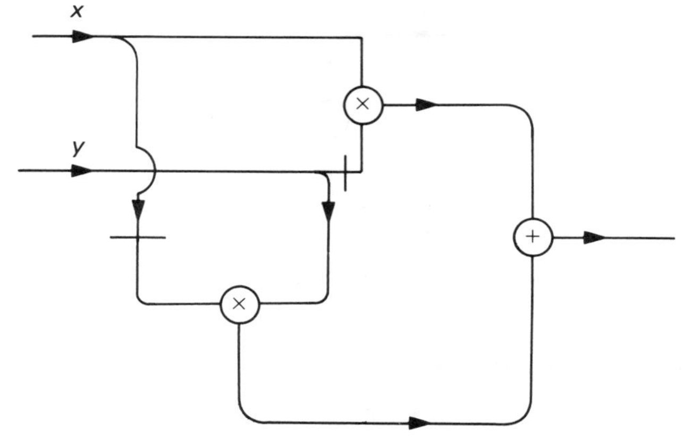

Figure 5.25

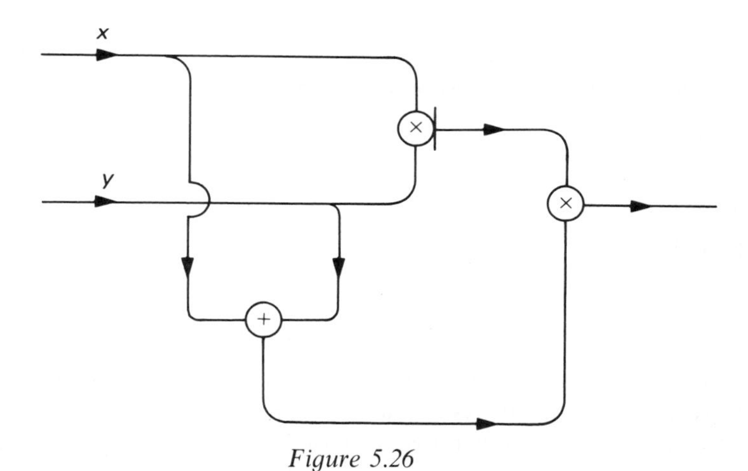

Figure 5.26

they will always result in identical output signals, that is, they both have the same input/output table.

Rather than compute the input/output tables directly, one way of proving two circuits to be equivalent is to show that their corresponding Boolean expressions are equal.

Example 5.5.5

The circuit in figure 5.25 computes $xy' + x'y$ for binary inputs x, y and the circuit in figure 5.26 computes $(xy)'(x + y)$. Since

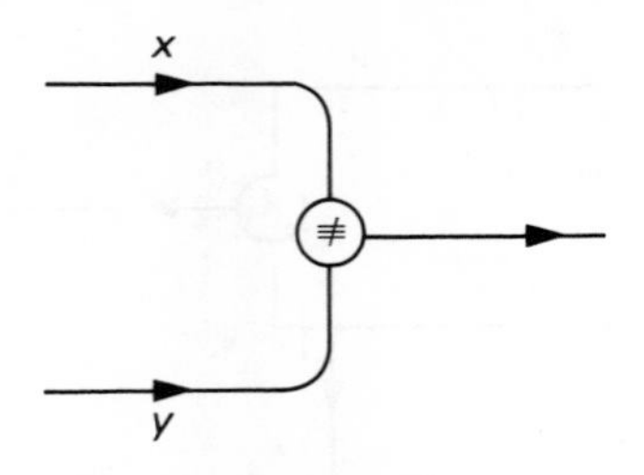

Figure 5.27

$$(xy)'(x+y)=(x'+y')(x+y)$$
$$=x'x+y'x+x'y+y'y$$
$$=y'x+x'y$$
$$=xy'+x'y$$

these two circuits are equivalent.

The input/output table corresponding to both circuits is

x	y	
1	1	0
0	1	1
1	0	1
0	0	0

A circuit which computes this value will be denoted as in figure 5.27.

The use of logic circuits will be demonstrated by constructing a circuit to add two binary numbers. As a first step, consider the problem of adding two binary digits. When two such digits are added together, the result can be one of three outcomes, that is, 0, 1 or 10. A circuit to add two binary digits will thus not only have two inputs but also requires two outputs. The first of these output wires will deliver c (the carry digit) and the second s (for sum). The circuit required will thus have input/output table

x	y	c	s
1	1	1	0
0	1	0	1
1	0	0	1
0	0	0	0

The carry digit is simply obtained from an and-element and the sum from the circuit given in figure 5.27. The resulting circuit (figure 5.28) is

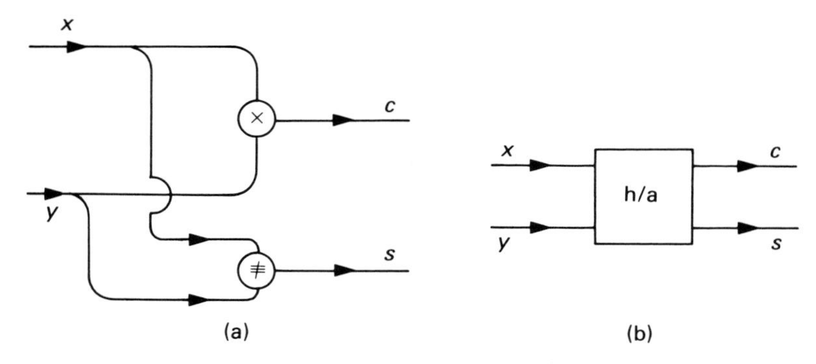

Figure 5.28 (a) A half-adder; (b) abbreviated form

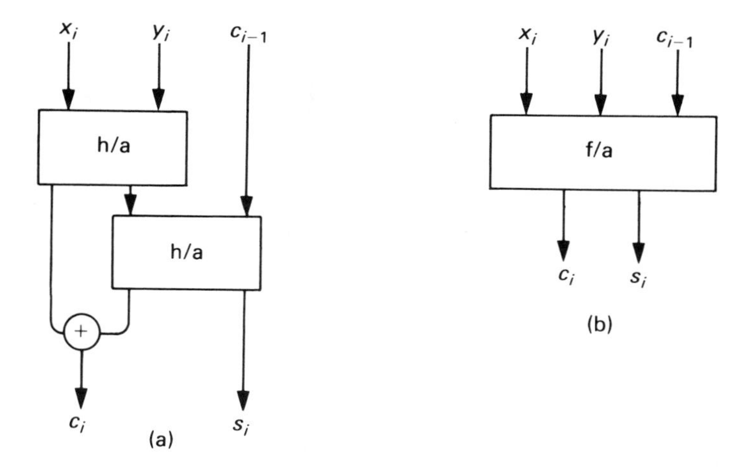

Figure 5.29 (a) A full-adder; (b) abbreviated form

called a *half-adder*. This circuit is used in the construction of a circuit that will add two binary numbers $\boldsymbol{x}$, $\boldsymbol{y}$ in the form of a sequence of binary digits

$$\boldsymbol{x} = x_{n-1}x_{n-2}\ldots x_2x_1x_0$$

$$\boldsymbol{y} = y_{n-1}y_{n-2}\ldots y_2y_1y_0$$

passing along two input wires. To add $\boldsymbol{x}$ and $\boldsymbol{y}$, x_0 is added to y_0 to get c_0, s_0; s_0 is output and c_0 is added to the sum of x_1 and y_1, giving c_1, s_1; s_1 is output and c_1 is then added to the sum of x_2 and y_2 giving c_2, s_2, etc. The *full-adder* (figure 5.29) performs the task of adding c_{i-1}, x_i and y_i to yield c_i and s_i.

A cheap but slow network for adding two binary numbers is the *serial*

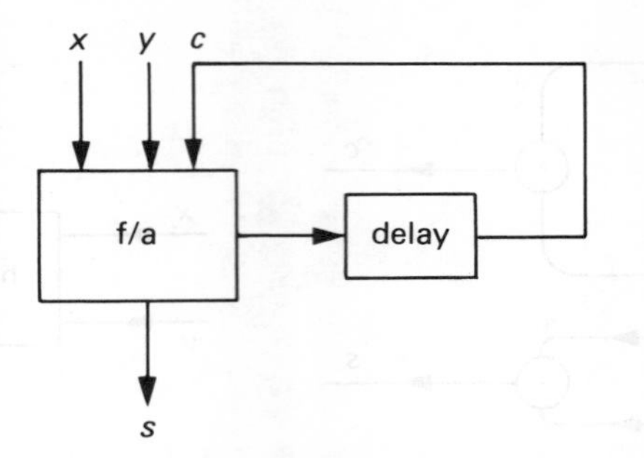

Figure 5.30 A serial adder

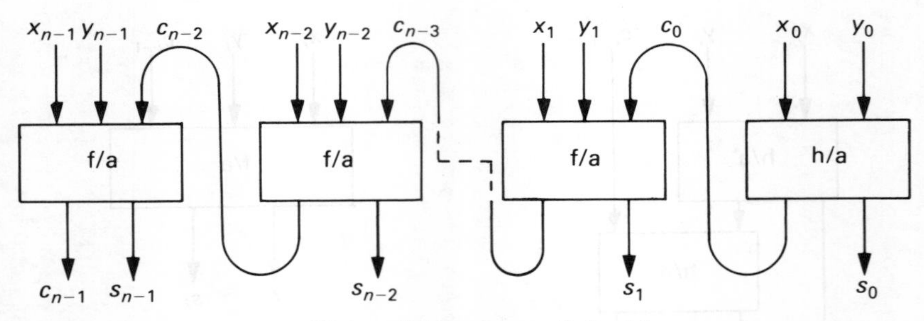

Figure 5.31 A parallel adder

adder (figure 5.30), which incorporates a delay device to delay a signal for a one-digit period. The binary numbers to be added are to be seen as a stream of binary digits arriving in the input wires **x**, **y**. The output wire, **s**, delivers a stream of binary digits equal to the sum of **x** and **y**. It is far more efficient, however, to use a *parallel adder*, as in figure 5.31. In this case there are $2n$ input wires, n for $x_{n-1}x_{n-2}\ldots x_0$ and n for $y_{n-1}y_{n-2}\ldots y_0$. The sum is output on $n+1$ wires in the form $c_{n-1}s_{n-1}s_{n-2}\ldots s_0$.

Exercise 5.5

1. Let a, b be any two elements of a Boolean algebra, B. Write $a \leqslant b$ iff $a+b=b$. Prove

(a) that $\leqslant$ is an ordering relation;

(b) that $0 \leqslant a \leqslant 1$, $\forall a \in B$;

(c) that $a \leqslant b'$ if and only if $b \leqslant a'$.

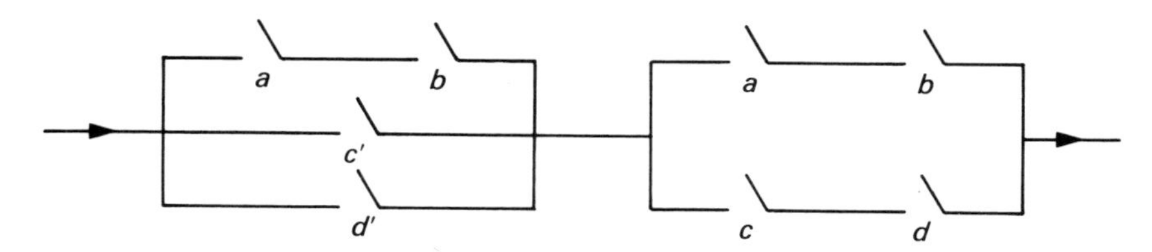

Figure 5.32 Exercise 5.5.4

2. Simplify the following Boolean expressions

(a) $(a+b)(a'c+b')+(c'+ab)'+(ac)'$;

(b) $(a(b'+c)+a'b')'$;

(c) $((a+b'c)(a'+b'))'$ [Hint: Look at example 5.5.3(3)];

(d) $((a+0)'(b+1)')'$.

3. Translate the results of exercise 5.5(1) to the equivalent results in set theory.

4. Use Boolean algebra to simplify the switching circuit in figure 5.32.

5. Three members of a committee can vote yes on a proposal by pressing a button. Devise a simple circuit that will allow current to flow only when at least two members vote in the affirmative.

6. A light is to be controlled by three different switches in a room such that flicking any of these switches turns the light on if it is off and off it is on. Construct a simple circuit to perform this task.

7. Design the circuit for a half-subtractor, that is, a device that subtracts a bit y from a bit x. (Hint: instead of using a carry output, c, use a borrow output, b.) Hence design the circuit for a full subtractor to be used in subtracting a binary number $\boldsymbol{y}$ from a binary number $\boldsymbol{x}$.

References

Allen, A. O., *Probability, Statistics and Queueing Theory, with Computer Science Applications* (Academic Press, New York, 1978).
Arthurs, A. M., *Probability Theory* (Routledge & Kegan Paul, London, 1965).
Bondy, J. A., and Murty, U. S. R., *Graph Theory with Applications* (Macmillan, London, 1977).
Burkill, J. C., *A First Course in Mathematical Analysis* (Cambridge University Press, 1962).
Davis, P. J., and Rabonowitz, P., *Methods of Numerical Integration* (Academic Press, New York, 1975).
Forsythe, G., and Moler, C. B., *Computer Solution of Linear Algebraic Systems* (Prentice-Hall, Englewood Cliffs, N.J., 1967) section 21.
Hall, M., *Theory of Groups* (Macmillan, New York, 1959).
Hamming, R. W., *Coding and Information Theory* (Prentice-Hall, Englewood Cliffs, N.J., 1980).
Knuth, D. E., *The Art of Computer Programming: Semi-numerical Algorithms* (Addison-Wesley, Reading, Mass., 1969).
Knuth, D. E., *The Art of Computer Programming*, vol. 3 (Addison-Wesley, Reading, Mass., 1973).
Minieka, E., *Optimization Algorithms for Networks and Graphs* (Marcel Dekker, New York, 1978).
Murdoch, J., *Queueing Theory* (Macmillan, London, 1978).
Page, E. S., and Wilson, L. B., *Information Representation and Manipulation in a Computer*, 2nd edn (Cambridge University Press, 1978).
Parzen, E., *Modern Probability Theory and Its Applications* (Wiley, New York, 1960).
Stone, H. S., *Introduction to Computer Organization and Data Structures* (McGraw-Hill, New York, 1972).

Solutions to Selected Exercises

CHAPTER 1

1.1

1. $J \equiv$ John is good, $M \equiv$ Mark is good: (a) $J \supset \sim M) \wedge (\sim M \supset J)$; (b) $\sim M \supset J$; (c) $J \supset \sim M$

2. (a) Neither; (b) Neither; (c) Contradiction; (d) Tautology; (e) Tautology

4. $(A \wedge B \wedge \sim C) \vee (A \wedge \sim B \wedge \sim C) \vee (\sim A \wedge B \wedge \sim C)$

5. $\sim A = A|A$; $A \vee B = (A|A)|(B|B)$

1.2

1. $\mathscr{U} = \{1, 2, 3, 4, 5, 6, 7, 8, 9, 10, 11, 12, 13, 14, 15, 16, 17, 18, 19\}$

(a) $\{1, 2, 3, 4, 5, 6, 7\}$; (b) $\{1, 2, 3, 5, 7, 11, 13, 17, 19\}$; (c) $\{1\}$; (d) $\emptyset$ or $\{\}$

2. (a) $\{2, 6, 10\}$; (b) $\{2, 4, 8\}$; (c) $\emptyset$; (d) $\{4, 8\}$; (e) $\{4, 8, 10\}$

3. If $A = \{1, 2\}$, $B = \{1\}$ then $A \backslash B = \{2\}$ and $B \backslash A = \emptyset$. If $A \backslash B = B \backslash A$ then $A \backslash B = \emptyset = B \backslash A$, so $A = B$

4. $\#(\emptyset) \leqslant \#(A \cap B) \leqslant \#(A) \leqslant \#(A \cup B) \leqslant \#(\mathscr{U})$

6. 14 play tennis. (a) No more than 4. (b) At least 17 but no more than 21

7. $P = \emptyset$

8. (a) C; (b) B; (c) $A' \cup B'$

1.3

2. (a) 1647; (b) $1356_8 = 750$; (c) 298

4. (a) 65.31; (b) 3.454

5. Four significant figures

6. (a) $x > -15$; (b) $-6 < x < 6$; (c) $x \geqslant 1$

7. (a) $x \in (-15, \infty)$; (b) $x \in (-6, 6)$; (c) $x \in [1, \infty)$

8. (a) $(-\infty, -15) \cup (14, \infty)$; (b) $[-15, 6]$; (c) $[-15, -2] \cup [1, 6]$

1.4

1. (a) $1 + i7$; (b) $\frac{-59}{74} + i\frac{95}{74}$; (c) $-\frac{1}{2} + i2$; (d) $\sqrt{2}\ \text{cis}(\pi/12) \approx 1.366 + 0.366i$

2. (a) $4 - 48i$; (b) $\sqrt{58} + \sqrt{40}$; (c) $56 + 90i$

4. (a) The roots are $2 \pm \sqrt{(5)}i$

5. $\sin 4\theta = 4 \sin\theta \cos\theta\ (2\cos^2\theta - 1)$ and $\cos 4\theta = \cos^4\theta + \sin^4\theta - 6\cos^2\theta \sin^2\theta$

6. α lies on the imaginary axis. Hence α is $-i$ or $3i$

1.5

1. (a) -3; (b) -1; (c) $\mathbb{R}$; (d) $\mathbb{R}$

2. (a) $x \mapsto x(1-x)$ $(x \in \mathbb{R})$; (b) $x \mapsto x^2 + (1-x)^2$ $(x \in \mathbb{R})$; (c) $x \mapsto x/(1-x)$ $(x \in \mathbb{R}, x \neq 1)$

3. (a) neither; (b) onto; (c) 1–1; (d) both

4. $h^{-1}: x \mapsto \sqrt[3]{[(x-3)/2]}$ $(x \in \mathbb{R})$

7. There are nine functions, six injections but no surjections

10. $(x - \frac{1}{2})^2 + (y + \frac{1}{2})^2 = 29/2$

11. $e = \sqrt{(15)}/4$, the focus is $(\sqrt{30}, 0)$ and the directrix is $x = 16\sqrt{(2/15)}$

13. $A \times B = \{(a, c), (a, d), (b, c), (b, d)\}$; $B \times A = \{(c, a)\ (c, b), (d, a), (d, b)\}$

CHAPTER 2

2.1

1. (a) $\begin{bmatrix} 3 \\ 2 \\ 9 \end{bmatrix}$; (b) $\begin{bmatrix} 5 \\ 6 \\ 21 \end{bmatrix}$; (c) $\sqrt{70}$. The vectors are linearly independent

2. $\boldsymbol{x} = [7/5 \;\; 2/5]$

3. $$\begin{bmatrix} 1 \\ 4 \\ 2 \end{bmatrix} = \frac{3}{4}\begin{bmatrix} 3 \\ 2 \\ 1 \end{bmatrix} + \frac{25}{12}\begin{bmatrix} 1 \\ 2 \\ 3 \end{bmatrix} - \frac{5}{3}\begin{bmatrix} 2 \\ 1 \\ 3 \end{bmatrix}$$

4. One such vector is $\begin{bmatrix} 4/5 \\ 3/5 \end{bmatrix}$

7. No two must be parallel and they must not be *coplanar*, that is, they must not all lie in the same plane

8. (a) linearly dependent; (b) linearly independent; (c) linearly independent; (d) linearly independent

2.2

1. Only (d) is defined which is equal to $\begin{bmatrix} 4 \\ 1 \end{bmatrix}$

2. (a) $\begin{bmatrix} 1 & 2 \\ 0 & -1 \end{bmatrix}$; (b) $\begin{bmatrix} 5 & -6 \\ 42 & -39 \end{bmatrix}$; (c) $\begin{bmatrix} -12 & 5 \\ 5 & -2 \end{bmatrix}$

6. 21, 105

2.3

1. $$\begin{bmatrix} 2 & 3 & 4 \\ 1 & 3 & 7 \\ 3 & 7 & 11 \end{bmatrix} \overset{E_{12}}{\sim} \begin{bmatrix} 1 & 3 & 7 \\ 2 & 3 & 4 \\ 3 & 7 & 11 \end{bmatrix} \overset{E_1(-2)}{\sim} \begin{bmatrix} -2 & -6 & -14 \\ 2 & 3 & 4 \\ 3 & 7 & 11 \end{bmatrix}$$

$$\overset{E_{23}(3)}{\sim} \begin{bmatrix} -2 & -6 & -14 \\ 11 & 24 & 37 \\ 3 & 7 & 11 \end{bmatrix} \overset{E_{13}(1)}{\sim} \begin{bmatrix} 1 & 1 & -3 \\ 11 & 24 & 37 \\ 3 & 7 & 11 \end{bmatrix}$$

$$\overset{E_{31}(2)}{\sim} \begin{bmatrix} 1 & 1 & -3 \\ 11 & 24 & 37 \\ 5 & 9 & 5 \end{bmatrix}$$

3. (i) $\begin{bmatrix} 1 & 0 & 0 & 0 & 0 \\ 0 & -6 & 0 & 0 & 0 \\ 0 & 0 & 1 & 0 & 0 \\ 0 & 0 & 0 & 1 & 0 \\ 0 & 0 & 0 & 0 & 1 \end{bmatrix}$ (ii) $\begin{bmatrix} 1 & 0 & 0 & 0 & 0 \\ 0 & 1 & 0 & 0 & 0 \\ 0 & 0 & 1 & 0 & 0 \\ 0 & 0 & 0 & 1 & 0 \\ 0 & 0 & 7 & 0 & 1 \end{bmatrix}$

(iii) $\begin{bmatrix} 0 & 0 & 0 & 1 & 0 \\ 0 & 1 & 0 & 0 & 0 \\ 0 & 0 & 1 & 0 & 0 \\ 1 & 0 & 0 & 0 & 0 \\ 0 & 0 & 0 & 0 & 1 \end{bmatrix}$

4. Add $-a_{21}/a_{11} \times$ row 1 to row 2, add $-a_{31}/a_{11} \times$ row 1 to row 3. Not valid if $a_{11}=0$.

$$\begin{bmatrix} 1 & 0 & 0 \\ -a_{21}/a_{11} & 1 & 0 \\ 0 & 0 & 1 \end{bmatrix}, \quad \begin{bmatrix} 1 & 0 & 0 \\ 0 & 1 & 0 \\ -a_{31}/a_{11} & 0 & 1 \end{bmatrix}$$

6. $\begin{bmatrix} 1 & 2 & 2 & 3 & 2 \\ 2 & 5 & 3 & 10 & 7 \\ 3 & 5 & 7 & 10 & 4 \end{bmatrix} \sim \begin{bmatrix} 1 & 0 & 4 & 15 & 0 \\ 0 & 1 & -1 & -11 & 0 \\ 0 & 0 & 0 & 5 & 1 \end{bmatrix}$

which has rank 3

2.4

1. $x_1 = x_2 = x_3 = 1$

$$\begin{bmatrix} 3 & -5 & 4 \\ -8 & 4 & 1 \\ 5 & -6 & 2 \end{bmatrix} = \begin{bmatrix} 1 & 0 & 0 \\ -\frac{8}{3} & 1 & 0 \\ \frac{5}{3} & -\frac{1}{4} & 1 \end{bmatrix} \begin{bmatrix} 3 & -5 & 4 \\ 0 & -\frac{28}{3} & \frac{35}{3} \\ 0 & 0 & -\frac{7}{4} \end{bmatrix}$$

2. $\begin{bmatrix} 6 & 13 & -17 \\ 13 & 29 & -38 \\ -17 & -38 & 50 \end{bmatrix}$

4. (i) $x_1 \approx 5.14$, $x_2 \approx 4.63$, $x_3 \approx 3.38$; (ii) $x_1 \approx 5.27$, $x_2 \approx 4.71$, $x_3 \approx 3.38$ (this solution is correct to the number of significant figures given)

5. $\alpha_1 \leftarrow a_1$;

for i **from** 2 **in steps of** 1 **to** n **do**

$\beta_i \leftarrow b_i/\alpha_{i-1}$;

$\alpha_i \leftarrow a_i - \beta_i * c_{i-1}$

endfor

$$\begin{bmatrix} 1 & & & & \\ 1 & 1 & & \bigcirc & \\ 0 & 1 & 1 & & \\ 0 & 0 & \frac{1}{2} & 1 & \\ 0 & 0 & 0 & \frac{2}{7} & 1 \end{bmatrix} \begin{bmatrix} 1 & 1 & 0 & 0 & 0 \\ & 1 & 1 & 0 & 0 \\ & & 2 & 1 & 0 \\ & \bigcirc & & \frac{7}{2} & 1 \\ & & & & \frac{33}{7} \end{bmatrix}$$

6. The exact solutions to the two systems of equations are $[1.20\ 0.60\ 0.60]^T$ and $[6.12\ -18.00\ 15.60]^T$, respectively

2.5

1. (a) 11; (b) 0; (c) 108; (d) $(a+b+c)(a^2+b^2+c^2-ab-bc-ca)$;
(e) $(a+b+c)(ab+bc+ca-a^2-b^2-c^2)$

2. 60

10. 0, -2, -2 and 4

11. $\text{Adj}(A) = \begin{bmatrix} -10 & 13 & 4 \\ 1 & 1 & -5 \\ -5 & -5 & 2 \end{bmatrix}$ and $A^{-1} = -\dfrac{1}{23}\,\text{Adj}(A)$

12. $$\frac{x}{\begin{vmatrix} 11 & -2 & 1 \\ -2 & 1 & -1 \\ 9 & -5 & 0 \end{vmatrix}} = \frac{y}{\begin{vmatrix} 3 & 11 & 1 \\ 1 & -2 & -1 \\ 2 & 9 & 0 \end{vmatrix}} = \frac{z}{\begin{vmatrix} 3 & -2 & 11 \\ 1 & 1 & -2 \\ 2 & -5 & 9 \end{vmatrix}}$$

$$= \frac{1}{\begin{vmatrix} 3 & -2 & 1 \\ 1 & 1 & -1 \\ 2 & -5 & 0 \end{vmatrix}}$$

that is $x/-36 = y/18 = z/-54 = 1/-18$. Hence $x=2$, $y=-1$, $z=3$.

13. The rank is 2.

CHAPTER 3

3.1

1. (a) $a_5 = 0.0016$, $a_{10} = 0.0001$; (b) $a_5 = 2.80092$, $a_{10} = 2.99357$;
(c) $a_5 = 0.21342$, $a_{10} = 0.15435$

6. (a) Convergent, with limit 4. (b) For any $K > 0$, there is an N such that $n! > K^n$, $\forall n \geqslant N$. Hence, $[n!]^{1/n} > K$, $\forall n \geqslant N$, that is, the sequence is divergent. (c) Convergent, with limit 3

8. 5.385165.

9. The scheme is convergent iff $r_n \to 0$ as $n \to \infty$. Now $r_{n+1} = r_1^{2n}$, $n \geqslant 1$. Thus, $r_n \to 0$ as $n \to \infty$ provided $|r_1| < 1$, that is, provided $-1 < 1 - ca_1 < 1$

10. (a) 0.081766; (b) divergent, since $0.5 \notin (0, 2/12.23)$

3.2

1. 2583

2. 2

3. 35/16

4. 5/28

6. $\sum_{n=1}^{p} (3n-2)(2n+1)(n+3) = d_1 p + d_2 p^2 + d_3 p^3 + d_4 p^4$, where $d_1 = -17/3$, $d_2 = 15/2$, $d_3 = 26/3$, $d_4 = 3/2$

7. (i) convergent; (ii) divergent; (iii) convergent

9. $s_9 = 0.9718918$, $s_{10} = 0.9817928$

$$\sum_{n=1}^{\infty} \frac{1}{n^2+1} \approx 1.07667$$

3.3

3. 1.77852

4. $64+576x+2160x^2+4320x^3+4860x^4+2916x^5+729x^6$

5. 1024

6. (i) 1; (ii) 1; (iii) ∞; (iv) ∞

7. $\frac{1}{2}\sum_{n=0}^{\infty}(n+1)(n+2)x^n$

8. $a_n=(-1)^{n-1}\left[1-\frac{1}{3}\left(\frac{2}{3}\right)^{n-1}\right]$

10. 10

3.4

1. (a) $6/(3-x)^2$; (b) $1/2\sqrt{x}$; (c) $-\sin x$

2. (a) $20x^3+21x^2+2$; (b) $4x^3+3x^2-14x-1$; (c) $[(1+x^2)\sin x+(1-x^2)x \cos x]/(1-x^2)^2$; (d) $2\tan x \sec^2 x$; (e) $-x/(1-x^2)^{1/2}$; (f) $1/(1+x^2)$; [$y=\tan^{-1} x$, $z=\tan y$; then $dy/dx=1/(dz/dy)=1/\sec^2 y=1/(1+\tan^2 y)$]

3. $-(2\sin 2x+y(e^{xy}+1/x))/(xe^{xy}+\log_e x)$

4. Let $y=x^{1/q}$. Then $x=y^q$ and $dy/dx=1/(dz/dy)=1/(qy^{q-1})$

5. (a) minimum turning point $x=2+\sqrt{3}/3$, minimum value $-2\sqrt{3}/9$; maximum turning point $x=2-\sqrt{(3)}/3$, maximum value $2\sqrt{(3)}/9$; (b) minimum turning point $x=-1$, minimum value $-1/2$; maximum turning point $x=1$, maximum value $1/2$; (c) maximum turning point $x=1$, maximum value $1/e$

6. $1/\pi$ cm/s

7. $x=\sqrt{0.2}$, $c=£536.66$.

10. $x-\frac{x^3}{3}+\frac{x^5}{5}-\frac{x^7}{7}+\ldots+\frac{(-1)^{n-1}x^{2n-1}}{2n-1}$

3.5

1. (a) $x^5+\frac{7}{2}x^2$; (b) $\frac{3}{4}x^{4/3}$; (c) $-\frac{1}{2}e^{-x^2}$

2. (a) $e^{cx}(cx-1)/c^2$; (b) $\frac{1}{2}e^x(\cos x+\sin x)$; (c) $x \tan x+\log_e \cos x$

3. $e^x(x^3-3x^2+6x-6)$

4. (a) $\log_e\frac{(2x-1)^2}{(x+5)^3}$; (b) $2\log_e(x+2)-1/(x+2)$;
(c) $\frac{1}{10}\log_e\frac{x^2+4}{(x+1)^2}+\frac{2}{5}\tan^{-1}\frac{x}{2}$

5. (a) $\frac{1}{4}\log_e\frac{(x^2+4)^{1/2}-2}{(x^2+4)^{1/2}+2}$ [substitution: $x=(t^2-4)^{1/2}$]; (b) $\tan^{-1}x^2$ (substitution: $x=t^{1/2}$); (c) $\frac{1}{2}(\log_e x)^2$ (substitution: $x=e^t$);
(d) $\frac{1}{5}\left(\frac{5}{3}\right)^{1/2}\tan^{-1}\left[\left(\frac{5}{3}\right)^{1/2}\tan x\right]$ (substitution: $x=\tan^{-1} t$)

6. (a) -1; (b) $-2/9$; (c) 0.31606

7. π square units

8. 8/3 square units

9. (a) -12.070346

10. (a) 2/27; (b) 4; (c) $\frac{2\pi}{3\sqrt{3}}$

11. 241/60

CHAPTER 4

4.1

1. $\mathscr{P}(E\cup F)=\frac{2}{3}$; $\mathscr{P}(E\cap F)=\frac{5}{36}$; $\mathscr{P}(E\cap F')=\frac{13}{36}$

3. (a) $\frac{1}{10}$; (b) $\frac{9}{10}$

4. $\frac{5}{6}$

5. $\frac{1}{9}$

6. $\frac{1}{425}$

4.2

2. $\frac{625}{1296}$

3. $\frac{1}{120}$

4. (a) $\frac{13}{52} \times \frac{12}{51} \times \ldots \times \frac{1}{40} \approx 1.6 \times 10^{-12}$; (b) 6.4×10^{-12}

5. $\frac{24}{43}$

6. $\frac{1}{3}$

7. (a) 0.12; (b) 0.32; (c) 0.88; (d) 0.64

8. (a) $\frac{5}{12}$; (b) $\frac{29}{72}$

9. $\begin{bmatrix} \frac{4}{11} & \frac{4}{11} & \frac{3}{11} \end{bmatrix}$

10. In the long run, for every 14 passes, A, B and C receive the ball 6, 3 and 5 times, respectively.

4.3

1. 0.2679

2.

x_i	0	1	2	3
p_i	$\frac{7}{24}$	$\frac{21}{40}$	$\frac{7}{40}$	$\frac{1}{120}$
$\mathscr{U}_i$	$\frac{7}{24}$	$\frac{98}{120}$	$\frac{119}{120}$	1

4. $\mu=\frac{9}{10}$, $\operatorname{Var}(X)=\frac{49}{100}$

10. (a) 0.368; (b) 0.176

4.4

1. 20

2.

$$\mathscr{U}: x \mapsto \begin{cases} 0, & x<10 \\ -20/x+2, & 10\leqslant x\leqslant 20 \\ 1, & x>20 \end{cases}$$

3. $\mu\approx 13.8629$; $\operatorname{Var}(X)\approx 7.819$

5. (a) $e^{-10}(1-e^{-10})$; (b) p: $x\mapsto 2e^{-2x}$, $x\geqslant 0$

8. 0.0769

4.5

1.

x_i	1	2	3
p_X	0.2	0.5	0.3

y_j	1	2
$p_{\mathscr{Y}}$	0.3	0.7

Since $p(x_i, y_j)=p_X(x_i)p_{\mathscr{Y}}(y_j)$ for each x_i and each y_j, X and $\mathscr{Y}$ are independent.

2. Let $\mathscr{Y}$ denote the random variable X^2.

(a)

y_j	1	4
$p_{\mathscr{Y}}$	0.5	0.5

(Note: $\mathscr{P}(y=4)=\mathscr{P}(x=-2 \vee x=2)=\mathscr{P}(x=-2)+\mathscr{P}(x=2)$, etc.).
(b) $\mathscr{P}((1, 4))=0$, $\mathscr{P}((1, 1))=\mathscr{P}(1)=0.25$, etc.

x_i \ y_j	1	4	p_X
−2	0	0.25	0.25
−1	0.25	0	0.25
1	0.25	0	0.25
2	0	0.25	0.25
$p_{\mathscr{Y}}$	0.5	0.5	

X and $\mathscr{Y}$ are not independent since $p(x_i, y_j) \neq p_X(x_i) p_{\mathscr{Y}}(y_j)$. (c) $E(X)=0$, $E(X^2)=2.5$, $E(X^3)=0$. (d) $\mathrm{Cov}(X, \mathscr{Y}) = E(X\mathscr{Y}) - E(X)E(\mathscr{Y}) = E(X^3) - E(X)E(X^2) = 0$, hence X and $\mathscr{Y}$ are uncorrelated.

3. (b) $p_X(x) = p_{\mathscr{Y}}(y) = 1$; (c) $\mathscr{U}(x, y) = xy(x+y-xy)$; (d) not independent; (e) $\dfrac{5}{64}$

4.6

1. 0.0009, 0.0910, 0.1911, 0.3012, 0.4213, 0.5513, 0.6814, 0.8215, 0.9716, 0.1317

2. 1, 1, 2, 2, 2, 2, 3, 3, 4, 2

3. 0.0018, 0.1908, 0.4242, 0.7168, 1.094, 1.603, 2.288, 3.446, 7.123, 0.2824

5.

x_i	1	2	3	4	5	6	7	8	9	10	11	12
$\mathscr{U}_i$	$\frac{31}{365}$	$\frac{59}{365}$	$\frac{90}{365}$	$\frac{120}{365}$	$\frac{151}{265}$	$\frac{181}{365}$	$\frac{212}{365}$	$\frac{243}{365}$	$\frac{273}{365}$	$\frac{304}{365}$	$\frac{334}{365}$	1

random values: 1, 2, 3, 4, 6, 7, 9, 10, 12, 2

CHAPTER 5

5.1

1. (a) $\{(2, 2), (3, 3), (4, 4), (6, 6), (8, 8), (9, 9)\}$. Domain = range = X.
(b) $\{(3, 2), (4, 2), (4, 3), (6, 2), (6, 3), (6, 4), (8, 2), (8, 3), (8, 4), (8, 6), (9, 2), (9, 3), (9, 4), (9, 6), (9, 8)\}$. Domain = $\{3, 4, 6, 8, 9\}$. Range = $\{2, 3, 4, 6, 8\}$.
(c) $\{(2, 2), (2, 4), (2, 6), (2, 8), (3, 3), (3, 6), (3, 9), (4, 4), (4, 8), (6, 6), (8, 8), (9, 9)\}$. Domain = range = X. (d) $\{(2, 6), (4, 8)\}$. Domain = $\{2, 4\}$. Range = $\{6, 8\}$.
(e) $\{(6, 2), (8, 2), (9, 2), (6, 3), (8, 3), (9, 3), (8, 4), (9, 4), (9, 6)\}$. Domain = $\{6, 8, 9\}$. Range = $\{2, 3, 4, 6\}$

2. (a) a, c; (b) a; (c) a, b, c, d, e (!); (d) a, b, c, d (when defined over X), e; The only equivalence relation is (a)

4. (a) and (b) are convex, (c) is not

7. $x_1=7/2$, $x_2=9/4$ gives $5x_1+2x_2=22$

5.2

3. Yes, No

4. $\sum_{r=1}^{n} 2^{r^2}$

6. The shortest route is a–d–b–e–f–g at a cost of 9

7. The minimal spanning tree comprises arcs a–d, d–b, d–c, b–e, e–f and $f-g$.

5.3

1. (a) Yes; No, $(0\pm 1)\pm -1\neq 0\pm(1\pm -1)$; No, $1\pm 0\neq 0\pm 1$; (c) No; (d) No

2. For ↑: (a) Yes; (b) Yes; (c) Yes (1); (d) No. For ↓: (a) Yes; (b) Yes; (c) No; (d) Yes (1)

5.5

2. (a) $(abc)'$; (b) $b(ac)'$; (c) $b+a'c'$; (d) 1

Index

203745

LINKED